Classic Texts Series

PROBLEMS IN ELEMENTARY MATHEMATICS

Classic Texts Series

PROBLEMS IN ELEMENTARY MATHEMATICS

V. Lidsky,
L. Ovsyannikov,
A. Tulaikov,
M. Shabunin

arihant
ARIHANT PRAKASHAN (Series), MEERUT

arihant

ARIHANT PRAKASHAN (Series), MEERUT

Administrative & Production Offices

Regd. Office

'Ramchhaya' 4577/15, Agarwal Road, Darya Ganj, New Delhi -110002

Tele: 011- 47630600, 43518550

Head Office

Kalindi, TP Nagar, Meerut (UP) - 250002

Tel: 0121-7156203, 7156204

Sales & Support Offices

Agra, Ahmedabad, Bengaluru, Bareilly, Chennai, Delhi, Guwahati, Hyderabad, Jaipur, Jhansi, Kolkata, Lucknow, Nagpur & Pune.

ISBN 978-93-24191-74-8

Published by Arihant Publications (India) Ltd.

For further information about the books published by Arihant, log on to www.arihantbooks.com or e-mail at info@arihantbooks.com

Follow us on

Contents

ALGEBRA

PROBLEMS

ALGEBRA

1. Arithmetic and Geometric Progressions

Preliminaries

Let a_n, d and S_n be, respectively, the nth term, the common difference and the sum of the first n terms of an *arithmetic progression*. Then

$$a_n = a_1 + d(n-1) \tag{1}$$

and

$$S_n = \frac{(a_1 + a_n)\,n}{2} = \frac{[2a_1 + d(n-1)]\,n}{2}. \tag{2}$$

If u_n, q and S_n are the nth term, the common ratio and the sum of the first n terms of a *geometric progression*, then

$$u_n = u_1 q^{n-1} \tag{3}$$

and

$$S_n = \frac{u_n q - u_1}{q-1} = \frac{u_1(q^n - 1)}{q-1}. \tag{4}$$

Finally, if S is the sum of an infinite geometric series with $|q| < 1$ then

$$S = \frac{u_1}{1-q}. \tag{5}$$

1. Prove that if positive numbers a, b and c form an arithmetic progression then the numbers

$$\frac{1}{\sqrt{b}+\sqrt{c}}, \frac{1}{\sqrt{c}+\sqrt{a}}, \frac{1}{\sqrt{a}+\sqrt{b}}$$

also form an arithmetic progression.

2. Positive numbers $a_1, a_2 \ldots, a_n$ form an arithmetic progression. Prove that

$$\frac{1}{\sqrt{a_1}+\sqrt{a_2}} + \frac{1}{\sqrt{a_2}+\sqrt{a_3}} + \ldots + \frac{1}{\sqrt{a_{n-1}}+\sqrt{a_n}} = \frac{n-1}{\sqrt{a_1}+\sqrt{a_n}}.$$

3. Prove that if numbers $a_1, a_2, \ldots, a_n$ are different from zero and form an arithmetic progression then

$$\frac{1}{a_1a_2}+\frac{1}{a_2a_3}+\frac{1}{a_3a_4}+\ldots+\frac{1}{a_{n-1}a_n}=\frac{n-1}{a_1a_n}.$$

4. Prove that any sequence of numbers $a_1, a_2, \ldots, a_n$ satisfying the condition

$$\frac{1}{a_1a_2}+\frac{1}{a_2a_3}+\frac{1}{a_3a_4}+\ldots+\frac{1}{a_{n-1}a_n}=\frac{n-1}{a_1a_n}$$

for every $n \geq 3$ is an arithmetic progression.

5. Prove that for every arithmetic progression $a_1, a_2, a_3, \ldots, a_n$ we have the equalities

$$a_1-2a_2+a_3=0,$$
$$a_1-3a_2+3a_3-a_4=0,$$
$$a_1-4a_2+6a_3-4a_4+a_5=0;$$

and, generally,

$$a_1-C_n^1a_2+C_n^2a_3-\ldots+(-1)^{n-1}C_n^{n-1}a_n+(-1)^nC_n^na_{n+1}=0$$

(where $n > 2$).

Hint. Here and in the problem below it is advisable to apply the identity $C_n^k=C_{n-1}^k+C_{n-1}^{k-1}$ which can be readily verified.

6. Given an arithmetic progression $a_1, \ldots, a_n, a_{n+1}, \ldots$ prove that the equalities

$$a_1^2-C_n^1a_2^2+\ldots+(-1)^nC_n^na_{n+1}^2=0$$

hold for $n \geq 3$.

7. Prove that if the numbers $\log_k x$, $\log_m x$ and $\log_n x\ (x \neq 1)$ form an arithmetic progression then

$$n^2=(kn)^{\log_k m}.$$

8. Find an arithmetic progression if it is known that the ratio of the sum of the first n terms to the sum of the kn subsequent terms is independent of n.

9. The numbers $x_1, x_2, \ldots, x_n$ form an arithmetic progression.

Find this progression if

$$x_1+x_2+\ldots+x_n=a,\ x_1^2+x_2^2+\ldots+x_n^2=b^2.$$

Hint. Here and in the problem below use the equality

$$1^2+2^2+3^2+\ldots+n^2=\frac{n(n+1)(2n+1)}{6}.$$

10. The number sequence 1, 4, 10, 19, ... satisfies the condition that the differences of two subsequent terms form an arithmetic progression. Find the nth term and the sum of the first n terms of this sequence.

11. Consider the table

1
2, 3, 4
3, 4, 5, 6, 7
4, 5, 6, 7, 8, 9, 10
. . . .

Prove that the sum of the terms in each row is equal to the square of an odd number.

12. Given the terms $a_{m+n} = A$ and $a_{m-n} = B$ of a geometric progression $a_1, a_2, a_3, \ldots$, find a_m and $a_n (A \neq 0)$.

13. Let S_n be the sum of the first n terms of a geometric progression $(S_n \neq 0,\ q \neq 0)$. Prove that

$$\frac{S_n}{S_{2n} - S_n} = \frac{S_{2n} - S_n}{S_{3n} - S_{2n}}.$$

14. Knowing the sum S_n of the first n terms of a geometric progression and the sum $\overline{S}_n$ of the reciprocals of these terms find the product Π_n of the first n terms of the progression.

15. Find the sum $1 + 2x + 3x^2 + 4x^3 + \ldots + (n+1)x^n$.

16. Find the sum

$$1 + 11 + 111 + \ldots + 111 \ldots 1$$

if the last summand is an n-digit number.

17. Find the sum

$$nx + (n-1)x^2 + \ldots + 2x^{n-1} + 1x^n.$$

18. Find the sum

$$\frac{1}{2} + \frac{3}{2^2} + \frac{5}{2^3} + \ldots + \frac{2n-1}{2^n}.$$

19. Prove that the numbers 49, 4489, 444889, ... obtained by inserting 48 into the middle of the preceding number are squares of integers.

20. Construct a geometric progression

$$1, q, q^2, \ldots, q^n, \ldots$$

with $|q| < 1$ whose every term differs from the sum of all subsequent terms by a given constant factor k. For what values of k is the problem solvable?

21. An infinite number sequence $x_1, x_2, x_3, \ldots, x_n, \ldots (x_1 \neq 0)$ satisfies the condition
$(x_1^2 + x_2^2 + \ldots + x_{n-1}^2)(x_2^2 + x_3^2 + \ldots + x_n^2) = (x_1x_2 + x_2x_3 + \ldots + x_{n-1}x_n)^2$ for any $n \geq 3$. Prove that the numbers x_1, x_2, ..., x_n,... form an infinite geometric progression.

Hint. Use the method of complete induction.

22. Given an arithmetic progression with general term a_n and a geometric progression with general term b_n. Prove that $a_n < b_n$ for $n > 2$ if $a_1 = b_1,\ a_2 = b_2,\ a_1 \neq a_2$ and $a_n > 0$ for all natural numbers n.

23. Prove that if the terms of a geometric progression $a_1, a_2, \dots, a_n, \dots$ and of an arithmetic progression $b_1, b_2, \dots, b_n, \dots$ satisfy the inequalities

$$a_1 > 0,\ \frac{a_2}{a_1} > 0,\ b_2 - b_1 > 0$$

then there exists a number α such that the difference $\log_\alpha a_n - b_n$ is independent of n.

2. Algebraic Equations and Systems of Equations

Preliminaries

In the problems below the original systems of equations should be simplified and reduced to equivalent systems whose all solutions either are known or can readily be found. In some cases it is necessary to introduce redundant equations which are a priori satisfied by the solutions of the original systems but may have, in the general case, some extraneous solutions. Then the values of the unknowns thus obtained must be tested by substituting them into the original systems.

In some problems one should use Vieta's theorem for the equation of the third degree

$$x^3 + px^2 + qx + r = 0. \tag{1}$$

The theorem establishes the following relations between the coefficients p, q and r of the equation and its roots x_1, x_2 and x_3:

$$x_1 + x_2 + x_3 = -p,\quad x_1x_2 + x_2x_3 + x_3x_1 = q,\quad x_1x_2x_3 = -r. \tag{2}$$

Formulas (2) are derived by equating the coefficients in the equal powers of x on both sides of the identity $x^3 + px^2 + qx + r \equiv (x - x_1)(x - x_2)(x - x_3)$.

24. Find all real solutions of the system of equations

$$\left.\begin{aligned} x^3 + y^3 &= 1, \\ x^2y + 2xy^2 + y^3 &= 2. \end{aligned}\right\}$$

25. Solve the system of equations

$$\left.\begin{aligned} x^2 + xy + y^2 &= 4, \\ x + xy + y &= 2. \end{aligned}\right\}$$

26. Find the real solutions of the system of equations

$$\left.\begin{aligned} x^3 + y^3 &= 5a^3, \\ x^2y + xy^2 &= a^3. \end{aligned}\right\}$$

provided a is real and different from zero.

27. Solve the system of equations

$$\left.\begin{aligned} \frac{x^2}{y} + \frac{y^2}{x} &= 12, \\ \frac{1}{x} + \frac{1}{y} &= \frac{1}{3}. \end{aligned}\right\}$$

28. Solve the system of equations

$$\left.\begin{aligned} x^4 + x^2y^2 + y^4 &= 91, \\ x^2 - xy + y^2 &= 7. \end{aligned}\right\}$$

29. Solve the system of equations

$$\left.\begin{aligned} x^3 - y^3 &= 19(x - y), \\ x^3 + y^3 &= 7(x + y). \end{aligned}\right\}$$

30. Find all real solutions of the system of equations

$$\left.\begin{aligned} 2(x + y) &= 5xy, \\ 8(x^3 + y^3) &= 65. \end{aligned}\right\}$$

31. Find the real solutions of the system of equations

$$\left.\begin{aligned} (x + y)(x^2 - y^2) &= 9, \\ (x - y)(x^2 + y^2) &= 5. \end{aligned}\right\}$$

32. Find all real solutions of the system of equations

$$\left.\begin{aligned} x + y &= 1, \\ x^4 + y^4 &= 7. \end{aligned}\right\}$$

33. Solve the system of equations

$$\left.\begin{aligned} x + y &= 1, \\ x^5 + y^5 &= 31. \end{aligned}\right\}$$

34. Find the real solutions of the system of equations

$$\left.\begin{aligned} x^4 + y^4 - x^2y^2 &= 13, \\ x^2 - y^2 + 2xy &= 1, \end{aligned}\right\}.$$

satisfying the condition $xy \geq 0$.

35. Solve the system of equations

$$\left.\begin{aligned} (x^2 + 1)(y^2 + 1) &= 10, \\ (x + y)(xy - 1) &= 3. \end{aligned}\right\}$$

Hint. Put $xy = v$ and $x + y = u$.

36. Solve the system of equations

$$\left.\begin{aligned}(x^2+y^2)\frac{x}{y}=6,\\(x^2-y^2)\frac{y}{x}=1.\end{aligned}\right\}$$

37. Solve the system of equations

$$\left.\begin{aligned}x^2+y^2=axy,\\x^4+y^4=bx^2y^2.\end{aligned}\right\}$$

38. Solve the equation

$$\left(\frac{x+a}{x+b}\right)^2+\left(\frac{x-a}{x-b}\right)^2-\left(\frac{a}{b}+\frac{b}{a}\right)\frac{x^2-a^2}{x^2-b^2}=0$$

by factorizing its left member.

39. Solve the equation

$$\frac{x^2}{3}+\frac{48}{x^2}=10\left(\frac{x}{3}-\frac{4}{x}\right).$$

40. Solve the system of equations

$$\left.\begin{aligned}\frac{x+y}{xy}+\frac{xy}{x+y}=a+\frac{1}{a},\\\frac{x-y}{xy}+\frac{xy}{x-y}=b+\frac{1}{b}.\end{aligned}\right\}$$

41. Find all the solutions of the equation

$$(x-4.5)^4+(x-5.5)^4=1.$$

42. Solve the system of equations

$$\left.\begin{aligned}|x-1|+|y-5|=1\\y=5+|x-1|^*.\end{aligned}\right\}$$

43. For what real x and y does the equality

$5x^2+5y^2+8xy+2y-2x+2=0$ hold?

44. Find all real values of x and y satisfying the equation

$$x^2+4x\cos(xy)+4=0.$$

hold?

45. Find the real solutions of the system

$$\left.\begin{aligned}x+y+z=2,\\2xy-z^2=4.\end{aligned}\right\}$$

---

* The absolute value of a number x (denoted as $|x|$ is the non -negative number determined by the conditions

$$|x|=\begin{cases}-x \text{ for } x<0,\\ x \text{ for } x\geq 0.\end{cases}$$

46. For what value of a does the system

$$\left.\begin{aligned} x^2+y^2&=z,\\ x+y+z&=a.\end{aligned}\right\}$$

possess a single real solution? Find this solution.

47. Prove that for every (complex, in the general case) solution of the system

$$\left.\begin{aligned} x^2+y^2+xy+\frac{1}{xy}&=a,\\ x^4+y^4+x^2y^2-\frac{1}{x^2y^2}-2&=b^2\end{aligned}\right\}$$

the sum x^2+y^2 is real for any real a and $b, a\neq 0$.

48. Solve the system of equations

$$\left.\begin{aligned} ax+by+cz&=a+b+c,\\ bx+cy+az&=a+b+c,\\ cx+ay+bz&=a+b+c,\end{aligned}\right\}$$

on condition that a, b and c are real and $a+b+c\neq 0$.

49. Solve the system of equations

$$\left.\begin{aligned} ax+y+z&=1,\\ x+ay+z&=a,\\ x+y+az&=a^2.\end{aligned}\right\}$$

50. What relationship must connect the numbers a_1, a_2, a_3 for the system

$$\left.\begin{aligned} (1+a_1)\,x+y+z&=1,\\ x+(1+a_2)\,y+z&=1,\\ x+y+(1+a_3)\,z&=1\end{aligned}\right\}$$

to be solvable and have a unique solution?

51. Solve the system of equations

$$\left.\begin{aligned} ax+by+cz+dt&=p,\\ -bx+ay+dz-ct&=q,\\ -cx-dy+az+bt&=r,\\ -dx+cy-bz+at&=s,\end{aligned}\right\}$$

where the coefficients a, b, c and d satisfy the condition

$$a^2+b^2+c^2+d^2\neq 0.$$

52. Solve the system of equations

$$\left.\begin{aligned} x_1+2x_2+3x_3+4x_4+\ldots+nx_n&=a_1,\\ nx_1+x_2+2x_3+3x_4+\ldots+(n-1)\,x_n&=a_2,\\ (n-1)\,x_1+nx_2+x_3+2x_4+\ldots+(n-2)\,x_n&=a_3,\\ \ldots\ldots\ldots\ldots\ldots\ldots\ldots\ldots\ldots\ldots&\\ 2x_1+3x_2+4x_3+5x_4+\ldots+1x_n&=a_n.\end{aligned}\right\}$$

53. Prove that if

$$\left.\begin{aligned} x_1+x_2+x_3&=0,\\ x_2+x_3+x_4&=0,\\ &\cdots\cdots\\ x_{99}+x_{100}+x_1&=0,\\ x_{100}+x_1+x_2&=0, \end{aligned}\right\}$$

then $x_1+x_2=\ldots=x_{99}=x_{100}=0.$

54. Solve the system of equations

$$\left.\begin{aligned} x^2+xy+xz-x&=2,\\ y^2+xy+yz-y&=4,\\ z^2+xz+yz-z&=6. \end{aligned}\right\}$$

55. Solve the system of equations

$$\left.\begin{aligned} x+y-z&=7,\\ x^2+y^2-z^2&=37,\\ x^3+y^3-z^3&=1. \end{aligned}\right\}$$

56. Solve the system of equations

$$\left.\begin{aligned} \frac{xyz}{x+y}&=2,\\ \frac{xyz}{y+z}&=\frac{6}{5},\\ \frac{xyz}{z+x}&=\frac{3}{2}. \end{aligned}\right\}$$

57. Solve the system of equations

$$\left.\begin{aligned} u^2+v^2+w&=2,\\ v^2+w^2+u&=2,\\ w^2+u^2+v&=2. \end{aligned}\right\}$$

58. Solve the system of equations

$$\left.\begin{aligned} x^2+xy+y^2&=1,\\ x^2+xz+z^2&=4,\\ y^2+yz+z^2&=7. \end{aligned}\right\}$$

59. Find the solutions of the system of equations

$$\left.\begin{aligned} \frac{x_2x_3\ldots x_n}{x_1}&=a_1,\\ \frac{x_1x_3\ldots x_n}{x_2}&=a_2,\\ &\cdots\cdots\\ \frac{x_1x_2\ldots x_{n-1}}{x_n}&=a_n, \end{aligned}\right\}.$$

if the numbers $a_1,\ldots,a_n$ *and* $x_1,\ldots,x_n$ are positive.

60. Solve the system of equations

$$\left.\begin{aligned}(x+y+z)(ax+y+z)&=k^2,\\(x+y+z)(x+ay+z)&=l^2,\\(x+y+z)(x+y+az)&=m^2,\end{aligned}\right\}$$

where a, k, l and m are positive numbers and $k^2+l^2+m^2>0$.

61. Find the real solutions of the system of equations

$$\left.\begin{aligned}x+y+z&=6,\\x^2+y^2+z^2&=14,\\xz+yz&=(xy+1)^2.\end{aligned}\right\}$$

62. Solve the system of equations

$$\left.\begin{aligned}x^2+xy+xz+yz&=a,\\y^2+xy+xz+yz&=b,\\z^2+xy+xz+yz&=c,\end{aligned}\right\}$$

assuming that $abc\neq 0$.

63. Solve the system of equations

$$\left.\begin{aligned}x(y+z)&=a^2,\\y(z+x)&=b^2,\\z(x+y)&=c^2,\end{aligned}\right\}$$

where $abc\neq 0$.

64. Find the real solution of the system of equations

$$\left.\begin{aligned}y^3+z^3&=2a(yz+zx+xy),\\z^3+x^3&=2b(yz+zx+xy),\\x^3+y^3&=2c(yz+zx+xy).\end{aligned}\right\}$$

65. Solve the system of equations

$$\left.\begin{aligned}y+2x+z&=a(x+y)(z+x),\\z+2y+x&=b(y+z)(x+y),\\x+2z+y&=c(z+x)(y+z).\end{aligned}\right\}$$

66. Solve the system of equations

$$\left.\begin{aligned}x+y+z&=9,\\\frac{1}{x}+\frac{1}{y}+\frac{1}{z}&=1,\\xy+xz+yz&=27.\end{aligned}\right\}$$

67. Solve the system of equations

$$\left.\begin{aligned}x+y+z&=a,\\xy+yz+xz&=a^2,\\xyz&=a^3.\end{aligned}\right\}$$

68. Show that the system of equations

$$\left.\begin{array}{c} 2x+y+z=0, \\ yz+zx+xy-y^2=0, \\ xy+z^2=0 \end{array}\right\}$$

has only the trivial solution $x=y=z=0$.

69. Solve the system of equations

$$\left.\begin{array}{c} x+y+z=a, \\ x^2+y^2+z^2=a^2, \\ x^3+y^3+z^3=a^3. \end{array}\right\}$$

70. Let (x, y, z) be a solution of the system of equations

$$\left.\begin{array}{c} x+y+z=a, \\ x^2+y^2+z^2=b^2, \\ \dfrac{1}{x}+\dfrac{1}{y}+\dfrac{1}{z}=\dfrac{1}{c}. \end{array}\right\}$$

Find the sum

$$x^3+y^3+z^3.$$

71. Solve the system of equations

$$\left.\begin{array}{c} x+y+z=2, \\ (x+y)(y+z)+(y+z)(z+x)+(z+x)(x+y)=1, \\ x^2(y+z)+y^2(z+x)+z^2(x+y)=-6. \end{array}\right\}$$

72. Solve the system of equations

$$\left.\begin{array}{c} x^2+(y-z)^2=a, \\ y^2+(x-z)^2=b, \\ z^2+(x-y)^2=c. \end{array}\right\}$$

73. Solve the system of equations

$$\left.\begin{array}{c} xy+yz+zx=47, \\ x^2+y^2=z^2, \\ (z-x)(z-y)=2. \end{array}\right\}$$

74. Find all real solutions of the system of equations

$$\left.\begin{array}{c} x=\dfrac{2z^2}{1+z^2}, \\ y=\dfrac{2x^2}{1+x^2}, \\ z=\dfrac{2y^2}{1+y^2}. \end{array}\right\}$$

75. Find the real solutions of the system of equations

$$\left.\begin{aligned} 2x_2 &= x_1 + \frac{2}{x_1}, \\ 2x_3 &= x_2 + \frac{2}{x_2}, \\ &\cdots\cdots\cdots\cdots \\ 2x_n &= x_{n-1} + \frac{2}{x_{n-1}}, \\ 2x_1 &= x_n + \frac{2}{x_n}. \end{aligned}\right\}$$

76. Show that if a, b, c and d are pairwise unequal real numbers and x, y, z is a solution of the system of equations

$$\left.\begin{aligned} 1 + x + y + z &= 0, \\ a + bx + cy + dz &= 0, \\ a^2 + b^2x + c^2y + d^2z &= 0, \end{aligned}\right\}$$

then the product xyz is positive.

In the equations below, if the index of a radical is even, consider only the values of the unknowns for which the radicand is non-negative and take only the non-negative value of the root.

When the index is odd the radicand can be any real number (in this case the sign of the root coincides with the sign of the radicand).

77. Solve the equation

$$\sqrt[3]{(a+x)^2} + 4\sqrt[3]{(a-x)^2} = 5\sqrt[3]{a^2-x^2}.$$

78. Solve the equation

$$\sqrt[m]{(1+x)^2} - \sqrt[m]{(1-x)^2} = \sqrt[m]{1-x^2}.$$

79. Solve the equation

$$\sqrt{y-2+\sqrt{2y-5}} + \sqrt{y+2+3\sqrt{2y-5}} = 7\sqrt{2}.$$

80. Solve the equation

$$\sqrt{x+\sqrt{x}} - \sqrt{x-\sqrt{x}} = \frac{3}{2}\sqrt{\frac{x}{x+\sqrt{x}}}.$$

81. Solve the equation

$$\frac{\sqrt{x^2+8x}}{\sqrt{x+1}} + \sqrt{x+7} = \frac{7}{\sqrt{x+1}}.$$

82. Find all real roots of the equation

$$\sqrt[3]{x-1} + \sqrt[3]{x+1} = x\sqrt[3]{2}.$$

83. Solve the equation $\sqrt{x-4a+16} = 2\sqrt{x-2a+4} - \sqrt{x}$.
For what real values of a is the equation solvable?

84. Solve the system of equations

$$\left.\begin{aligned}\sqrt{1-16y^2}-\sqrt{1-16x^2}&=2(x+y),\\ x^2+y^2+4xy&=\frac{1}{5}.\end{aligned}\right\}$$

85. Solve the system of equations

$$\left.\begin{aligned}x-y&=\frac{7}{2}\left(\sqrt[3]{x^2y}-\sqrt[3]{xy^2}\right),\\ \sqrt[3]{x}-\sqrt[3]{y}&=3.\end{aligned}\right\}$$

86. Solve the system of equations

$$\left.\begin{aligned}\sqrt{\frac{x}{y}}-\sqrt{\frac{y}{x}}&=\frac{3}{2},\\ x+yx+y&=9.\end{aligned}\right\}$$

87. Solve the system of equations

$$\left.\begin{aligned}\sqrt{\frac{y+1}{x-y}}+2\sqrt{\frac{x-y}{y+1}}&=3,\\ x+xy+y&=7.\end{aligned}\right\}$$

88. Find all real solutions of the system

$$\left.\begin{aligned}x+y-\sqrt{\frac{x+y}{x-y}}&=\frac{12}{x-y},\\ xy&=15.\end{aligned}\right\}$$

89. Solve the system of equations

$$\left.\begin{aligned}y+\frac{2\sqrt{x^2-12y+1}}{3}&=\frac{x^2+17}{12},\\ \frac{x}{8y}+\frac{2}{3}&=\sqrt{\frac{x}{3y}+\frac{1}{4}}-\frac{y}{2x}.\end{aligned}\right\}$$

90. Solve the system of equations

$$\left.\begin{aligned}\frac{x+\sqrt{x^2-y^2}}{x-\sqrt{x^2-y^2}}+\frac{x-\sqrt{x^2-y^2}}{x+\sqrt{x^2-y^2}}&=\frac{17}{4},\\ x(x+y)+\sqrt{x^2+xy+4}&=52.\end{aligned}\right\}$$

91. Solve the system of equations

$$\left.\begin{aligned}y^2+\sqrt{3y^2-2x+3}&=\frac{2}{3}x+5,\\ 3x-2y&=5.\end{aligned}\right\}$$

92. Find the real solutions of the system of equations

$$\left.\begin{aligned} y+\frac{4}{3}\sqrt{x^2-6y+1}=\frac{x^2+17}{6},\\ \frac{x^2y-5}{49}=\frac{2}{y}-\frac{12}{x^2}+\frac{4}{9}.\end{aligned}\right\}$$

93. Solve the system of equations

$$\left.\begin{aligned} (x-y)\sqrt{y}=\frac{\sqrt{x}}{2},\\ (x+y)\sqrt{x}=3\sqrt{y}.\end{aligned}\right\}$$

94. Solve the system of equations

$$\left.\begin{aligned} \sqrt{x+y}-\sqrt{x-y}=a,\\ \sqrt{x^2+y^2}+\sqrt{x^2-y^2}=a^2\end{aligned}\right\}(a>0).$$

95. Solve the system of equations

$$\left.\begin{aligned} x\sqrt{x}-y\sqrt{y}=a(\sqrt{x}-\sqrt{y}),\\ x^2+xy+y^2=b^2\end{aligned}\right\}(a>0, b>0).$$

3. Algebraic Inequalities

Preliminaries

Here are some inequalities which are used for solving the problems below.

For any real a and b we have

$$a^2+b^2\geq 2\,|ab|. \tag{1}$$

Inequality (1) is a consequence of the obvious inequality $(a\pm b)^2\geq 0$.

Relation (1) turns into an equality only if $|a|=|b|$.

If $ab>0$, then dividing both sides of inequality (1) by ab we obtain

$$\frac{a}{b}+\frac{b}{a}\geq 2. \tag{2}$$

If $u\geq 0$ and $\upsilon\geq 0$, then, putting $u=a^2$ and $\upsilon=b^2$ in (1) we obtain

$$\frac{u+\upsilon}{2}\geq\sqrt{u\upsilon}. \tag{3}$$

In inequalities (2) and (3) the sign of equality appears only for

$$a=b \text{ and } (u=\upsilon).$$

In addition, let us indicate some properties of the *quadratic trinomial*

$$y=ax^2+bx+c \tag{4}$$

which are used in some problems below.

The representation of trinomial (4) in the form

$$y = a\left(x + \frac{b}{2a}\right)^2 - \frac{b^2 - 4ac}{4a} \qquad (5)$$

implies that if the discriminant of the trinomial satisfies the condition

$$D = b^2 - 4ac < 0$$

(in this case the roots of the trinomial are nonreal), then, for all x, the trinomial takes on values of the same sign which coincides with the sign of the coefficient a in the second power of x.)

If $D = 0$ the trinomial vanishes only for $x = -\frac{b}{2a}$ and retains its sign for all the other values of x.

Finally, if $D > 0$ (in this case the trinomial has real distinct roots x_1 and x_2), it follows from the factorization

$$y = a\,(x - x_1)(x - x_2),$$

that the trinomial attains the values whose sign is opposite to that of a only for x satisfying the condition

$$x_1 < x < x_2.$$

For all the other values of x different from x_1 and x_2 the trinomial has the same sign as a.

Thus, a trinomial always retains the sign of the coefficient in x^2 except for the case when its roots x_1 and x_2 are real and $x_1 \le x \le x_2$.

96. Find all real values of r for which the polynomial

$$(r^2 - 1)\,x^2 + 2\,(r - 1)\,x + 1$$

is positive for all real x.

97. Prove that the expression

$$3\left(\frac{x^2}{y^2} + \frac{y^2}{x^2}\right) - 8\left(\frac{x}{y} + \frac{y}{x}\right) + 10$$

is non-negative for any real x and y different from zero.

98. For what values of a is the system of inequalities

$$-3 < \frac{x^2 + ax - 2}{x^2 - x + 1} < 2$$

fulfilled for all x ?

99. Prove that for any real numbers a, b, c and d the inequality $a^4 + b^4 + c^4 + d^4 \ge 4abcd$ is valid.

100. Find all the values of a for which the system

$$\left.\begin{array}{r} x^2 + y^2 + 2x \le 1, \\ x - y + a = 0 \end{array}\right\}$$

has a unique solution. Find the corresponding solutions.

101. Find the pairs of integers x and y satisfying the system of inequalities

$$\left.\begin{array}{l} y-\left|x^2-2x\right|+\frac{1}{2}>0, \\ y+|x-1|<2. \end{array}\right\}$$

102. Prove that the inequality

$$\frac{1}{n+1}+\frac{1}{n+2}+\ldots+\frac{1}{2n}>\frac{1}{2}$$

holds for every integer $n>1$.

103. Prove that the inequality

$$\frac{1}{m+1}+\frac{1}{m+2}+\ldots+\frac{1}{m+(2m+1)}>1$$

is valid for every positive integer m.

104. Show that for any natural n we have

$$\frac{1}{2^2}+\frac{1}{3^2}+\ldots+\frac{1}{n^2}<\frac{n-1}{n}.$$

105. Prove that

$$(n!)^2>n^n$$

for $n>2$.

106. Prove that, given three line segments of length $a>0$, $b>0$ and $c>0$, a triangle with these segments as sides can be constructed if and only if $pa^2+qb^2>pqc^2$ for any numbers p and q satisfying the condition $p+q=1$.

107. Prove that for any real x, y and z we have the inequality

$$4x(x+y)(x+z)(x+y+z)+y^2z^2\geq 0.$$

108. Prove that the inequality

$x^2+2xy+3y^2+2x+6y+4\geq 1$ holds for any real x and y.

109. Prove that if $2x+4y=1$, the inequality

$x^2+y^2\geq\frac{1}{20}$ is fulfilled.

110. What conditions must be imposed on the number $d>0$ for the inequality

$0<\frac{d^2+R^2-r^2}{2dR}\leq 1$ to be valid for $R\geq r>0$?

111. Prove the inequality

$$\frac{1}{a}+\frac{1}{b}+\frac{1}{c}\geq\frac{9}{a+b+c},$$

where a, b and c are positive.

112. Prove that if a, b and c are numbers of the same sign and $a < b < c$, then
$a^3(b^2-c^2)+b^3(c^2-a^2)+c^3(a^2-b^2)<0.$

113. Prove that if $a_1, a_2, a_3, \ldots, a_n$ are positive numbers and $a_1a_2a_3\ldots a_n=1$, then
$(1+a_1)(1+a_2)(1+a_3)\ldots(1+a_n)\geq 2^n.$

114. Prove that if $a+b=1$ then

$$a^4+b^4\geq\frac{1}{8}.$$

115. Prove that the polynomial

$$x^8-x^5+x^2-x+1$$

is positive for all real x.

116. Prove that if $|x|<1$ the inequality

$$(1-x)^n+(1+x)^n<2^n$$

is fulfilled for any integer $n\geq 2$.

117. Prove that

$$|x_1a_1+x_2a_2+\ldots+x_na_n|\leq\frac{1}{\varepsilon}(x_1^2+x_2^2+\ldots+x_n^2)+\frac{\varepsilon}{4}(a_1^2+a_2^2+\ldots+a_n^2),$$

where $x_1, x_2, \ldots, x_n$ and $a_1, a_2, \ldots, a_n$ and ε are arbitrary real numbers and $\varepsilon>0$.

118. For what real values of x is the inequality

$$\frac{1-\sqrt{1-4x^2}}{x}<3$$

fulfilled?

119. Prove that for all positive x and y and positive integers m and n $(n\geq m)$ we have the inequality

$$\sqrt[m]{x^m+y^m}\geq\sqrt[n]{x^n+y^n}.$$

120. Prove the inequality

$$\sqrt{a+\sqrt{a+\ldots+\sqrt{a}}}<\frac{1+\sqrt{4a+1}}{2},\ a>0.$$

121. Prove the inequality

$$\frac{2-\sqrt{2+\sqrt{2+\sqrt{2+\ldots+\sqrt{2}}}}}{2-\sqrt{2+\sqrt{2+\ldots+\sqrt{2}}}}>\frac{1}{4}$$

provided the numerator of the left member of the inequality contains n radical signs and the denominator contains $n-1$ radical signs.

122. Prove that for any real numbers $a_1, a_2, \dots, a_n$ and $b_1, b_2, \dots, b_n$ satisfying the relations

$$\left.\begin{aligned} a_1^2 + a_2^2 + \dots + a_n^2 = 1, \\ b_1^2 + b_2^2 + \dots + b_n^2 = 1, \end{aligned}\right\}$$

the inequality $|a_1 b_1 + a_2 b_2 + \dots + a_n b_n| \le 1$ is valid.

123. Prove that if the numbers $x_1, x_2, \dots, x_n$ are positive and satisfy the relation

$$x_1 x_2 \dots x_n = 1,$$

then

$$x_1 + x_2 + \dots + x_n \ge n.$$

4. Logarithmic and Exponential Equations, Identities and Inequalities

Preliminaries

The definition of the logarithm of a number N to a base a states that

$$a^{\log_a N} = N. \qquad (1)$$

Here N is any positive number, a is an arbitrary base and $a > 0$, $a \ne 1$..

The solution of some problems below is based on the following formula for converting from logarithms to a base a to the logarithms to a base b:

$$\log_a N = \frac{\log_b N}{\log_b a}. \qquad (2)$$

The formula is proved by taking the logarithms to the base b of the both sides of identity (1). In particular, for N=b formula (1) implies

$$\log_a b = \frac{1}{\log_b a}. \qquad (3)$$

124. Solve the equation

$$\frac{\log_2 x}{\log_2^2 a} - \frac{2 \log_a x}{\log_{\frac{1}{b}} a} = \log_{3\sqrt{a}} x \log_a x.$$

125. Solve the equation

$$\log_x 2 \log_{\frac{x}{16}} 2 = \log_{\frac{x}{64}} 2.$$

126. Solve the equation

$$\log_2(9^{x-1} + 7) = 2 + \log_2(3^{x-1} + 1).$$

127. Solve the equation

$$\log_{3x}\left(\frac{3}{x}\right)+\log_3^2 x=1.$$

128. Prove that the equation

$$\log_{2x}\left(\frac{2}{x}\right)\log_2^2 x+\log_2^4 x=1$$

has only one root satisfying the inequality $x > 1$. Find this root.

129. Solve the equation

$$\frac{\log_{a^2\sqrt{xa}}}{\log_{2x} a}+\log_{ax} a\ \log_{\frac{1}{a}} 2x=0.$$

130. What conditions must be imposed on the numbers a and b for the equation

$$1+\log_b(2\log a-x)\log_x b=\frac{2}{\log_b x}$$

to have at least one solution? Find all the solutions of this equation.

131. Solve the equation*

$$\sqrt{\log_a\sqrt[4]{ax}+\log_x\sqrt[4]{ax}}+\sqrt{\log_a\sqrt[4]{\frac{x}{a}}+\log_x\sqrt[4]{\frac{a}{x}}}=a.$$

132. Solve the equation

$$\frac{\log(\sqrt{x+1}+1)}{\log\sqrt[3]{x-40}}=3.$$

133. Solve the equation

$$1+\frac{\log_a(p-x)}{\log_a(x+q)}=\frac{2-\log_{p-q}4}{\log_{p-q}(x+q)}\qquad(p>q>0).$$

134. Solve the equation

$$\log_{\sqrt{5}} x\sqrt{\log_x 5\sqrt{5}+\log_{\sqrt{5}} 5\sqrt{5}}=-\sqrt{6}.$$

135. Solve the equation

$$(0.4)^{\log^2 x+1}=(6.25)^{2-\log x^2}.$$

136. Solve the equation

$$1+\log_x\frac{4-x}{10}=(\log\log n-1)\log_x 10.$$

How many roots has the equation for a given value of n?

137. Solve the equation

$$\log_{\sin x}2\cdot\log_{\sin^2 x}a+1=0.$$

---

* Here and henceforward the roots are understood as mentioned on page 11.

138. Solve the system of equations

$$\left.\begin{array}{c}\log_2(x+y)-\log_3(x-y)=1,\\ x^2-y^2=2.\end{array}\right\}$$

139. Solve the system of equations

$$\left.\begin{array}{c}x^a=y^b,\\ \log_c\dfrac{x}{y}=\dfrac{\log_c x}{\log_c y}\end{array}\right\}(a\neq b,\ ab\neq 0).$$

140. Solve the system of equations

$$\left.\begin{array}{c}\log_5 x+3^{\log_3 y}=7,\\ x^y=5^{12}.\end{array}\right\}$$

141. Solve the system of equations

$$\left.\begin{array}{c}yx^{\log_y x}=x^{\frac{5}{2}},\\ \log_4 y\log_y(y-3x)=1.\end{array}\right\}$$

142. Solve the system of equations

$$\left.\begin{array}{c}a^x b^y=ab,\\ 2\log_a x=\log_{\frac{1}{b}} y\log_{\sqrt{a}} b.\end{array}\right\}$$

143. Solve the system of equations

$$\left.\begin{array}{c}3\,(2\log_{y^2} x-\log_{\frac{1}{x}} y)=10,\\ xy=81.\end{array}\right\}$$

144. Solve the system of equations

$$\left.\begin{array}{c}\log_{12} x\left(\dfrac{1}{\log_x 2}+\log_2 y\right)=\log_2 x,\\ \log_2 x\log_3(x+y)=3\log_3 x.\end{array}\right\}$$

145. Solve the system of equations

$$\left.\begin{array}{c}x\log_2 y\log_{\frac{1}{x}} 2=y\sqrt{y}(1-\log_x 2),\\ \log_{y^3} 2\log_{\sqrt{2}} x=1.\end{array}\right\}$$

146. Solve the system of equations

$$\left.\begin{array}{c}\log_2 x+\log_4 y+\log_4 z=2,\\ \log_3 y+\log_9 z+\log_9 x=2,\\ \log_4 z+\log_{10} x+\log_{19} y=2.\end{array}\right\}$$

147. Solve the system of equations

$$\left.\begin{array}{c}\log_{0.5}(y-x)+\log_2\dfrac{1}{y}=-2,\\ x^2+y^2=25.\end{array}\right\}$$

148. Solve the equation

$$4^x - 3^{x-\frac{1}{2}} = 3^{x+\frac{1}{2}} - 2^{2x-1}.$$

149. Find the positive roots of the system of equations

$$\left.\begin{aligned} x^{x+y} &= y^{x-y}, \\ x^2 y &= 1. \end{aligned}\right\}$$

150. Solve the system of equations

$$\left.\begin{aligned} a^{2x} + a^{2y} &= 2b, \\ a^{x+y} &= c \end{aligned}\right\} \qquad (a > 0).$$

Under what conditions on b and c is the system solvable?

151. Find the positive solutions of the system of equations

$$\left.\begin{aligned} x^{x+y} &= y^n, \\ y^{x+y} &= x^{2n} y^n, \end{aligned}\right\}$$

where $n > 0$.

152. Solve the system of equations

$$\left.\begin{aligned} (3x+y)^{x-y} &= 9, \\ \sqrt[x-y]{324} &= 18x^2 + 12xy + 2y^2. \end{aligned}\right\}$$

153. Find the positive roots of the system of equations

$$\left.\begin{aligned} x^y &= y^x, \\ x^p &= y^q, \end{aligned}\right\}$$

where $pq > 0$.

154. Solve the system of equations

$$\left.\begin{aligned} x^y &= y^x, \\ p^x &= q^y, \end{aligned}\right\}$$

assuming that $x > 0$, $y > 0$, $p > 0$ and $q > 0$.

155. Prove that

$$\log_{c+b} a + \log_{c-b} a = 2 \log_{c+b} a \log_{c-b} a,$$
if $a^2 + b^2 = c^2$ and $a > 0, b > 0, c > 0$.

156. Simplify the expression

$$(\log_b a - \log_a b)^2 + (\log_{b^{\frac{1}{2}}} a - \log_{a^2} b)^2 + \ldots + \left(\log_{b^{\frac{1}{2^n}}} a - \log_{a^{2^n}} b\right)^2.$$

157. Simplify the expression $a^{\frac{\log \log a}{\log a}}$ where all the logarithms are taken to the same base b.

158. Let $\log_a b = A$ and $\log_q b = B$. Compute $\log_c b$ where c is the product of n terms of a geometric progression with common ratio q and the first term a.

159. Prove that if the relation

$$\frac{\log_a N}{\log_c N} = \frac{\log_a N - \log_b N}{\log_b N - \log_c N}$$

is fulfilled for a given positive $N \neq 1$ and three positive numbers a, b and c, then b is the mean proportional between a and c, and the relation is fulfilled for any positive $N \neq 1$.

160. Prove the identity

$$\log_a N \log_b N + \log_b N \log_c N + \log_c N \log_a N = \frac{\log_a N \log_b N \log_c N}{\log_{abc} N}.$$

161. Prove the identity

$$\frac{\log_a x}{\log_{ab} x} = 1 + \log_a b.$$

162. Solve the inequality

$$\log_{\frac{1}{2}} x + \log_3 x > 1.$$

163. Solve the inequality

$$x^{\log_a x + 1} > a^2 x \qquad (a > 1).$$

164. Solve the inequality

$$\log_a x + \log_a (x+1) < \log_a (2x+6) \ (a > 1).$$

165. Solve the inequality

$$\log_3 (x^2 - 5x + 6) < 0.$$

166. Solve the inequality

$$\frac{1}{\log_2 x} - \frac{1}{\log_2 x - 1} < 1.$$

167. Solve the inequality

$$x^{2 - \log_2^2 x - \log_2 x^2} - \frac{1}{x} > 0.$$

168. For what real x and α is the inequality $\log_2 x + \log_x 2 + 2\cos\alpha \leq 0$ valid?

169. Solve the inequality

$$\log_{\frac{1}{3}} [\log_4 (x^2 - 5)] > 0.$$

5. Combinatorial Analysis and Newton's Binomial Theorem

Preliminaries

The number of *permutations* of n things taken m at a time is given by the formula

$$P(n,\ m) = n(n-1)\ldots(n-m+1). \tag{1}$$

The number of *permutations* of n things taken all at a time is equal to factorial n:

$$n! = 1\cdot 2\cdot 3 \ldots n. \tag{2}$$

The number of *combinations* of n elements, m at a time, is defined by the formula

$$C(n,\ m) = \frac{n(n-1)(n-2)\ \ldots (n-m+1)}{1\cdot 2\cdot 3\ldots m} = \frac{p\,(n,m)}{m!}. \tag{3}$$

There is a relation of the form

$$C(n,m) = C(n,\ n-m).$$

For positive integers n and any x and a we have binomial formula

$$(x+a)^n = x^n + C(n,1)ax^{n-1} + C(n,2)a^2x^{n-2}$$
$$+\ldots + C(n,\ n-2)a^{n-2}x^2 + C(n,\ n-1)a^{n-1}x + a^n, \tag{4}$$

whose general term is equal to

$$C\,(n,\ k)a^k x^{n-k}. \tag{5}$$

Formula (4) implies the equalities

$$1 + C(n,1) + C(n,2) + \ldots. + C\,(n, n-2) + C\,(n, n-1) + (1)^n = 2^n.$$

and

$$1 - C(n,1) + C(n,2) - C(n,3) + \ldots. + (-1)^n = 0.$$

170. Find m and n knowing that

$$C(n+1,\ m+1) : C(n+1,\ m) : C(n+1,\ m-1) = 5 : 5 : 3.$$

171. Find the coefficient in x^8 in the binomial expansion of

$$(1 + x^2 - x^3)^9.$$

172. Find the coefficient in x^m in the expansion of the expression

$$(1+x)^k + (1+x)^{k+1} + \ldots + (1+x)^n$$

in powers of x. Consider the cases $m < k$ and $m \geq k$.

173. In the expansion, by the binomial formula, of the expression $\left(x\sqrt{x} + \frac{1}{x^4}\right)^n$ the binomial coefficient in the third term is by 44 larger than that in the second term. Find the term not containing x.

174. In the expansion of the expression

$$\left(1+x+\frac{6}{x}\right)^{10}$$

find the term not containing x.

175. Find out for what value of k the $(k+1)$th term of the expansion, by the binomial formula, of the expression $(1+\sqrt{3})^{100}$

is simultaneously greater than the preceding and the subsequent terms of the expansion?

176. Find the condition under which the expansion of $(1+a)^n$ in powers of a (where n is an integer and $a \neq 0$) contains two equal consecutive terms. Can this expansion contain three equal consecutive terms?

177. Find the total number of dissimilar terms obtained after the expression $x_1 + x_2 + x_3 + \ldots + x_n$ has been cubed.

178. Let $p_1, p_2, \ldots, p_n$ be different prime numbers and $q = p_1 p_2 \ldots p_n$.

Determine the number of the divisors (including 1 and q) of q.

179. Prove that if each coefficient in the expansion of the expression $x(1+x)^n$ in powers of x divided by the exponent of the corresponding power, then the sum of the quotients thus obtained is equal to

$$\frac{2^{n+1}-1}{n+1}.$$

180. Prove that

$$C(n,1)x(1-x)^{n-1} + 2C(n,2)x^2(1-x)^{n-2} + \ldots\ldots +$$
$$+ kC(n,k)x^k(1-x)^{n-k} + \ldots\ldots + nC(n,n)x^n = nx,$$

where $n > 0$ is an arbitrary integer.

181. In how many ways can a pack of 36 cards be split in two so that each portion contains two aces?

182. How many five-digit telephone numbers with pairwise distinct digits can be composed?

183. Given a set of $2n$ elements. Consider all the possible partitions of the set into the pairs of elements on condition that the partitions solely differing in the order of elements within the pairs and in the order of the pairs are regarded as coincident. What is the total number of these partitions?

184. Determine the number of permutations of n elements taken all at a time in which two given elements a and b are not adjacent.

185. Eight prizes are distributed by a lottery. The first participant takes 5 tickets from the urn containing 50 tickets. In how many ways can be extract them so that (1) exactly two tickets are winning, (2) at least two tickets are winning.

186. m points are taken on one of two given parallel lines and n points on the other. Join with line segments each of the m points on the former line to each of the n points on the latter. What is the number of points of intersection of the segments if it is known that there are no points in which three or more segments intersect.

187. n parallel lines in a plane are intersected by a family of m parallel lines. How many parallelograms are formed in the network thus formed?

188. An alphabet consists of six letters which are coded in Morse code as

.; —; —;— —; — ; —.

A word was transmitted without spaces between the letters so that the resultant continuous line of dots and dashes contained 12 characters. In how many ways can that word be read?

6. Problems in Forming Equations

189. In multiplying two numbers one of which exceeds the other by 10 the pupil reduced, by mistake, the tens digit in the product by 4. When checking the answer by dividing the product thus obtained by the smaller of the factors he obtained the quotient 39 and the remainder 22. Determine the factors.

190. Two cyclists simultaneously start out from a point A and proceed with different but constant speeds to a point B and then return without stopping. One of them overtakes the other and meets him on the way back at a point a kilometres from B. Having reached A he starts for B and again meets the second cyclist after covering $\frac{1}{k}$th the distance between A and B. Find the distance from A to B.

191. Two cars simultaneously start out from a point and proceed in the same direction, one of them going at a speed of 50 km/hr and the other at 40 km/hr. In half an hour a third car starts out from the same point and overtakes the first car 1.5 hours after catching up with the second car. Determine the speed of the third car.

192. A pedestrian and a cyclist start out from points A and B towards one another. After they meet the pedestrian continues to go in the direction from A to B while the cyclist turns and also goes towards B. The pedestrian reaches B t hours later than the cyclist. Find the time period between the start and meeting if the speed of the cyclist is k times that of the pedestrian.

193. Walking without stopping a postman went from a point A through a point B to a point C. The distance from A to B was covered with a speed of 3.5 km/hr and from B to C of 4 km/hr. To get back from C to A in the same time following the same route with a constant speed he was to walk 3.75 km per hour. However, after walking at that speed and reaching B he stopped for 14 minutes and then, in order to reach A at the appointed time he had to move from B to A walking 4 km per hour. Find the distances between A and B and between B and C.

194. The distance from a point A to a point B is 11.5 km. The road between A and B first goes uphill, then horizontally and then downhill. A pedestrian went from A to B in 2 hours and 54 minutes but it took him 3 hours and 6 minutes to get back from B to A. His speeds were 3 km/hr uphill, 4 km/hr on the horizontal part of the road and 5 km/hr downhill. Determine the length of the horizontal part.

195. In a motorcycle test two motorcyclists simultaneously start out from A to B and from B to A, each driving at a constant speed. After arriving at their terminal points they turn back without stopping. They meet at a distance of p km from B and then, in t hours, at q km from A. Find the distance between A and B and the speeds of the motorcyclists.

196. An airplane was in flight from A to B in a straight line. Due to a head wind, after a certain time, it reduced its speed to v km/hr and therefore was t_1 minutes late. During a second flight from A to B the airplane for the same reason reduced its speed to the same level but this time d km farther from A than in the first flight and was t_2 minutes late. Find the original speed of the airplane.

197. There are two pieces of an alloy weighing m kg and n kg with different percentages of copper. A piece of the same weight is cut from either alloy. Each of the cut-off pieces is alloyed with the rest of the other piece which results into two new alloys with the same percentage of copper. Find the weights of the cut-off pieces.

198. Given two pieces of alloys of silver and copper. One of them contains p% of copper and the other contains q % of copper. In what ratio are the weights of portions of the alloys if the new alloy made up of these portions contains r% of copper? For what relationships between p, q and r is the problem solvable? What is the greatest weight of the new alloy that can be obtained if the first piece weighs P grams and the second Q grams?

199. Workers A and B have been working the same number of days. If A worked one day less and B 7 days less then A would earn 72 roubles and B 64 roubles 80 kopecks. If, conversely, A worked 7 days less and B one day less B would earn 32 roubles and 40 kopecks more than A. How much did in fact either worker earn?

200. Two bodies move in a circle in opposite directions, one of them being in a uniform motion with linear speed v and the other in a uniformly accelerated motion with linear acceleration a. At the initial moment of time the bodies are at the same point A, and the velocity of the second one is equal to zero. In what time does their first meeting take place if the second meeting occurs at the point A?

201. A tank was being filled with water from two taps. One of the taps was first open during one third of the time required for filling the tank by the other tap alone. Then, conversely, the second tap was kept open for one third of the time required to fill the tank by using the first tap alone, after which the tank was $\frac{13}{18}$ full. Compute the time needed to fill the tank by each tap separately if both taps, when open together, fill the tank in 3 hours and 36 minutes.

202. A cylindrical pipe with a piston is placed vertically into a tank of water so that there is a column of air h metres high between the piston and the water (at the atmospheric pressure). The piston is then elevated b metres above the water level in the tank. Compute the height of the column of water in the pipe if it is known that the column of liquid in a water barometer is c metres high at the atmospheric pressure.

203. A cylindrical pipe with a moving piston is placed vertically into a cup of mercury. The mercury level in the pipe is 12 cm above that in the cup, and the column of air in the pipe between the mercury and the piston is $29\frac{3}{4}$ cm high. The piston is then moved 6 cm downward. What is the resultant height of the column of mercury if the external air pressure is 760 mm Hg?

204. At a certain moment a watch shows a 2-minutes lag although it is fast. If it showed a 3-minutes lag at that moment but gained half a minute more a day than it does it would show true time one day sooner than it actually does. How many minutes a day does the watch gain?

205. Two persons deposited equal sums of money in a savings bank. One of them withdrew his money after m months and received p roubles, and the other withdrew the money after n months and received q roubles. How much money did either person deposit and what interest does the savings bank pay?

206. In a circle of radius R two points uniformly move in the same direction. One of them describes one circuit t seconds faster than the other. The time period between two consecutive meetings of the points is equal to T. Determine the speeds of the points.

207. A flask contains a solution of sodium chloride. $\frac{1}{n}$th part of the solution is poured into a test tube and evaporated until the percentage of sodium chloride in the test tube is doubled. The evaporated solution is then poured back into the flask. This increases the percentage of sodium chloride in the flask by p%. Determine the original percentage of sodium chloride.

208. Two identical vessels, each of 30 litres, contain a total of only 30 litres of alcohol. Water is added to the top of one vessel, the resulting mixture is added to the top of the other vessel and then 12 litres of the new mixture are poured from the second vessel into the first. How much alcohol did each vessel contain originally if after the above procedure the second vessel contains 2 litres of alcohol less than the first?

209. Three travellers A, B and C are crossing a water obstacle s km wide. A is swimming at a speed of v km/hr, and B and C are in a motor boat going at v_1 km/hr. Some time after the start C decides to swim the rest of the distance, his speed being equal to that of A. At this moment B decides to pick up A and turns back. A then takes the motor boat and continues his way with B. All the three travellers simultaneously arrive at the opposite bank. How long did the crossing take?

210. A train left a station A for B at 13:00. At 19:00 the train was brought to a halt by a snow drift. Two hours later the railway line was cleared and to make up for the lost time the train proceeded at a speed exceeding the original speed by 20% and arrived at B only one hour later. The next day a train going from A to B according to the same timetable was stopped by a snow drift 150 km farther from A than the former train. Likewise, after a two-hour halt it went with a 20% increase of speed but failed to make up for the lost time and arrived at B 1 hour 30 minutes late. Find the distance between A and B.

211. A landing stage B is a kilometres up the river from A. A motor boat makes trips going from A to B and returning to A without stopping in T hours. Find the speed of the boat in still water and the speed of the current if it is known that once, when returning from B to A, the motor boat had an accident at a distance of b km from A which delayed it for T_0 hours and reduced its speed twice so that it went from B to A during the same time as from A to B.

212. A tank of a volume of 425 m^3 was filled with water from two taps. One of the taps was open 5 hours longer than the other. If the first tap had been kept open as long as the second and the second tap as long as the first, then the first tap would have released one half the amount of water flowed out from the second. If both taps had been opened simultaneously the tank would have been filled in 17 hours.

Taking into account all these conditions determine how long the second tap was open.

213. According to the timetable, a train is to cover the distance of 20 km between A and B at a constant speed. The train covered half the distance at that speed and then stopped for three minutes; in order to arrive at B on schedule it had to increase the speed by 10 km/hr on the remaining half of the trip. Another time the train was delayed for 5 minutes after passing half the way. At what speed must the train go after the stop in order to arrive at B on schedule?

214. Two airplanes simultaneously take off from A and B. Flying towards each other, they meet at a distance of a kilometres from the midpoint of AB. If the first airplane took off b hours later than the second, they would meet after passing half the distance from A to B. If, conversely, the second airplane took off b hours after the first, they would meet at a point lying at the quarter of that distance from B. Find the distance between A and B and the speeds of the airplanes.

215. A motor boat and a raft simultaneously start out downstream from A. The motor boat covers 96 km, turns back and arrives at A in 14 hours. Find the speed of the motor boat in still water and the speed of the current if it is known that the two craft met at a distance of 24 km from A when the motor boat was returning.

216. Two bodies simultaneously start out in the same direction from two points 20 metres apart. The one behind is in uniformly accelerated motion and covers 25 metres during the first second and $\frac{1}{3}$ of a metre more in the next second. The other body is in uniformly decelerated motion and passes 30 metres in the first second and half a metre less in the next second. How many seconds will it take the first body to catch up with the second?

217. A boat moves 10 km downstream and then 6 km upstream. The river current is 1 km/hr. Within what limits must the relative speed of the boat lie for the entire trip to take from 3 to 4 hours?

218. The volumes of three cubic vessels A, B and C are in the ratio $1:8:27$ while the amounts of water in them are in the ratio $1:2:3$. After water has been poured from A into B and from B into C, the water level in the vessels is the same. $128\frac{4}{7}$ litres of water are then poured out from C into B after which a certain amount is poured from B into A so that the depth of water in A becomes twice that in B. This results in the amount of water in A being by 100 litres less than the original amount. How much water did each vessel contain originally?

219. Find a four-digit number using the following conditions: the sum of the squares of the extreme digits equals 13; the sum of the squares of the middle digits is 85; if 1089 is subtracted from the desired number, the result is a number expressed by the same digits as the sought-for number but written in reverse order.

220. Two points move in a circle whose circumference is l metres at the speeds v and $w < v$. At what moments of time reckoned from the start of the first point will successive meetings of the points occur if they move in the same direction, and the first point starts t seconds before the second and is a metres behind the second point at the initial moment ($a < l$)?

221. A piece of an alloy of two metals weighs P kg and loses A kg in weight when immersed in water. A portion of P kg of one of the metals loses B kg in water and a portion of the same weight of the other metal loses C kg. Find the weights of the components of the alloy and test the solvability of the problem depending on the magnitudes of the quantities P, A, B and C.

222. Log rafts floated downstream from a point A to the mouth of a river where they were picked up by a towboat and towed across a lake to a point B $17\frac{1}{8}$ days after the departure from A. How long did it take the towboat to bring the log rafts to B across the lake if it is known that, alone, the towboat goes from A to B in 61 hours and from B to A in 79 hours and that in towing the relative speed of the towboat is reduced twice?

223. The current of a river between A and B is negligibly small but between B and C it is rather strong. A boat goes downstream from A to C in 6 hours and upstream from C to A in 7 hours. If between A and B the current were the same as between B and C the whole distance from A to C would be covered in 5.5 hours. How long would it take to go upstream from C to A in the latter case?

224. A vessel contains a p% solution of an acid. a litres of the solution are then poured out and the same quantity of a q % solution of the acid is added ($q < p$). After mixing this operation is repeated $k-1$ times which results in a r% solution. Find the volume of the vessel.

225. A roubles are invested in a savings bank which pays an interest of p%. At the end of every year the depositor takes out B roubles. In how many years will the rest be three times the original sum? Under what conditions is the problem solvable?

226. A forestry has a p% annual growth rate of wood. Every winter an amount x of wood is obtained. What must x be so that in n years the amount of wood in the forestry becomes q times the original amount a?

227. One of n identical cylindrical vessels is full of alcohol and the others are half-full with a mixture of water and alcohol, the concentration of alcohol in each vessel being $\frac{1}{k}$th that in the preceding one. Then the second vessel is filled to the top from the first one after which the third is filled from the second and so on to the last vessel. Find the resultant concentration of alcohol in the last vessel.

228. Consider a quotient of two integers in which the divisor is less by unity than the square of the dividend. If 2 is added to the dividend and to the divisor the value of the quotient will exceed $\frac{1}{3}$ but if 3 is subtracted from the numerator and denominator, the quotient will remain positive but less than $\frac{1}{10}$. Find the quotient.

7. Miscellaneous Problems

Algebraic Transformations

229. Compute the sum

$$\frac{1}{n(n+1)}+\frac{1}{(n+1)(n+2)}+\ldots+\frac{1}{(n+k-1)(n+k)}.$$

230. Simplify the expression

$$(x+a)(x^2+a^2)\ldots(x^{2^{n-1}}+a^{2^{n-1}}).$$

231. Simplify the expression

$$(x^2-ax+a^2)(x^4-a^2x^2+a^4)\ldots(x^{2^n}-a^{2^{n-1}}x^{2^{n-1}}+a^{2^n}).$$

232. Given two sequences of numbers

$$a_1, a_2, \ldots, a_n,$$
$$b_1, b_2, \ldots, b_n,$$

prove that

$$a_1b_1+a_2b_2+\ldots+a_nb_n=(a_1-a_2)S_1+(a_2-a_3)S_2+\ldots$$
$$\ldots+(a_{n-1}-a_n)S_{n-1}+a_nS_n,$$

where $S_k=b_1+b_2+\ldots+b_k$.

233. Show that the equality

$$a^2+b^2+c^2=bc+ac+ab,$$

where a, b and c are real numbers, implies $a=b=c$.

234. Prove that if $a^3+b^3+c^3=3abc$ then either $a^2+b^2+c^2=bc+ca+ab$ or $a+b+c=0$.

235. Show that if

$$a_1^2 + a_2^2 + \ldots + a_n^2 = p^2,$$
$$b_1^2 + b_2^2 + \ldots + b_n^2 = q^2,$$
$$a_1 b_1 + a_2 b_2 + \ldots + a_n b_n = pq$$

and $pq \neq 0$, then $a_1 = \lambda b_1, a_2 = \lambda b_2, \ldots, a_n = \lambda b_n$, where $\lambda = \frac{p}{q}$. (All the quantities are supposed to be real.)

236. It is known that the number sequence $a_1, a_2, a_3, \ldots$ satisfies, for any n, the relation

$$a_{n+1} - 2a_n + a_{n-1} = 1.$$

Express a_n in terms of a_1, a_2 and n.

237. The sequence of numbers $a_1, a_2, a_3, \ldots, a_n, \ldots$ satisfies for $n > 2$ the relation

$$a_n = (\alpha + \beta)\, a_{n-1} - \alpha\beta a_{n-2},$$

where α and β ($\alpha \neq \beta$) are given numbers. Express a_n in terms of α, β, a_1 and a_2.

BEZOUT'S THEOREM. PROPERTIES OF ROOTS OF POLYNOMIALS

238. The roots x_1 and x_2 of the equation $x^2 - 3ax + a^2 = 0$ satisfy the condition $x_1^2 + x_2^2 = 1.75$. Determine a.

239. Given the equation $x^2 + px + q = 0$, form a quadratic equation whose roots are

$$y_1 = x_1^2 + x_2^2 \text{ and } y_2 = x_1^3 + x_2^3.$$

240. Let x_1 and x_2 be the roots of the equation $ax^2 + bx + c = 0$ $(ac \neq 0)$. Without solving the equation express the quantities

1) $\frac{1}{x_1^2} + \frac{1}{x_2^2}$ and 2) $x_1^4 + x_1^2 x_2^2 + x_2^4$

in terms of the coefficients a, b and c.

241. What conditions must be imposed on the real coefficients a_1, b_1, a_2, b_2, a_3, and b_3 for the expression

$(a_1 + b_1 x)^2 + (a_2 + b_2 x)^2 + (a_3 + b_3 x)^2$

to be the square of a polynomial of the first degree in x with real coefficients?

242. Prove that the roots of the quadratic equation $x^2 + px + q = 0$ with real coefficients are negative or have a negative real part if and only if $p > 0$ and $q > 0$.

243. Prove that if both roots of the equation

$$x^2 + px + q = 0$$

are positive, then the roots of the equation $qy^2 + (p - 2rq)y + 1 - pr = 0$ are positive for all $r \geq 0$. Is this assertion true for $r < 0$?

244. Find all real values of p for which the roots of the equation $(p-3)x^2 - 2px + 6p = 0$ are real and positive.

245. For any positive λ all the roots of the equation

$$ax^2 + bx + c + \lambda = 0$$

are real and positive. Prove that in this case $a = 0$ (the coefficients a, b and c are real).

246. Prove that both roots of the equation $x^2 + x + 1 = 0$ satisfy the equation

$$x^{3m} + x^{3n+1} + x^{3p+2} = 0,$$

where m, n and p are arbitrary integers.

247. The system of equations

$$\left.\begin{aligned} a(x^2 + y^2) + x + y - \lambda &= 0, \\ x - y + \lambda &= 0 \end{aligned}\right\}$$

has real solutions for any λ. Prove that $a = 0$.

248. Prove that for any real values of a, p and q the equation

$$\frac{1}{x-p} + \frac{1}{x-q} = \frac{1}{a^2}$$

has real roots.

249. Prove that the quadratic equation

$$a^2x^2 + (b^2 + a^2 - c^2)x + b^2 = 0$$

cannot have real roots if $a + b > c$ and $|a - b| < c$.

250. It is known that x_1, x_2 and x_3 are the roots of the equation

$$x^3 - 2x^2 + x + 1 = 0.$$

Form a new algebraic equation whose roots are the numbers

$$y_1 = x_2x_3, \; y_2 = x_3x_1, \; y_3 = x_1x_2.$$

251. It is known that x_1, x_2 and x_3 are the roots of the equation

$$x^3 - x^2 - 1 = 0.$$

Form a new equation whose roots are the numbers

$$y_1 = x_2 + x_3, \; y_2 = x_3 + x_1, \; y_3 = x_1 + x_2.$$

252. Express the constant term c of the cubic equation

$$x^3 + ax^2 + bx + c = 0$$

in terms of the coefficients a and b, knowing that the roots of the equations form an arithmetic progression.

253. Let it be known that all roots of an equation

$$x^3 + px^2 + qx + r = 0$$

are positive. What additional condition must be imposed on its coefficients p, q and r so that the line segments of lengths equal to the roots are the sides of a triangle?

Hint. Consider the expression

$$(x_1 + x_2 - x_3)(x_2 + x_3 - x_1)(x_3 + x_1 - x_2).$$

254. The equations

$$x^3 + p_1 x + q_1 = 0$$

and

$$x^3 + p_2 x + q_2 = 0$$

$(p_1 \neq p_2, q_1 \neq q_2)$ have a common root. Find this root and also the other roots of both equations.

255. Find all the values of λ for which two equations

$$\lambda x^3 - x^2 - x - (\lambda + 1) = 0$$

and

$$\lambda x^2 - x - (\lambda + 1) = 0$$

have a common root. Determine this root.

256. All the roots of the polynomial

$$P(x) = x^3 + px + q$$

with real coefficients p and q $(q \neq 0)$ are real. Prove that $p < 0$.

257. Prove that the equation

$$x^3 + ax^2 - b = 0$$

where a and b $(b > 0)$ are real has one and only one positive root.

258. Find all the real values of a and b for which the equations

$$x^3 + ax^2 + 18 = 0$$

and

$$x^3 + bx + 12 = 0$$

have two common roots and determine these roots.

259. Prove that

$$\sqrt[3]{20 + 14\sqrt{2}} + \sqrt[3]{20 - 14\sqrt{2}} = 4.$$

260. Let a, b and c be pairwise different numbers.

Prove that the expression

$$a^2(c - b) + b^2(a - c) + c^2(b - a)$$

is not equal to zero.

261. Factorize the expression

$$(x + y + z)^3 - x^3 - y^3 - z^3.$$

262. Prove that if three real numbers a, b and c satisfy the relationship

$$\frac{1}{a}+\frac{1}{b}+\frac{1}{c}=\frac{1}{a+b+c},$$

then two of them are necessarily equal in their absolute values and have opposite signs.

263. Find out for what complex values of p and q the binomial x^4-1 is divisible by the quadratic trinomial x^2+px+q.

264. For what values of a and n is the polynomial $x^n-ax^{n-1}+ax-1$ divisible by $(x-1)^2$?

265. The division of the polynomial $p(x)$ by $x-a$ gives the remainder A, the division by $x-b$ gives the remainder B and the division by $x-c$ gives the remainder C. Find the remainder polynomial obtained by dividing $p(x)$ by $(x-a)(x-b)(x-c)$ on condition that the numbers a, b and c are pairwise different.

MATHEMATICAL INDUCTION

The following problems are solved by the method of complete mathematical induction. To prove that an assertion is true for every natural n it is sufficient to prove that (a) this assertion is true for $n=1$ and (b) if this assertion is true for a natural number n then it is also true for $n+1$.

266. Prove that

$$1+3+6+10+\ldots+\frac{(n-1)n}{2}+\frac{n(n+1)}{2}=\frac{n(n+1)(n+2)}{6}.$$

267. Prove that

$$1^2+2^2+3^2+\ldots+n^2=\frac{n(n+1)(2n+1)}{6}.$$

268. Prove that

$$\frac{1}{1\times2\times3}+\frac{1}{2\times3\times4}+\ldots+\frac{1}{n(n+1)(n+2)}=\frac{n(n+3)}{4(n+1)(n+2)}.$$

269. Prove De Moivre's formula

$$(\cos\varphi+i\sin\varphi)^n=\cos n\varphi+i\sin n\varphi.$$

270. Prove that for any positive integer n the quantity $a_n=\dfrac{a^n-b^n}{\sqrt5}$ where $a=\dfrac{1+\sqrt5}{2}$ and $b=\dfrac{1-\sqrt5}{2}$ is a positive integer.

271. Prove that if real numbers $a_1, a_2, \ldots, a_n, \ldots$ satisfy the condition $-1<a_i\le 0$, $i=1,2,\ldots$, then for any n we have the inequality

$$(1+a_1)(1+a_2)\ldots(1+a_n)\ge 1+a_1+a_2+\ldots+a_n.$$

272. The generalized nth power of an arbitrary number a (denoted by $(a)_n$) is defined for non-negative integers n as follows: if $n=0$ then $(a)_n = 1$ and if $n > 0$ then $(a)_n = a(a-1)\dots(a-n+1)$.

Prove that for the generalized power of a sum of two numbers we have the formula

$$(a+b)_n = C_n^0 (a)_0 (b)_n - C_n^1 (a)_1 (b)_{n-1} + \dots + C_n^n (a)_n (b)_0$$

which generalizes Newton's binomial theorem to this case.

THE GREATEST AND LEAST VALUES

To find the *least* value of a quadratic trinomial

$$y = ax^2 + bx + c \quad (1)$$

for $a > 0$ it is represented in the form

$$y = a\left(x + \frac{b}{2a}\right)^2 - \frac{b^2 - 4ac}{4a}. \quad (2)$$

The first summand on the right-hand side being non-negative for any x and the second summand being independent of x, the trinomial attains its least value when the first summand vanishes. Thus, the least value of the trinomial is

$$y_0 = -\frac{b^2 - 4ac}{4a}. \quad (3)$$

It is assumed for

$$x = x_0 = -\frac{b}{2a}. \quad (4)$$

A similar technique yields the *greatest* value of a trinomial $y = ax^2 + bx + c$ for $a < 0$.

273. Two rectilinear railway lines AA' and BB' are mutually perpendicular and intersect at a point C, the distances AC and BC being equal to a and b. Two trains whose speeds are, respectively, v_1 and v_2 start simultaneously from the points A and B toward C. In what time after the departure will the distance between the trains be the least? Find this least distance.

274. Two stations A and B are on a rectilinear highway passing from west to east, B lying 9 km to the east of A. A car starts from A and moves uniformly eastwards at a speed of 40 km/hr. A motorcycle simultaneously starts from B in the same direction and moves with a constant acceleration of 32 km/hr^2. Determine the greatest distance between the car and motorcycle during the first two hours of motion.

Hint. It is advisable to plot the graph of the distance between the car and motorcycle against the time of motion.

275. Find the greatest value of the expression

$$\log_2^4 x + 12 \log_2^2 x \log_2 \frac{8}{x}$$

when x varies between 1 and 64.

276. Find the greatest value of the function

$$y = \frac{x}{ax^2 + b} \quad (a > 0, b > 0).$$

277. Find the least value of the expression

$$\frac{1 + x^2}{1 + x}$$

for $x \geq 0$.

278. Find the least value of the function

$$\varphi(x) = |x - a| + |x - b| + |x - c| + |x - d|,$$

where $a < b < c < d$ are fixed real numbers and x takes arbitrary real values.

Hint. Mark a, b, c, and d on a number scale.

COMPLEX NUMBERS

279. Find all the values of z satisfying the equality

$$z^2 + |z| = 0$$

where $|z|$ denotes the modulus of the complex number z.

280. Find the complex number z satisfying the equalities

$$\left|\frac{z - 12}{z - 8i}\right| = \frac{5}{3} \text{ and } \left|\frac{z - 4}{z - 8}\right| = 1.$$

281. Compute the product

$$\left[1 + \left(\frac{1+i}{2}\right)\right]\left[1 + \left(\frac{1+i}{2}\right)^2\right]\left[1 + \left(\frac{1+i}{2}\right)^{2^2}\right] \ldots \left[1 + \left(\frac{1+i}{2}\right)^{2^n}\right].$$

282. Among the complex numbers z satisfying the condition,

$$|z - 25i| \leq 15,$$

find the number having the least argument. Make a drawing.

283. Find the condition for a complex number $a + bi$ to be representable in the form

$$a + bi = \frac{1 - ix}{1 + ix},$$

where x is a real number?

284. Find the greatest value of the moduli of complex numbers z satisfying the equation

$$\left|z+\frac{1}{z}\right|=1.$$

285. Through a point A n rays are drawn which form the angles $\frac{2\pi}{n}$ with each other. From a point B lying on one of the rays at a distance d from A a perpendicular is drawn to the next ray. Then from the foot of this perpendicular a new perpendicular is drawn to the neighbouring ray and so on, unlimitedly. Determine the length L of the broken line thus obtained which sweeps out an infinity of circuits round the point A. Also investigate the variation of L as the number n is increased and, in particular, the case when n approaches infinity.

286. A six-digit number begins with 1. If this digit is carried from the extreme left decimal place to the extreme right without changing the order of the other digits the new number thus obtained is three times the original number. Find the original number.

287. Prove that if a natural number $p = abc$ where a, b and c are the decimal digits is divisible by 37 then the numbers $q = bca$ and $r = cab$ are also divisible by 37.

288. Prove that the sum of the cubes of three successive integers is divisible by 9.

289. Prove that the sum

$$S_n = n^3 + 3n^2 + 5n + 3$$

is divisible by 3 for any positive integer n.

290. 120 identical balls are tightly stacked in the form of a regular triangular pyramid. How many balls lie at the base of the pyramid?

291. k smaller boxes are put in a box. Then in each of the smaller boxes either k still smaller boxes are put or no boxes and so on. Determine the number of empty boxes if it is known that here are m filled boxes.

GEOMETRY

A. PLANE GEOMETRY

B. SOLID GEOMETRY

GEOMETRY

A. PLANE GEOMETRY

Preliminaries

Here are some basic relations between the elements of a triangle with sides a, b and c and the respective opposite angles A, B and C.

1. Law of sines:

$$\frac{a}{\sin A} = \frac{b}{\sin B} = \frac{c}{\sin C} = 2R,$$

where R is the radius of the circumscribed circle.

2. Law of cosines:

$$a^2 = b^2 + c^2 - 2bc\cos A.$$

For computing the area S of a triangle use the following formulas:

$$S = \frac{1}{2}ah_a,$$

where a is a side of the triangle and h_a is the altitude drawn to this side;

$$S = \sqrt{p(p-a)(p-b)(p-c)} \quad \text{(Heron's formula)}$$

where

$$p = \frac{a+b+c}{2};$$

$$S = \frac{1}{2}ab\sin C;$$

$$S = rp,$$

where r is the radius of the inscribed circle.

1. Computation Problems

292. In a triangle ABC the angle A is twice as large as the angle B. Given the sides b and c, find a.

293. The legs of a right triangle are equal to b and c. Find the length of the bisector of the right angle.

294. Given two sides a and b of a triangle, find its third side if it is known that the medians drawn to the given sides intersect at a right angle. What are the conditions for the triangle to exist?

295. The vertex angle of a triangle with lateral sides of lengths a and b ($a < b$) is trisected by straight lines whose segments inside the triangle form the ratio $m : n$ ($m < n$). Find the lengths of the segments.

296. Intersect a given triangle ABC by a straight line DE parallel to BC so that the area of the triangle BDE is of a given magnitude k^2. What relationship between k^2 and the area of the triangle ABC guarantees the solvability of the problem and how many solutions has the problem?

297. Through a point lying inside a triangle three straight lines parallel to its sides are drawn. The lines divide the triangle into six parts three of which are triangles with areas S_1, S_2 and S_3, respectively. Find the area of the given triangle.

298. Given the sides b and c of a triangle. Find the third side x knowing that it is equal to the altitude drawn to it. Under what condition connecting b and c does the triangle exist?

299. In a triangle ABC the altitudes AA_1, BB_1 and CC_1 are drawn, and the points A_1, B_1, and C_1 are joined. Determine the ratio of the area of the triangle $A_1B_1C_1$ to that of the triangle ABC if the angles of the triangle ABC are given.

300. In a triangle ABC through the point of intersection of the bisectors of the angles B and C a straight line parallel to BC is drawn. This line intersects the sides AB and AC at points M and N respectively. Find the relationship between the line segments MN, BM and CN.

Consider the following cases:

(1) both bisectors divide interior angles of the triangle;

(2) both bisectors divide exterior angles of the triangle;

(3) one of the bisectors cuts an interior angle and the other cuts an exterior angle.

When do the points M and N coincide?

301. Inside an equilateral triangle ABC an arbitrary point P is taken from which the perpendiculars PD, PE and PF are dropped onto BC, CA and AB respectively. Compute

$$\frac{PD + PE + PF}{BD + CE + AF}.$$

302. Find the ratio of the area of a triangle ABC to the area of a triangle whose sides are equal to the medians of the triangle ABC.

303. In a triangle with sides a, b and c a semicircle is inscribed whose diameter lies on the side c. Find the radius of the semicircle.

304. Determine the acute angles of a right triangle knowing that the ratio of the radius of the circumscribed circle to the radius of the inscribed circle is $5 : 2$.

305. About a given rectangle circumscribe a new one with given area m^2. For what m is the problem solvable?

306. On the side AB of the rectangle $ABCD$ find a point E from which the sides AD and DC are seen at equal angles. What relationship between the sides guarantees the solvability of the problem?

307. Find the area of an isosceles trapezoid with altitude h if its nonparallel sides are seen from the centre of the circumscribed circle at angles α.

308. Given the upper and lower bases a and b of a trapezoid. Find the length of the line segment joining the midpoints of the diagonals of the trapezoid.

309. Each vertex of a parallelogram is connected with the midpoints of two opposite sides by straight lines. What portion of the area of the parallelogram is the area of the figure bounded by these lines?

310. P, Q, R and S are respectively the midpoints of the sides AB, BC, CD, and DA of a parallelogram $ABCD$. Find the area of the figure bounded by the straight lines AQ, BR, CS and DP knowing that the area of the parallelogram is equal to a^2.

311. Given the chords of two arcs of a circle of radius R, find the chord of an arc equal to the sum of these arcs or to their difference.

312. The distance between the centres of two intersecting circles of radii R and r is equal to d. Find the area of their common portion.

313. Three circles of radii r, r_1 and R are pairwise externally tangent. Find the length of the chord cut off by the third circle from the internal common tangent of the first two circles.

314. Two circles of radii R and r ($R > r$) are internally tangent.Find the radius of the third circle tangent to the two given circles and to their common diameter.

315. Three equal circles are externally tangent to a circle of radius r and pairwise tangent to one another. Find the areas of the three curvilinear triangles formed by these circles.

316. On a line segment of length $2a + 2b$ and on its parts of lengths $2a$ and $2b$ as diameters semicircles lying on one side of the line segment are constructed. Find the radius of the circle tangent to the three semicircles.

317. Given two parallel straight lines and a point A between them. Find the sides of a right triangle with vertex of the right angle at the point A and vertices of the acute angles on the given parallel lines if it is known that the area of the triangle is of a given magnitude k^2.

318. n equal circles are inscribed in a regular n-gon with side a so that each circle is tangent to two adjacent sides of the polygon and to two other circles. Find the area of the star-shaped figure formed in the centre of the polygon.

319. Through a point C of an arc AB of a circle two arbitrary straight lines are drawn which intersect the chord AB at points D and E and the circle at points F and G. What position does the point C occupy on the arc AB if it is possible to circumscribe a circle about the quadrilateral $DEGF$?

320. Circles are inscribed in an acute angle so that every two neighbouring circles are tangent. Show that the radii of the circles form a geometric progression. Find the relationship between the common ratio of the progression and the magnitude of the acute angle.

321. A light source is located at a point A of a plane P. A hemispherical mirror of unit radius is placed above the plane so that its reflecting inner side faces the plane and its axis of symmetry passes through the point A and is perpendicular to the plane P. Knowing that the least angle between the rays reflected by the mirror and the plane P is equal to 15° determine the distance from the mirror to the plane and the radius of the illuminated circle of the plane P.

322. The centres of four circles of radius r are at the vertices of a square with side a. Find the area S of the common part of all circles contained inside the square.

323. A trapezoid is divided into four triangles by its diagonals.

Find the area of the trapezoid if the areas of the triangles adjacent to the bases of the trapezoid are equal to S_1 and S_2.

324. Express the diagonals of an inscribed quadrilateral of a circle in terms of its sides. Based on this result, deduce the Ptolemy theorem which states that the product of the diagonals of a quadrilateral inscribed in a circle is equal to the sum of the products of the two pairs of opposite sides.

2. Construction Problems

325. Given two circles of different radii with no points in common and a point A on one of them. Draw a third circle tangent to the two given circles and passing through the point A. Consider various possible cases of location of the point A on the circle.

326. Given a circle and a straight line with point A on it. Construct a new circle tangent to the given line and circle and passing through the point A. Consider in detail how many solutions the problem has in various particular cases.

327. Given a straight line and a circle with point A on it. Construct a new circle tangent to the given line and circle and passing through the point A. Consider in detail how many solutions the problem has in various particular cases.

328. Construct a right triangle, given the hypotenuse c and the altitude h drawn to it. Determine the lengths of the legs of the triangle and find the relationship between h and c for which the problem is solvable.

329. Given the lengths of the sides AB, BC, CD and DA of a plane quadrilateral. Construct this quadrilateral if it is known that the diagonal AC bisects the angle A.

330. Reconstruct the triangle from the points at which the extended bisector, median and altitude drawn from a common vertex intersect the circumscribed circle.

331. Draw three pairwise tangent circles with centres at the vertices of a given triangle. Consider the cases when the circles are externally and internally tangent.

332. Inscribe a triangle ABC in a given circle if the positions of the vertex A and of the point of intersection of the altitude h_B with the circle and the direction of the altitude h_A are known.

333. Intersect a trapezoid by a straight line parallel to its base so that the segment of this line inside the trapezoid is trisected by the diagonals.

334. Construct a square, given a vertex and two points lying on two sides not passing through this vertex or on their extensions.

335. Through a point M lying on the side AC of a triangle ABC draw a straight line MN cutting from the triangle a part whose area is $\frac{1}{k}$ that of the whole triangle. How many solutions has the problem?

336. Make a ruler and compass construction of a rectangle with given diagonal inscribed in a given triangle.

337. About a given circle circumscribe a triangle with given angle and given side opposite this angle. Find the solvability condition for the problem.

338. Given a straight line CD and two points A and B not lying on it. Find a point M on the line such that
$\angle AMC = 2\angle BMD$.

3. Proof Problems

339. Prove that a median of a triangle is less than half-sum of the sides it lies between and greater than the difference of this half-sun and half the third side.

340. Prove that in any triangle ABC the distance from the centre of the circumscribed circle to the side BC is half the distance between the point of intersection of the altitudes and the vertex A.

341. Prove that the sum of the distances from any point lying inside an equilateral triangle to the sides of the triangle is a constant independent of the position of the point.

342. Prove that in any triangle a shorter bisector of an interior angle corresponds to a longer side.

343. Prove that if P, Q and R are respectively the points of intersection of the sides BC, CA and AB (or their extensions) of a triangle ABC and a straight line then

$$\frac{PB}{PC}\frac{QC}{QA}\frac{RA}{RB}=1.$$

344. In a right triangle ABC the length of the leg AC is three times that of the leg AB. The leg AC is trisected by points K and F. Prove that

$$\angle AKB + \angle AFB + \angle ACB = \frac{\pi}{2}.$$

345. Let a, b, c and h be respectively the two legs of a right triangle, the hypotenuse and the altitude drawn from the vertex of the right angle to the hypotenuse. Prove that a triangle with sides $h, c+h$ and $a+b$ is right.

346. In an isosceles triangle with base a and congruent side b the vertex angle is equal to 20°. Prove that $a^3+b^3=3ab^2$.

347. Prove that an angle of a triangle is acute, right or obtuse depending on whether the side opposite this angle is less than, equal to, or greater than the doubled length of the corresponding median.

348. In an isosceles triangle ABC the vertex angle B is equal to 20° and points Q and P are taken respectively on the sides AB and BC so that $\angle ACQ=60°$ and $\angle CAP=50°$. Prove that $\angle APQ=80°$.

349. Prove that if the sides a, b and c of a triangle are connected by the relation $a^2=b^2+bc$ then the angles A and B subtended by the sides a and b satisfy the equality $\angle A=2\angle B$.

350. A triangle AOB is turned in its plane about the vertex O by 90°, the new positions of the vertices A and B being, respectively, A_1 and B_1. Prove that in the triangle OAB_1 the median of the side AB_1 is an altitude of the triangle OA_1B (analogously, the median of the side A_1B in the triangle OA_1B is an altitude of the triangle OAB_1).

351. Prove that the sum of the products of the altitudes of an acute triangle by their segments from the orthocentre to the corresponding vertices equals half-sum of the squares of the sides. Generalize this assertion to the case of an obtuse triangle.

352. Let the lengths a, b and c of the sides of a triangle satisfy the condition $a < b < c$ and form an arithmetic progression. Prove that $ac = 6\ Rr$ where R is the radius of the circumscribed circle of the triangle and r is the radius of the inscribed circle.

353. Prove that the square of the bisector of an angle in a triangle is equal to the difference of the product of the sides including this angle and the product of the segments of the base. What is the meaning of this equality for the case of an isosceles triangle?

354. In a triangle ABC two equal line segments $BD = CE$ are set off in opposite directions on the sides AB and AC. Prove that the ratio in which the segment DE is divided by the side BC is the reciprocal of the ratio of the side AB to the side AC.

355. From a vertex of a triangle the median, the bisector of the interior angle and the altitude are drawn. Prove that the bisector lies between the median and the altitude.

356. Prove that the straight line which is the reflection of a median through the concurrent bisector of an interior angle of a triangle divides the opposite side into parts proportional to the squares of the adjacent sides.

357. On the sides of a triangle ABC points P, Q and R are taken so that the three straight lines AP, BQ and CR are concurrent. Prove that
$AR \cdot BP \cdot CQ = RB \cdot PC \cdot QA$.

358. Prove that the radius R of the circumscribed circle of a triangle and the radius r of the inscribed circle satisfy the relation $l^2 = R^2 - 2Rr$ where l is the distance between the centres of these circles.

359. Prove that in any triangle the ratio of the radius of the inscribed circle to the radius of the circumscribed circle does not exceed $\frac{1}{2}$.

360. Prove that for any right triangle we have the inequality $0.4 < \frac{r}{h} < 0.5$ where r is the radius of the inscribed circle and h is the altitude drawn to the hypotenuse.

361. Prove that for any acute triangle we have the relation $k_a + k_b + k_c = r + R$ where k_a, k_b and k_c are the perpendiculars drawn from the centre of the circumscribed circle to the corresponding sides and r (R) is the radius of the inscribed (circumscribed) circle.

Hint. Express the left-hand and right-hand sides of the required equality in terms of the sides and the angles of the triangle.

362. The vertices A, B and C of a triangle are connected by straight lines with points A_1, B_1, and C_1, arbitrarily placed on the opposite sides (but not at the vertices). Prove that the midpoints of the segments AA_1, BB_1, and CC_1 do not lie in a common straight line.

363. Straight lines DE, FK and MN parallel to the sides AB, AC and BC of a triangle ABC are drawn through an arbitrary point O lying inside the triangle so that the points F and M are on AB, the points E and K are on BC and the points N and D on AC. Prove that

$$\frac{AF}{AB}+\frac{BE}{BC}+\frac{CN}{CA}=1.$$

364. A square is inscribed in a triangle so that one of its sides lies on the longest side of the triangle. Derive the inequality $\sqrt{2}\,r < x < 2\,r$ where x is the length of the side of the square and r is the radius of the inscribed circle of the triangle.

365. Prove that the midpoints of the sides of a triangle, the feet of the altitudes and the midpoints of the segments of the altitudes from the vertices to the orthocentre are nine points of a circle. Show that the centre of this circle lies at the midpoint of the line segment joining the orthocentre of the triangle with the centre of the circumscribed circle and its radius equals half the radius of the circumscribed circle.

366. From the foot of each altitude of a triangle perpendiculars are dropped on the other two sides. Prove the following assertions:

(1) the feet of these perpendiculars are the vertices of a hexagon whose three sides are parallel to the sides of the triangle; (2) it is possible to circumscribe a circle about this hexagon.

367. Prove that in a right triangle the sum of the legs is equal to the sum of the diameters of the inscribed and circumscribed circles.

368. Prove that in a right triangle the bisector of the right angle is simultaneously the bisector of the angle between the median and altitude drawn to the hypotenuse.

369. Two triangles ABC and A_1, B_1 C_1 are symmetric about the centre of their common inscribed circle of radius r. Prove that the product of the areas of the triangles ABC, A_1 B_1 C_1 and of the six other triangles formed by the intersecting sides of the triangles ABC and A_1 B_1 C_1 is equal to r^{16}.

370. Prove that the difference of the sum of the squares of the distances from an arbitrary point M of a plane to two opposite vertices of a parallelogram $ABCD$ in the plane and the sum of the squares of the distances from the same point to the other two vertices is a constant quantity.

371. On the sides of a triangle ABC equilateral triangles ABC_1, BCA_1, and CAB_1, are constructed which do not overlap the triangle ABC. Prove that the straight lines AA_1, BB_1, and CC_1 are concurrent.

372. On the sides *AB*, *AC* and *BC* of a triangle *ABC* as bases three similar isosceles triangles *ABP*, *ACQ* and *BCR* are constructed, the first two triangles lying outside the given triangle and the third being on the same side of *BC* as the triangle *ABC*. Prove that either the figure *APRQ* is a parallelogram or the points *A*, *P*, *R*, *Q* are in a straight line.

373. A point *O* of a plane is connected by straight lines with the vertices of a parallelogram *ABCD* lying in the plane. Prove that the area of the triangle *AOC* is equal to the sum or difference of the areas of two adjacent triangles each of which is formed by two of the straight lines *OA*, *OB*, *OC* and *OD* and the corresponding side of the parallelogram. Consider the cases when the point *O* is inside and outside the parallelogram.

374. In a trapezoid *ABCD* the sum of the base angles *A* and *D* is equal to $\frac{\pi}{2}$. Prove that the line segment connecting the midpoints of the bases equals half the difference of the bases.

375. Prove that the sum of the squares of the diagonals of a trapezoid is equal to the sum of the squares of its sides plus twice the product of the bases.

376. Prove that the straight line joining the midpoints of the bases of a trapezoid passes through the point of intersection of the diagonals.

377. Prove that if the line segment connecting the midpoints of opposite sides of a quadrilateral equals half-sum of the other two sides, then the quadrilateral is a trapezoid.

378. Prove that if the diagonals of two quadrilaterals are respectively equal and intersect at equal angles, then these quadrilaterals have the same area.

379. Prove that at least one of the feet of the perpendiculars drawn from an arbitrary interior point of a convex polygon to its sides lies on the side itself but not on its extension.

380. Prove that the bisectors of the interior angles of a parallelogram form a rectangle whose diagonals are equal to the difference of two adjacent sides of the parallelogram.

381. Given a parallelogram, prove that the straight lines consecutively joining the centres of the squares constructed outside the parallelogram on its sides also form a square.

382. Prove that if in an arbitrary quadrilateral *ABCD* the bisectors of the interior angles are drawn, then the four points at which the bisectors of the angles *A* and *C* intersect the bisectors of the angles *B* and *D* lie on a circle.

383. Two tangent lines are drawn to a circle. Prove that the length of the perpendicular drawn from an arbitrary point of the circle to the chord joining the points of tangency is the mean proportional between the lengths of the perpendiculars drawn from the same point to the tangent lines.

384. Prove that the feet of the perpendiculars dropped from an arbitrary point of a circle onto the sides of the inscribed triangle lie in a straight line.

385. Three equal circles intersect in a point. The other point of intersection of every two of the circles and the centre of the third circle lie on a straight line. Prove that the three straight lines thus specified are concurrent.

386. Two circles are internally tangent at a point A, the segment AB being the diameter of the larger circle. The chord BK of the larger circle is tangent to the smaller circle at a point C. Prove that AC is the bisector of the angle A of the triangle ABK.

387. A circle of radius r is inscribed in a sector of a circle of radius R. The length of the chord of the sector is equal to $2a$. Prove that $\frac{1}{r} = \frac{1}{R} + \frac{1}{a}$.

388. Two tangent lines are drawn to a circle. They intersect a straight line passing through the centre of the circle at points A and B and form equal angles with it. Prove that the product of the line segments AC and BD which are cut off from the given (fixed) tangent lines by any (moving) tangent line is a constant quantity.

389. Prove that the sum of the squares of the lengths of two chords of a circle intersecting at a right angle is greater than the square of the diameter of the circle and the sum of the squares of the four line segments into which the chords are divided by the point of intersection is equal to the square of the diameter.

390. Prove that if a chord of a circle is trisected and the endpoints of the chord and the points of division are joined with the centre of the circle, then the corresponding central angle is divided into three parts one of which is greater than the other two.

391. Prove that if two intersecting chords are drawn from the endpoints of a diameter of a circle, then the sum of the products of each chord by its segment from the endpoint of the diameter to the point of intersection is a constant quantity.

392. From each of two points of a straight line two tangent lines are drawn to a circle. Circles of equal radii are inscribed in the angles thus formed with the vertices at these points. Prove that the centre line of the circles is parallel to the given line.

393. The diameter of a semicircle is divided into two arbitrary parts, and on each part as diameter a semicircle lying inside the given semicircle is constructed. Prove that the area contained between the three semicircular arcs is equal to the area of a circle whose diameter is equal to the length of the perpendicular erected to the diameter of the original semicircle at the point of division.

394. Prove that if two points lie outside a circle and the straight line passing through them does not intersect the circle, then the distance between these two points is greater than the difference between the lengths of the tangent lines drawn from the given points to the circle and less than their sum. Show that either the former or the latter inequality is violated if the straight line intersects the circle.

395. Through the midpoint C of an arbitrary chord AB of a circle two chords KL and MN are drawn, the points K and M lying on one side of AB. Prove that $QC = CP$ where Q is the point of intersection of AB and KN and P is the point of intersection of AB and ML.

396. A circle is arbitrarily divided into four parts, and the midpoints of the arcs thus obtained are connected by line segments.

Show that two of these segments are mutually perpendicular.

397. Prove that for any closed plane polygonal line without self-intersection there exists a circle whose radius is $\frac{1}{4}$ the perimeter of the polygonal line such that none of the points of the polygonal line lies outside this circle.

398. Can a triangle be equilateral if the distances from its vertices to two given mutually perpendicular straight lines are expressed by integers?

399. On one side of a straight line at its points A and B two perpendiculars $AA_1 = a$ and $BB_1 = b$ are erected. Prove that for constant a and b the distance from the point of intersection of the straight lines AB_1 and A_1B to the straight line AB is also constant irrespective of the position of the points A and B.

400. A circle is inscribed in a right angle with point A as vertex, B and C being the points of tangency. Prove that if a tangent line intersecting the sides AB and AC at points M and N is drawn to this circle, then the sum of the lengths of the segments MB and NC is greater than $\frac{1}{3}(AB + AC)$ and less than $\frac{1}{2}(AB + AC)$.

401. Prove that if a circle of radius equal to the altitude of an isosceles triangle rolls upon the base of the triangle, then the length of the arc cut off from the circle by the congruent sides of the triangle remains constant. Is this assertion true for a scalene triangle?

402. Prove that the ratio of the diagonals of an inscribed quadrilateral of a circle is equal to the ratio of the sums of the products of the sides passing through the endpoints of the diagonals.

403. Prove that the sum of the squares of the distances from a point on a circle to the vertices of an equilateral inscribed triangle is a constant independent of the position of the point on the circle.

404. Prove that if a circle is internally tangent to three sides of a quadrilateral and intersects the fourth side, then the sum of the latter and the side opposite to it is greater than the sum of the other two sides of the quadrilateral.

405. Prove that if a circle is internally tangent to three sides of a quadrilateral whose fourth side does not intersect the circle, then the sum of the fourth side and the side opposite it is less than the sum of the other two sides of the quadrilateral.

406. Two equal semicircles whose diameters lie in a common straight line are tangent to each other. Draw a tangent line to them and inscribe a circle tangent to this line and to the two semicircles. Then inscribe another circle tangent to the first one and to the semicircles after which inscribe one more circle tangent to the second one and to the semicircles and so on, unlimitedly.

Using this construction prove that the sum of the fractions

$$\frac{1}{1\times 2}+\frac{1}{2\times 3}+\frac{1}{3\times 4}+\frac{1}{4\times 5}+\ldots+\frac{1}{n\,(n+1)}$$

tends to unity for $n \to \infty$, that is

$$\frac{1}{1\times 2}+\frac{1}{2\times 3}+\ldots+\frac{1}{n\,(n+1)}+\ldots=1.$$

407. An elastic ball of negligible dimensions rests at a point A at a distance a from the centre of circular billiards of radius R. To what point B of the cushion must the ball be directed so that it returns to the point A after being reflected twice from the cushion?

408. A ray of light is issued from a point A lying inside an angle with reflecting sides. Prove that the number of reflection of the ray from the sides is always finite. Determine this number if the angle is equal to α and the initial ray is directed at an angle β to one of the sides. Under what conditions does the reflected ray again pass through the point A?

4. Loci of Points

409. Two fixed points A and B and a moving point M are taken on a circle.On the extension of the line segment AM a segment $MN = MB$ is laid off outside the circle. Find the locus of points N.

410. Given two parallel straight lines and a point O between them. Through this point an arbitrary secant is drawn which intersects the parallel lines at points A and A'. Find the locus of the endpoints of the perpendicular of length OA erected to the secant at the point A'.

411. Find the locus of points for which the sum of their distances from two given straight lines m and l is equal to the length a of a given line segment. Consider the cases of intersecting and parallel lines.

412. Find the locus of points for which the difference of their distances from two given straight lines m and l is equal to a line segment of given length. Consider the cases of parallel and intersecting lines.

413. Two line segments AB and CD are taken in the plane. Find the locus of points M for which the sum of the areas of the triangles AMB and CMD is equal to a constant a^2.

414. Given a circle K and its chord AB. Consider all the inscribed triangles of the circle with given chord as base. Find the locus of orthocentres of these triangles.

415. Inside a given circle a point A not coincident with the centre is fixed. An arbitrary chord passing through the point A is taken, and through its endpoints two tangent lines to the circle intersecting at a point M are drawn. Find the locus of points M.

416. Prove that the locus of points M, for which the ratio of their distances from two given points A and B equals

$$\frac{p}{q} \neq 1,$$

is a circle with centre on the straight line AB.

Express the diameter of this circle in terms of the length a of the line segment AB. Also consider the case

$$\frac{p}{q} = 1.$$

417. Given a line segment AB and a point C on it. Each pair of equal circles one of which passes through the points A and C and the other through the points C and B has, besides C, another common point D. Find the locus of points D.

418. A polygon is deformed in such a way that its sides remain respectively parallel to given directions whereas all its vertices but one slide along given straight lines. Find the locus of positions of that vertex.

419. Given a circle K of radius r and its chord AB whose length is $2a$. Let CD be a moving chord of this circle with length $2b$. Find the locus of points of intersection of the straight lines AC and BD.

420. Through a point P lying in a given circle and a point Q belonging to a given straight line an arbitrary circle is drawn whose second point of intersection with the given circle is R and the point of intersection with the given straight line is S. Prove that all the straight lines RS thus specified have a common point lying on the given circle.

5. The Greatest and Least Values

421. Given two parallel straight lines and a point A between them at distances a and b from the lines. The point A is the vertex of the right angles of the right triangles whose other two vertices lie on either parallel line. Which of the triangles has the least area?

422. Given a right triangle with acute angle α. Find the ratio of the radii of the circumscribed and inscribed circles and determine the value of α for which this ratio attains its minimum.

423. A right triangle with legs a_1 and b_1 is cut off from a quadrilateral with sides a and b. How must the quadrilateral of maximum area with sides parallel to those of the initial quadrilateral be cut off from the remaining part of the quadrilateral?

424. Two points A and B are taken on a side of an acute angle. Find a point C on the other side of the angle such that the angle ACB attains its maximum value. Make a ruler and compass construction of the point C.

425. On a given straight line l find a point for which the difference of its distances from two given points A and B lying on one side of the straight line attains its minimum value, and also a point such that this difference attains the maximum value.

426. Through a point A inside an angle a straight line is drawn which cuts off from the angle a triangle with the least area. Prove that the segment of this line between the sides of the angle is bisected at the point A.

427. Prove that among all triangles with common vertex angle φ and given sum $a + b$ of the lengths of the sides including this angle the isosceles triangle has the least base.

428. Among all triangles with equal bases and the same vertex angle find the triangle having the greatest perimeter.

429. In a triangle ABC an arbitrary point D is taken on the base BC or on its extension, and circles are circumscribed about the triangles ACD and BCD. Prove that the ratio of the radii of these circles is a constant quantity. Find the position of the point D for which these radii attain their least values.

430. Cut off two equal circles having the greatest radius from a given triangle.

B. SOLID GEOMETRY

Preliminaries

Here is a number of formulas to be used for computing volumes and surface areas of polyhedrons and solids of revolution, the notation being as follows: V, volume; S_{lat}, lateral surface area; S, area of base; H, altitude.

Pyramid: $V = \frac{SH}{3}$.

Frustum of a pyramid:

$V = \frac{H}{3}(S_1 + S_2 + \sqrt{S_1 S_2})$, where S_1 and S_2 are the areas of the upper and lower bases.

Right circular cone: $V = \frac{\pi R^2 H}{3}$, where R is the radius of the base; $S_{lat} = \pi R l$, where l is the slant height.

Right circular cylinder : $V = \pi R^2 H$, where R is the radius of the base; $S_{lat} = 2\pi RH$.

Frustum of a cone : $V = \frac{\pi H}{3}(R_1^2 + R_2^2 + R_1 R_2)$, where R_1 and R_2 are the radii of the bases; $S_{lat} = \pi(R_1 + R_2)l$, where l is the slant height.

Sphere : $V = \frac{4}{3}\pi R^3$; $S = 4\pi R^2$, where R is the radius of the sphere.

Spherical sector : $V = \frac{2\pi R^2 h}{3}$,where R is the radius of the sphere and h is the altitude of the zone forming the base of the sector.

Spherical segment : $V = \frac{1}{3}\pi h^2(3R - h)$; $S_{lat} = 2\pi Rh$, where R is the radius of the sphere and h is the altitude of the segment.

1. Computation Problems

431. The volume of a regular triangular prism is equal to V and the angle between the diagonals of two faces drawn from one vertex is equal to α. Find the side of the base of the prism.

432. From the vertex S of a regular quadrangular pyramid the perpendicular SB is dropped on the base. From the midpoint O of the line segment SB the perpendicular OM of length h is drawn to a lateral edge and the perpendicular OK of length b is dropped on a lateral face. Compute the volume of the pyramid.

433. Find the lateral area of a regular n-gonal pyramid of volume V if the radius of the inscribed circle of its base is equal to the radius of the circumscribed circle of the parallel section drawn at a distance h from the base.

434. A regular pentagonal pyramid $SABCDE$ is intersected by the plane passing through the vertices A and C of the base and the midpoints of the lateral edges DS and ES. Find the area of the section if the length of the side of the base is equal to q and the length of the lateral edge is equal to b.

435. A regular triangular pyramid is cut by the plane passing through a vertex of the base and the midpoints of two lateral edges. Find the ratio of the lateral area of the pyramid to the area of the base if it is known that the cutting plane is perpendicular to the lateral face opposite that vertex.

436. A pyramid of total surface area S is cut off from a regular quadrangular prism by a plane passing through a diagonal of the lower base and a vertex of the upper base. Find the total surface area of the prism if the vertex angle of the triangle in the section is equal to α.

437. Compute the volume of a regular triangular pyramid knowing that the face angle at the vertex is equal to α and the radius of the circumscribed circle of the lateral face is equal to r.

438. A regular quadrangular pyramid with side of its base equal to a is cut by a plane bisecting its dihedral angle at the base which is equal to 2α. Find the area of the section.

439. Above the plane ceiling of a hall having the form of a square with side a a roof is made which is constructed in the following way: each pair of adjacent vertices of the square forming the ceiling is joined by straight lines with the midpoint of the opposite side and on each of the four triangles thus obtained a pyramid is constructed whose vertex is projected into the midpoint of the corresponding side of the square. The elevated parts of the faces of the four pyramids form the roof. Find the volume of the garret (i.e. the space between the ceiling and the roof) if the altitude of each pyramid is equal to h.

440. Find the dihedral angle formed by two lateral faces of a regular triangular pyramid if the dihedral angle formed by its lateral face and base is equal to α.

441. In a regular triangular pyramid $SABC$ the face angle at the vertex is equal to α and the shortest distance between a lateral edge and the opposite side of the base is equal to d. Find the volume of the pyramid.

442. The base of a pyramid is an isosceles trapezoid in which the lengths of the bases are equal to a and b ($a > b$) and the angle between the diagonals subtended by its lateral side is equal to φ. Find the volume of the pyramid if its altitude dropped from the vertex passes through the point of intersection of the diagonals of the base and the ratio of the dihedral angles whose edges are the parallel sides of the base is $2 : 1$.

443. An angle BAC of 60° is taken in a plane P. The distances from a point S to the vertex A, the side AB and the side AC are irrespectively 25 cm, 7 cm and 20 cm. Find the distance between the point S and the plane P.

444. A regular hexagonal pyramid with face angle at the vertex equal to α is intersected by a plane passing at an angle β to the base through its longest diagonal. Find the ratio of the area of the plane section to the area of the base.

445. All the three face angles of a trihedral angle are acute and one of them is equal to α. The dihedral angles whose edges are the sides of this face angle are equal to β and γ respectively. Find the other two face angles.

446. Compute the volume of a regular pyramid of altitude h knowing that its base is a polygon for which the sum of the interior angles is equal to $n\pi$ and the ratio of the lateral area of the pyramid to the area of the base is equal to k.

447. Consider a cube with edge a. Through the endpoints of each triple of concurrent edges a plane is drawn. Find the volume of the solid bounded by these planes.

448. A regular hexahedral pyramid is intersected by a plane parallel to its lateral face and passing through the centre of the base. Find the ratio of the area of the plane section to the area of the lateral face.

449. Through each edge of a tetrahedron a plane parallel to the opposite edge is drawn. Find the ratio of the volume of the parallelepiped thus formed to the volume of the tetrahedron.

450. On the lateral faces of a regular quadrangular pyramid as bases regular tetrahedrons are constructed. Find the distance between the exterior vertices of two adjacent tetrahedrons if the side of the base of the pyramid is equal to a.

451. Through a point on a diagonal of a cube with edge a a plane is drawn perpendicularly to this diagonal.

(1) What polygon is obtained in the section of the faces of the cube by the plane?

(2) Find the lengths of the sides of this polygon depending on the distance x from the centre of symmetry O of the cube to the cutting plane.

452. Consider the projection of a cube with edge a onto a plane perpendicular to a diagonal of the cube. What is the ratio of the area of this projection to the area of the section of the cube by the plane passing through the midpoint of the diagonal perpendicularly to it?

453. Given a regular quadrangular pyramid with altitude h and side of the base a. Through a side of the base of the pyramid and the midpoint of a lateral edge not intersecting this side the plane section is drawn. Determine the distance from the vertex of the pyramid to the cutting plane.

454. Given a regular tetrahedron $SABC$ with edge a. Through the vertices of the base ABC of the tetrahedron three planes are drawn each of which passes through the midpoints of two lateral edges. Find the volume of the portion of the tetrahedron lying above the three cutting planes.

455. A rhombus with diagonals $AC = a$ and $BD = b$ is the base of a pyramid $SABCD$. The lateral edge SA of length q is perpendicular to the base. Through the point A and the midpoint K of the edge SC a plane parallel to the diagonal BD of the base is drawn. Determine the area of the plane section thus obtained.

456. In a regular quadrangular prism two parallel plane sections are drawn. One of them passes through the midpoints of two adjacent sides of the base and the midpoint of the axis of the prism and the other divides the axis in the ratio 1 : 3. Knowing that the area of the former section is S, find the area of the latter.

457. A triangular pyramid is cut by a plane into two polyhedrons. Find the ratio of volumes of these polyhedrons if it is known that the cutting plane divides three concurrent lateral edges of the pyramid so that the ratios of the segments of these edges adjacent to the common vertex to the remaining parts of the edges are 1 : 2, 1 : 2 and 2 : 1.

458. Find the volume of a triangular pyramid if the areas of its faces are S_0, S_1, S_2 and S_3, and the dihedral angles adjacent to the face with area S_0 are equal.

459. In a cube with edge a through the midpoints of two parallel edges not lying in one face a straight line is drawn, and the cube is turned about it by 90°. Determine the volume of the common portion of the initial and turned cubes.

460. Through the vertex of a cone a plane is drawn at an angle α to the base of the cone. This plane intersects the base along the chord AB of length a subtending an arc of the base of the cone with central angle β. Find the volume of the cone.

461. A cone and a cylinder have a common base, and the vertex of the cone is in the centre of the other base of the cylinder. Find the angle between the axis of the cone and its element if the ratio of the total surface area of the cylinder to the total surface area of the cone is 7 : 4.

462. A cylinder is inscribed in a cone, the altitude of the cylinder being equal to the radius of the base of the cone. Find the angle between the axis of the cone and its element if the ratio of the total surface area of the cylinder to the area of the base of the cone is 3 : 2.

463. In a cone with slant height l and element inclined to the base at an angle α a regular n-gonal prism whose all edges are congruent is inscribed. Find the total surface area of the prism.

464. The four sides of an isosceles trapezoid are tangent to a cylinder whose axis is perpendicular to the bases of the trapezoid. Find the angle between the plane of the trapezoid and the axis of the cylinder if the lengths of the bases of the trapezoid are respectively equal to a and b and the altitude of the trapezoid is equal to h.

465. A sphere is inscribed in a right prism whose base is a right triangle. In this triangle a perpendicular of length h dropped from the vertex of the right angle on the hypotenuse forms an angle α with a leg of the triangle. Find the volume of the prism.

466. In a regular n-gonal pyramid with side of the base a and lateral edge b a sphere is inscribed. Find its radius.

467. A sphere is inscribed in a regular triangular pyramid. Determine the angle between its lateral edge and the base if the ratio of the volume of the pyramid to the volume of the sphere is equal to $\frac{27\sqrt{3}}{4\pi}$.

468. About a sphere of radius r a regular n-gonal pyramid with dihedral angle at the base α is circumscribed. Find the ratio of the volume of the sphere to that of the pyramid.

469. Find the ratio of the volume of a regular n-gonal pyramid to the volume of its inscribed sphere, knowing that the circumscribed circles of the base and lateral faces of the pyramid are of the same radius.

470. Find the altitude of a regular quadrangular pyramid if it is known that the volume of its circumscribed sphere is equal to V and the perpendicular drawn from the centre of the sphere to its lateral face forms with the altitude of the pyramid an angle α.

471. A sphere of radius R is inscribed in a pyramid whose base is a rhombus with acute angle α. The lateral faces of the pyramid are inclined to the plane of the base at an angle ψ. Find the volume of the pyramid.

472. The congruent bases of two regular n-gonal pyramids are made coincident. Find the radius of the inscribed sphere of the polyhedron thus obtained if the sides of the bases of the pyramids are equal to a and their altitudes are equal to h and H respectively.

473. The congruent bases of two regular n-gonal pyramids are made coincident, the altitudes of the pyramids being different. Determine these altitudes if the radius of the circumscribed sphere of the polyhedron thus formed is equal to R and the sides of the bases of the pyramids are equal to a. What is the relationship between the values of a and R for which the problem is solvable?

474. An inscribed sphere of a regular n-gonal prism touches all the faces of the prism. Another sphere is circumscribed about the prism. Find the ratio of the volume of the latter to that of the former.

475. A regular tetrahedron is inscribed in a sphere, and another sphere is inscribed in the tetrahedron. Find the ratio of the surface areas of the spheres.

476. A sphere is inscribed in a regular tetrahedron, and another regular tetrahedron is inscribed in the sphere. Find the ratio of the volumes of the tetrahedrons.

477. Given two concentric spheres of radii r and $R (R > r)$. What relationship connects R and r if it is possible to construct a regular tetrahedron inside the larger sphere so that the three vertices of its base lie on the larger sphere and the three lateral faces are tangent to the smaller sphere?

478. A plane dividing a cube into two parts passes through two opposite vertices of the cube and the midpoints of the six edges not containing these vertices. Into each part of the cube a sphere is placed so that it is tangent to three faces of the cube and the cutting plane. Find the ratios of the volume of the cube to the volumes of the spheres.

479. From a point on a sphere of radius R three equal chords are drawn at an angle α to one another. Find the length of these chords.

480. In a triangular pyramid $SABC$ the edges SA, SC and SB are pairwise perpendicular, $AB = BC = a$ and $BS = b$. Find the radius of the inscribed sphere of the pyramid.

481. Find the dihedral angle φ formed by the base of a regular quadrangular pyramid and its lateral face if the radius of the circumscribed sphere of the pyramid is three times that of the inscribed sphere.

482. In a sphere of radius R a regular tetrahedron is inscribed, and all its faces are extended to intersect the sphere. The lines of intersection of the faces of the tetrahedron with the sphere cut off from its surface four spherical triangles and several spherical lunes. Compute the areas of these spherical parts.

483. A sphere is inscribed in a cone. The ratio of the surface area of the sphere to the area of the base of the cone is 4 : 3. Find the vertex angle of the axial section of the cone.

484. A hemisphere is inscribed in a cone so that its great circle lies in the base of the cone. Determine the vertex angle of the axial section of the cone if the ratio of the total surface area of the cone to the surface area of the hemisphere is 18 : 5.

485. In a sphere of radius R a cone is inscribed whose lateral area is k times the area of its base. Find the volume of the cone.

486. The ratio of the altitude of a cone to the radius of its circumscribed sphere is equal to q. Find the ratio of the volumes of these solids. For what q is the problem solvable?

487. Find the ratio of the volume of a sphere to that of a right cone circumscribed about the sphere if the total surface of the cone is n times the surface area of the sphere.

488. Determine the radii of the bases of a frustum of a cone circumscribed about a sphere of radius R knowing that the ratio of the total surface area of the frustum to the surface area of the sphere is equal to m.

489. A sphere of radius r is inscribed in a cone. Find the volume of the cone knowing that the distance from the vertex of the cone to the tangent plane to the sphere which is perpendicular to an element of the cone is equal to d.

490. A sphere of radius R is inscribed in a cone with vertex angle of its axial section equal to α. Find the volume of the part of the cone above the sphere.

491. Determine the radii of two intersecting spheres forming biconvex lense with thickness $2a$, total surface area S and diameter $2R$.

492. A sphere is inscribed in a cone, the ratio of their volumes being equal to k. Find the ratio of the volumes of the spherical segments cut off from the sphere by the plane passing through the line of tangency of the sphere and cone.

493. In a sphere S of radius R eight equal spheres of smaller radius are inscribed so that each of them is tangent to two adjacent spheres and all the eight spheres touch the given sphere S along its great circle. Then in the space between the spheres a sphere S_1 is placed which touches all the spheres of smaller radius and the sphere S. Find the radius ρ of the sphere S_1.

494. In a sphere S of radius R eight equal spheres are inscribed each of which is tangent to three adjacent spheres and the given one. Find the radius of the inscribed spheres if their centres are at the vertices of a cube.

495. In a sphere two equal cones with coinciding axes are inscribed whose vertices are at the opposite endpoints of a diameter of the sphere. Find the ratio of the volume of the common portion of the cones to that of the sphere knowing that the ratio of the altitude h of each cone to the radius R of the sphere is equal to k.

496. The areas of two parallel plane sections of a sphere drawn on one side of its centre are equal to S_1 and S_2, and the distance between them is d. Find the area of the section parallel to the two given sections and equidistant from them.

497. Three equal spheres of radius R tangent to one another lie on a plane P. A right circular cone with its base in P is externally tangent to the spheres. Find the radius of the base of the cone if its altitude is equal to qR.

498. Given four equal spheres of radius R each of which is tangent to the other three. A fifth sphere is externally tangent to each given sphere, and one more sphere is internally tangent to them. Find the ratio of the volume V_6 of the sixth sphere to the volume V_5 of the fifth.

499. Three equal pairwise tangent spheres of radius R lie on a plane. A fourth sphere is tangent to the plane and to each given sphere. Find the radius of the fourth sphere.

500. Four equal spheres of radius R lie on a plane. Three of them are pairwise tangent, and the fourth sphere touches two of these three. Two equal tangent spheres of smaller radius are placed above these spheres so that each of them touches three larger spheres. Find the ratio of the radius of a larger sphere to that at a smaller.

2. Proof Problems

501. Given a frustum of a cone with lateral area equal to the area of a circle whose radius is equal to the slant height of the frustum. Prove that it is possible to inscribe a sphere in the frustum.

502. Given a frustum of a cone whose altitude is the mean proportional between the diameters of the bases. Prove that it is possible to inscribe a sphere in the given frustum.

503. Prove that the straight lines joining three vertices of a regular tetrahedron to the midpoint of the altitude dropped from the fourth vertex are pairwise perpendicular.

504. Let R be the radius of the circumscribed sphere of a regular quadrangular pyramid, and r be the radius of the inscribed sphere. Prove that

$$\frac{R}{r} \geq \sqrt{2} + 1.$$

Hint. Express $\frac{R}{r}$ in terms of $\tan \frac{\alpha}{2}$ where α is the dihedral angle between the base of the pyramid and its lateral face.

505. From a point O in the base ABC of a triangular pyramid $SABC$ are drawn the straight lines OA', OB' and OC' respectively parallel to the edges SA, SB and SC which intersect the faces SBC, SCA and SAB at points A', B' and C'. Prove that $\frac{OA'}{SA} + \frac{OB'}{SB} + \frac{OC'}{SC} = 1.$

506. Consider two triangles ABC and $A_1B_1C_1$ with pairwise nonparallel sides lying in intersecting planes. The straight lines joining the corresponding vertices of the triangles intersect in one point O. Prove that the extensions of the corresponding sides of the triangles are pairwise concurrent and the points of intersection lie in a straight line.

507. Show that the line segments joining the vertices of a triangular pyramid to the centroids of the opposite faces meet in one point and are divided by this point in the ratio 1 : 3.

508. Show that the area of any triangular section of an arbitrary triangular pyramid does not exceed the area of at least one of its faces.

509. One of two triangular pyramids with common base is inside the other. Prove that the sum of the face angles at the vertex of the interior pyramid is greater than that of the exterior one.

510. Four spheres with non-coplanar centres are pairwise tangent to one another. For every two spheres a common tangent plane is drawn perpendicularly to their centre line. Prove that the six planes thus constructed have a common point.

511. Prove that if the sums of the lengths of any pair of opposite edges of a triangular pyramid are equal, then the vertices of the pyramid are the centres of four pairwise tangent spheres.

512. What condition on the radii of three pairwise tangent spheres guarantees the existence of a common tangent plane to the spheres?

513. Prove that if a point moves inside the base of a regular pyramid in its plane, then the sum of the distances from this point to the lateral faces remains constant.

514. Prove that two planes drawn through the endpoints of two triples of edges of a parallelepiped meeting in the endpoints of a diagonal of the parallelepiped trisect this diagonal.

515. Show that if a plane drawn through the endpoints of three edges of a parallelepiped meeting in one vertex cuts off a regular tetrahedron from the parallelepiped, then the latter can be intersected by a plane so that the section is a regular hexagon.

516. Prove that every plane passing through the midpoints of two opposite edges of a tetrahedron divides this tetrahedron into two parts of equal volumes.

517. Prove that if all dihedral angles of a triangular pyramid are equal, all the edges of the pyramid are also equal.

518. The endpoints of two line segments AB and CD lying in two parallel planes are the vertices of a triangular pyramid. Prove that the volume of the pyramid does not change when the segments are translated in these planes.

519. Prove that a straight line intersecting the two faces of a dihedral angle forms equal angles with them if and only if the points of intersection are equidistant from the edge.

520. Consider two line segments AB and CD not lying in one plane. Let MN be the line segment joining their midpoints. Prove that

$$\frac{AD+BC}{2} > MN$$

where AD, BC and MN designate the lengths of the corresponding segments.

521. Prove that every face angle of an arbitrary tetrahedral angle is less than the sum of the other three face angles.

522. Prove that any convex tetrahedral angle can be intersected by a plane so that the section is a parallelogram.

523. Prove that if the faces of a triangular pyramid are of the same area, they are congruent.

3. Loci of Points

524. Find the locus of projections of a point in space on planes passing through another fixed point.

525. Find the locus of centres of the sections of a sphere by the planes passing through a given straight line l. Consider the cases when the line and the sphere intersect, are tangent or have no points in common.

526. Find the locus of centres of the sections of a sphere by the planes passing through a given point C. Consider the cases when the point is outside the sphere, on its surface or inside it.

527. Find the locus of points from which it is possible to draw three tangent lines to a given sphere of radius R which are the edges of a trihedral angle with three right face angles.

528. Find the locus of feet of the perpendiculars dropped from a given point in space on the straight lines lying in a given plane and intersecting in one point.

529. Given a plane P and two points A and B not lying in it. Consider all the possible spheres tangent to the plane P and passing through A and B. Find the locus of points of tangency.

530. A trihedral angle is intersected by a plane, a triangle ABC being the section. Find the locus of the centroids of triangles ABC on condition that

(a) vertices A and B are fixed; (b) vertex A is fixed.

4. The Greatest and Least Values

531. A cube is intersected by a plane passing through its diagonal. How must this plane be drawn to obtain the section of the least area?

532. A triangular pyramid is intersected by the planes parallel to two nonintersecting edges. Find the section having the greatest area.

TRIGONOMETRY

TRIGONOMETRY

Preliminaries

Here are some formulas to be used in the suggested problems.

1. Addition and subtraction formulas:

$$\sin(x+y) = \sin x \cos y + \cos x \sin y, \tag{1}$$

$$\sin(x-y) = \sin x \cos y - \cos x \sin y, \tag{2}$$

$$\cos(x+y) = \cos x \cos y - \sin x \sin y, \tag{3}$$

$$\cos(x-y) = \cos x \cos y + \sin x \sin y. \tag{4}$$

2. Double-angle and triple-angle formulas:

$$\sin 2x = 2\sin x \cos x, \tag{5}$$

$$\cos 2x = \cos^2 x - \sin^2 x, \tag{6}$$

$$\sin 3x = 3\sin x - 4\sin^3 x, \tag{7}$$

$$\cos 3x = 4\cos^3 x - 3\cos x. \tag{8}$$

3. Sum and difference of trigonometric functions:

$$\sin x + \sin y = 2\sin\frac{x+y}{2}\cos\frac{x-y}{2}, \tag{9}$$

$$\sin x - \sin y = 2\cos\frac{x+y}{2}\sin\frac{x-y}{2}, \tag{10}$$

$$\cos x + \cos y = 2\cos\frac{x+y}{2}\cos\frac{x-y}{2}, \tag{11}$$

$$\cos x - \cos y = 2\sin\frac{x+y}{2}\sin\frac{y-x}{2}. \tag{12}$$

4. Product formulas:

$$\sin x \sin y = \frac{1}{2}[\cos(x-y) - \cos(x+y)], \tag{13}$$

$$\cos x \cos y = \frac{1}{2}[\cos(x-y) + \cos(x+y)], \tag{14}$$

$$\sin x \cos y = \frac{1}{2}[\sin(x-y) + \sin(x+y)], \tag{15}$$

$$\sin^2 x = \frac{1-\cos 2x}{2}, \tag{16}$$

$$\cos^2 x = \frac{1+\cos 2x}{2}. \tag{17}$$

5. Expressing $\sin x$, $\cos x$ and $\tan x$ in terms of $\tan\frac{x}{2}$:

$$\sin x=\frac{2\tan\frac{x}{2}}{1+\tan^2\frac{x}{2}}, \tag{18}$$

$$\cos x=\frac{1-\tan^2\frac{x}{2}}{1+\tan^2\frac{x}{2}}, \tag{19}$$

$$\tan x=\frac{2\tan\frac{x}{2}}{1-\tan^2\frac{x}{2}}. \tag{20}$$

6. Inverse trigonometric functions.

(a) Principal values of inverse trigonometric functions:

$$y=\text{arc}\sin x,\ \text{if } x=\sin y \text{ and } -\frac{\pi}{2}\le y\le\frac{\pi}{2}, \tag{21}$$

$$y=\text{arc}\cos x \text{ if } x=\cos y \text{ and } 0\le y\le\pi, \tag{22}$$

$$y=\text{arc}\tan x \text{ if } x=\tan y \text{ and } -\frac{\pi}{2}< y<\frac{\pi}{2}, \tag{23}$$

$$y=\text{arc}\cot x \text{ if } x=\cot y \text{ and } 0< y<\pi. \tag{24}$$

(b) Multiple-valued functions:

$$\text{Arc}\sin x=(-1)^n\,\text{arc}\sin x+\pi n,\ n=0,\pm 1,\pm 2,\ldots, \tag{25}$$

$$\text{Arc}\cos x=\pm\,\text{arc}\cos x+2\pi n, \tag{26}$$

$$\text{Arc}\tan x=\text{arc}\tan x+\pi n, \tag{27}$$

$$\text{Arc}\cot x=\text{arc}\cot x+\pi n. \tag{28}$$

Formulas (25) to (28) determine the general expressions for the angles corresponding to given values of trigonometric functions.

1. Transforming Expressions Containing Trigonometric Functions

533. Prove the identity

$$\sin^6 x+\cos^6 x=1-\frac{3}{4}\sin^2 2x.$$

534. Prove the identity

$$\cos^2\alpha+\cos^2(\alpha+\beta)-2\cos\alpha\cos\beta\cos(\alpha+\beta)$$
$$=\sin^2\beta.$$

535. Prove that

$$\tan x + \tan 2x - \tan 3x = -\tan x \tan 2x \tan 3x$$

for all permissible values of x.

536. Prove that the equality

$$\tan 3x = \tan x \tan\left(\frac{\pi}{3} - x\right)\tan\left(\frac{\pi}{3} + x\right)$$

for all permissible values of x.

537. Prove the identity

$$\sin\alpha + \sin\beta + \sin\gamma - \sin(\alpha+\beta+\gamma) = 4\sin\frac{\alpha+\beta}{2}\sin\frac{\beta+\gamma}{2}\sin\frac{\gamma+\alpha}{2}.$$

538. Prove that

$$\sin\alpha + \sin\beta + \sin\gamma = 4\cos\frac{\alpha}{2}\cos\frac{\beta}{2}\cos\frac{\gamma}{2}$$

if $\alpha+\beta+\gamma=\pi$.

539. For $\alpha+\beta+\gamma=\pi$ prove the identity

$\sin 2n\alpha + \sin 2n\beta + \sin 2n\gamma = (-1)^{n+1}\, 4\sin n\alpha \sin n\beta \sin n\gamma$ where n is an integer.

540. Prove that if $\cos(\alpha+\beta)=0$ then

$$\sin(\alpha+2\beta) = \sin\alpha.$$

541. Prove that if $3\sin\beta = \sin(2\alpha+\beta)$ then

$$\tan(\alpha+\beta) = 2\tan\alpha$$

for all permissible values of α and β.

542. Prove that if $\sin\alpha = A\sin(\alpha+\beta)$ then

$$\tan(\alpha+\beta) = \frac{\sin\beta}{\cos\beta - A}$$

for all permissible values of α and β.

543. Prove that if the angles α and β satisfy the relation

$$\frac{\sin\beta}{\sin(2\alpha+\beta)} = \frac{n}{m}\ (|m|>|n|),$$

then

$$\frac{1+\dfrac{\tan\beta}{\tan\alpha}}{m+n} = \frac{1-\tan\alpha\cdot\tan\beta}{m-n}.$$

544. Prove that if $\cos x\cdot\cos y\cdot\cos z \neq 0$, the formula

$$\cos(x+y+z) = \cos x\cos y\cos z(1-\tan x\tan y - \tan y\tan z - \tan z\tan x)$$ holds true.

545. Prove that if α, β, γ are the angles of a triangle then

$$\tan\frac{\alpha}{2}\tan\frac{\beta}{2}+\tan\frac{\beta}{2}\tan\frac{\gamma}{2}+\tan\frac{\gamma}{2}\tan\frac{\alpha}{2}=1.$$

546. Let $x+y+z=\frac{\pi}{2}k$. For what integral k is the sum

$\tan y\tan z+\tan z\tan x+\tan x\tan y$ independent of x, y and z?

547. Find the algebraic relation between the angles α, β and γ if

$\tan\alpha+\tan\beta+\tan\gamma=\tan\alpha\tan\beta\tan\gamma.$

548. Rewrite as a product the expression

$$\cot^2 2x-\tan^2 2x-8\cos 4x\cot 4x.$$

549. Transform into a product the expression

$$\sin^2\alpha+\sin^2\beta+\sin^2\gamma+2\cos\alpha\cos\beta\cos\gamma-2.$$

550. Compute

$$\frac{1}{2\sin 10^\circ}-2\sin 70^\circ$$

without using tables.

551. Prove that

$$\cos\frac{\pi}{5}-\cos\frac{2\pi}{5}=\frac{1}{2}.$$

552. Prove that

$$\cos\frac{2\pi}{7}+\cos\frac{4\pi}{7}+\cos\frac{6\pi}{7}=-\frac{1}{2}.$$

553. Compute

$$\sin^4\frac{\pi}{16}+\sin^4\frac{3\pi}{16}+\sin^4\frac{5\pi}{16}+\sin^4\frac{7\pi}{16}$$

without using tables.

554. Prove that $\tan 20^\circ\tan 40^\circ\tan 80^\circ=\sqrt{3}.$

2. Trigonometric Equations and Systems of Equations

A. TRIGONOMETRIC EQUATIONS

555. Solve the equation

$$\sin^3 x\cos x-\sin x\cos^3 x=\frac{1}{4}.$$

556. Solve the equation

$$\frac{1-\tan x}{1+\tan x}=1+\sin 2x.$$

557. Solve the equation

$$1+\sin x+\cos x+\sin 2x+\cos 2x=0.$$

558. Solve the equation

$$1+\sin x+\cos 3x=\cos x+\sin 2x+\cos 2x.$$

559. Solve the equation

$$(\sin 2x+\sqrt{3}\cos 2x)^2-5=\cos\left(\frac{\pi}{6}-2x\right).$$

560. Solve the equation

$$2\sin 17x+\sqrt{3}\cos 5x+\sin 5x=0.$$

561. Solve the equation

$$\sin^2 x(\tan x+1)=3\sin x(\cos x-\sin x)+3.$$

562. Solve the equation

$$\sin^3 x+\cos^3 x=1-\frac{1}{2}\sin 2x.$$

563. Solve the equation

$$\frac{1}{\sin^2 x}-\frac{1}{\cos^2 x}-\frac{1}{\tan^2 x}-\frac{1}{\cot^2 x}-\frac{1}{\sec^2 x}-\frac{1}{\csc^2 x}=-3.$$

564. Solve the equation

$$\sin^4\frac{x}{3}+\cos^4\frac{x}{3}=\frac{5}{8}.$$

565. Solve the equation

$$\frac{1}{2}(\sin^4 x+\cos^4 x)=\sin^2 x\cos^2 x+\sin x\cos x.$$

566. Solve the equation

$$(1+k)\cos x\cos(2x-\alpha)=(1+k\cos 2x)\cos(x-\alpha).$$

567. Solve the equation

$\sin ax\sin bx=\sin cx\sin dx,$

where a, b, c and d are consecutive positive terms of an arithmetic progression.

568. Solve the equation

$$2+\cos x=2\tan\frac{x}{2}.$$

569. Solve the equation

$$\cot x-2\sin 2x=1.$$

570. Find $\tan x$ from the equation

$$2\cos x\cos(\beta-x)=\cos\beta.$$

571. Find $\cos\varphi$ if

$$\sin\alpha+\sin(\varphi-\alpha)+\sin(2\varphi+\alpha)=\sin(\varphi+\alpha)+\sin(2\varphi-\alpha)$$

and the angle φ is in the third quadrant.

572. Find $\cot x$ from the equation

$$\cos^2(\alpha+x)+\cos^2(\alpha-x)=a,$$

where $0<a<2$. For what α is the problem solvable?

573. Find $\tan\frac{\alpha}{2}$ if $\sin\alpha+\cos\alpha=\frac{\sqrt{7}}{2}$ and the angle α lies between 0° and 45°.

574. Solve the equation

$$\sin 2x-12(\sin x-\cos x)+12=0.$$

575. Solve the equation

$$1+2\csc x=-\frac{\sec^2\frac{x}{2}}{2}.$$

576. Solve the equation

$$\cot^2 x=\frac{1+\sin x}{1+\cos x}.$$

577. Solve the equation

$$2\tan 3x-3\tan 2x=\tan^2 2x\tan 3x.$$

578. Solve the equation

$$2\cot 2x-3\cot 3x=\tan 2x.$$

579. Solve the equation

$$6\tan x+5\cot 3x=\tan 2x.$$

580. Solve the equation

$$\sin^5 x-\cos^5 x=\frac{1}{\cos x}-\frac{1}{\sin x}.$$

581. Solve the equation

$$\tan\left(x-\frac{\pi}{4}\right)\tan x\tan\left(x+\frac{\pi}{4}\right)=\frac{4\cos^2 x}{\tan\frac{x}{2}-\cot\frac{x}{2}}.$$

582. For what a is the equation

$\sin^2 x-\sin x\cos x-2\cos^2 x=a$ solvable? Find the solutions.

583. Determine all the values of a for which the equation

$$\sin^4 x-2\cos^2 x+a^2=0$$

is solvable. Find the solutions.

584. Solve the equation

$$\cos\pi\frac{x}{31}\cos 2\pi\frac{x}{31}\cos 4\pi\frac{x}{31}\cos 8\pi\frac{x}{31}\cos 16\pi\frac{x}{31}=\frac{1}{32}.$$

585. Solve the equation

$$\cos 7x - \sin 5x = \sqrt{3}\,(\cos 5x - \sin 7x).$$

586. Solve the equation

$$2 - (7 + \sin 2x)\sin^2 x + (7 + \sin 2x)\sin^4 x = 0.$$

587. Find $\sin x$ and $\cos x$ if

$$a\cos x + b\sin x = c.$$

What condition connecting a, b and c guarantees the solvability of the problem?

588. Solve the equation

$$\frac{a\sin x + b}{b\cos x + a} = \frac{a\cos x + b}{b\sin x + a}\ (a^2 \neq 2b^2).$$

589. Solve the equation

$$32\cos^6 x - \cos 6x = 1.$$

590. Solve the equation

$$8\sin^6 x + 3\cos 2x + 2\cos 4x + 1 = 0.$$

591. Solve the equation

$$\cos 3x \cos^3 x + \sin 3x \sin^3 x = 0.$$

592. Solve the equation

$$\sin^8 x + \cos^8 x = \frac{17}{32}.$$

593. Solve the equation

$$\sin^{10} x + \cos^{10} x = \frac{29}{16}\cos^4 2x.$$

594. Solve the equation

$$\sin^3 x + \sin^3 2x + \sin^3 3x = (\sin x + \sin 2x + \sin 3x)^3.$$

595. Solve the equation

$$\sin^{2n} x + \cos^{2n} x = 1,$$

where n is a positive integer.

596. Solve the equation

$$\sin\left(\frac{\pi}{10} + \frac{3x}{2}\right) = 2\sin\left(\frac{3\pi}{10} - \frac{x}{2}\right).$$

597. Solve the equation

$$(\cos 4x - \cos 2x)^2 = \sin 3x + 5.$$

598. Solve the equation

$$(\sin x + \cos x)\sqrt{2} = \tan x + \cot x.$$

599. Prove that the equation $(\sin x + \sqrt{3}\cos x)\sin 4x = 2$ has no solutions.

600. Determine the range of the values of the parameter λ for which the equation

$$\sec x + \csc x = \lambda$$

possesses a root x satisfying the inequality $0 < x < \frac{\pi}{2}$.

B. SYSTEMS OF EQUATIONS

601. Find all solutions of the system of equations

$$\left.\begin{aligned} \sin(x+y) &= 0, \\ \sin(x-y) &= 0, \end{aligned}\right\}$$

satisfying the conditions $0 \le x \le \pi$ and $0 \le y \le \pi$.

602. Solve the system of equations

$$\left.\begin{aligned} \sin x &= \csc x + \sin y, \\ \cos x &= \sec x + \cos y. \end{aligned}\right\}$$

603. Solve the system of equations

$$\left.\begin{aligned} \sin^3 x &= \frac{1}{2}\sin y, \\ \cos^3 x &= \frac{1}{2}\cos y. \end{aligned}\right\}$$

604. Solve the system of equations

$$\left.\begin{aligned} \tan x + \tan y &= 1, \\ \cos x \cos y &= \frac{1}{\sqrt{2}}. \end{aligned}\right\}$$

605. Solve the system of equations

$$\left.\begin{aligned} \sin x \sin y &= \frac{1}{4\sqrt{2}}, \\ \tan x \tan y &= \frac{1}{3}. \end{aligned}\right\}$$

606. Solve the system of equations

$$\left.\begin{aligned} x + y &= \varphi \\ \cos x \cos y &= a. \end{aligned}\right\}$$

For what a is the system solvable?

607. Find all the values of a for which the system of equations

$$\left.\begin{aligned} \sin x \cos 2y &= a^2 + 1, \\ \cos x \sin 2y &= a \end{aligned}\right\}$$

is solvable and solve the system.

608. Solve the system of equations

$$\left.\begin{aligned}\cos(x-2y)&=a\cos^3 y,\\ \sin(x-2y)&=a\cos^3 y.\end{aligned}\right\}$$

For what values of a is the system solvable?

609. Find $\cos(x+y)$ if x and y satisfy the system of equations

$$\left.\begin{aligned}\sin x+\sin y&=a,\\ \cos x+\cos y&=b.\end{aligned}\right\}\text{ and } a^2+b^2\neq 0.$$

610. For what values of α is the system

$$\left.\begin{aligned}x-y&=\alpha,\\ 2(\cos 2x+\cos 2y)&=1+4\cos^2(x-y)\end{aligned}\right\}$$

solvable? Find the solutions.

611. Find all the solutions of the system

$$\left.\begin{aligned}8\cos x\cos y\cos(x-y)+1&=0,\\ x+y&=\alpha.\end{aligned}\right\}$$

For what α do the solutions exist?

612. Solve the system of equations

$$\left.\begin{aligned}\tan x+\frac{1}{\tan x}&=2\sin\left(y+\frac{\pi}{4}\right),\\ \tan y+\frac{1}{\tan y}&=2\sin\left(x-\frac{\pi}{4}\right).\end{aligned}\right\}$$

613. Eliminate x and y from the system of equations

$$\left.\begin{aligned}a\sin^2 x+b\cos^2 x&=1,\\ a\cos^2 y+b\sin^2 y&=1,\\ a\tan x&=b\tan y,\end{aligned}\right\}$$

under the assumption that the system is solvable and $a\neq b$.

614. Express $\cos\alpha$ and $\sin\beta$ in terms of A and B if

$$\sin\alpha=A\sin\beta,\ \tan\alpha=B\tan\beta.$$

615. Solve the system of equations

$$\left.\begin{aligned}\tan x&=\tan^3 y,\\ \sin x&=\cos 2y.\end{aligned}\right\}$$

616. Solve the system of equations

$$\left.\begin{aligned}\sin x+\sin y&=\sin(x+y),\\ |x|+|y|&=1.\end{aligned}\right\}$$

617. Solve the system of equations

$$\left.\begin{aligned}\sin(y-3x)&=2\sin^3 x,\\ \cos(y-3x)&=2\cos^3 x.\end{aligned}\right\}$$

618. What conditions must be satisfied by the numbers a, b and c for the system of equations

$$\left.\begin{aligned} \sin x+\sin y&=2a,\\ \cos x+\cos y&=2b,\\ \tan x\tan y&=c \end{aligned}\right\}$$

to have at least one solution?

3. Inverse Trigonometric Functions

619. Compute $\operatorname{arc\,cos}\left[\sin\left(-\frac{\pi}{7}\right)\right]$.

620. Compute $\operatorname{arc\,sin}\left(\cos\frac{33}{5}\pi\right)$.

621. Prove that $\operatorname{arc\,tan}\frac{1}{3}+\operatorname{arc\,tan}\frac{1}{5}+\operatorname{arc\,tan}\frac{1}{7}+\operatorname{arc\,tan}\frac{1}{8}=\frac{\pi}{4}$.

622. Derive the formula

$$\operatorname{arc\,sin} x+\operatorname{arc\,cos} x=\frac{\pi}{2}.$$

623. Show that for $\alpha<\frac{1}{32}$ the equation

$$(\operatorname{arc\,sin} x)^3+(\operatorname{arc\,cos} x)^3=\alpha\pi^3$$

has no roots.

624. Prove that

$$\operatorname{arc\,cos} x=\begin{cases}\operatorname{arc\,sin}\sqrt{1-x^2} & \text{if } 0\le x\le 1;\\ \pi-\operatorname{arc\,sin}\sqrt{1-x^2} & \text{if } -1\le x\le 0.\end{cases}$$

625. Prove the formulas

$$\operatorname{arc\,sin}(-x)=-\operatorname{arc\,sin} x \text{ and } \operatorname{arc\,cos}(-x)=\pi-\operatorname{arc\,cos} x.$$

626. Prove that if

$$-\frac{\pi}{2}+2k\pi\le x\le\frac{\pi}{2}+2k\pi \text{ then}$$

$$\operatorname{arc\,sin}(\sin x)=x-2k\pi.$$

627. Prove that if $0<x<1$ and

$$\alpha=2\operatorname{arc\,tan}\frac{1+x}{1-x},$$

$$\beta=\operatorname{arc\,sin}\frac{1-x^2}{1+x^2} \text{ then } \alpha+\beta=\pi.$$

628. Find the relationship between

$\operatorname{arc\,sin}\cos\operatorname{arc\,sin} x$ and $\operatorname{arc\,cos}\sin\operatorname{arc\,cos} x$.

4. Trigonometric Inequalities

629. Solve the inequality $\sin x > \cos^2 x$.

630. For what x is the inequality

$$4\sin^2 x + 3\tan x - 2\sec^2 x > 0 \text{ fulfilled?}$$

631. Solve the inequality $\sin x \sin 2x < \sin 3x \sin 4x$ if $0 < x < \frac{\pi}{2}$.

632. Solve the inequality

$$\frac{\sin^2 x - \frac{1}{4}}{\sqrt{3} - (\sin x + \cos x)} > 0.$$

633. Find all positive values of x not exceeding 2π for which the inequality $\cos x - \sin x - \cos 2x > 0$ is satisfied.

634. Solve the inequality

$$\tan\frac{x}{2} > \frac{\tan x - 2}{\tan x + 2}.$$

635. Solve the inequality

$$\cos^3 x \cos 3x - \sin^3 x \sin 3x > \frac{5}{8}.$$

636. For $0 < \varphi < \frac{\pi}{2}$ prove the inequality

$$\cot\frac{\varphi}{2} > 1 + \cot\varphi.$$

637. Prove that the inequality

$$(1 - \tan^2 x)(1 - 3\tan^2 x)(1 + \tan 2x \tan 3x) > 0$$

hold for all the values of x entering into the domain of definition of the left-hand side.

638. Prove that the inequality

$$(\cot^2 x - 1)(3\cot^2 x - 1)(\cot 3x \tan 2x - 1) \le -1$$

is valid for all the values of x belonging to the domain of definition of the left-hand side.

639. Putting $\tan\theta = n\tan\varphi\,(n > 0)$ prove that

$$\tan^2(\theta - \varphi) \le \frac{(n-1)^2}{4n}.$$

640. Prove the inequality

$$\frac{\sin x - 1}{\sin x - 2} + \frac{1}{2} \ge \frac{2 - \sin x}{3 - \sin x}.$$

For what values of x does it turn into an equality?

641. Prove that if $0 \le \varphi \le \frac{\pi}{2}$, the inequality

$\cos \sin \varphi > \sin \cos \varphi$ is fulfilled.

642. By the method of complete induction, prove that

$\tan n\alpha > n \tan \alpha$

n is a positive integer greater than unity and α is an angle satisfying the inequality $0 < \alpha < \frac{\pi}{4(n-1)}$.

643. Let $0 < \alpha_1 < \alpha_2 < \ldots < \alpha_n < \frac{\pi}{2}$. Prove that

$$\tan \alpha_1 < \frac{\sin \alpha_1 + \ldots + \sin \alpha_n}{\cos \alpha_1 + \ldots + \cos \alpha_n} < \tan \alpha_n.$$

644. Prove that if A, B and C are the angles of a triangle then

$$\sin \frac{A}{2} \sin \frac{B}{2} \sin \frac{C}{2} \le \frac{1}{8}.$$

645. Prove that if $0 < x < \frac{\pi}{4}$ then $\frac{\cos x}{\sin^2 x (\cos x - \sin x)} > 8$.

5. Miscellaneous Problems

646. Compute $\sin\left(2 \arctan \frac{1}{5} - \arctan \frac{5}{12}\right)$.

647. Prove that if $\tan \alpha = \frac{1}{7}$ and $\sin \beta = \frac{1}{\sqrt{10}}$ where the angles α and β are in the first quadrant then $\alpha + 2\beta = 45°$.

648. Prove that the expression $y = \frac{\sin x + \tan x}{\cos x + \cot x}$

assumes positive values for all permissible values of x.

649. Prove that the equality $\sin \alpha \sin 2\alpha \sin 3\alpha = \frac{4}{5}$ does not hold for all the values of α.

650. Express $\sin 5x$ in terms of $\sin x$. With the aid of the formula thus obtained compute $\sin 36°$ without using tables.

651. Find the greatest and the least values of the function

$$\varphi(x) = \sin^6 x + \cos^6 x.$$

652. Find the greatest and the least values of the function

$$y = 2\sin^2 x + 4\cos^2 x + 6 \sin x \cos x.$$

653. Find out for what integral values of n the number 3π is a period * of the function $\cos nx \sin \frac{5}{n} x$.

654. Prove that if the sum

$$a_1 \cos(\alpha_1 + x) + a_2 \cos(\alpha_2 + x) + \ldots + a_n \cos(\alpha_n + x)$$

vanishes for $x = 0$ and $x = x_1 \neq k\pi$ where k is an integer, then it is identically equal to zero for all x.

655. Prove that the function $\cos\sqrt{x}$ is nonperiodic (i.e. there is no constant number $T \neq 0$ such that $\cos\sqrt{x+T} = \cos\sqrt{x}$ for all x).

656. Prove the formula

$$\sin x + \sin 2x + \ldots + \sin nx = \frac{\sin\frac{nx}{2}\sin\frac{(n+1)x}{2}}{\sin\frac{x}{2}}.$$

Hint. Use De Moivre's formula $(\cos x + i\sin x)^n = \cos nx + i\sin nx$.

657. Compute the sum

$$\frac{\cos\frac{\pi}{4}}{2} + \frac{\cos\frac{2\pi}{4}}{2^2} + \ldots + \frac{\cos\frac{n\pi}{4}}{2^n}.$$

Hint. Apply De Moivre's formula.

658. Consider the function $f(x) = A\cos x + B\sin x$,

where A and B are constants.

Prove that if a function $f(x)$ vanishes for two values x_1 and x_2 such that $x_1 - x_2 \neq k\pi$,

where k an integer, then $f(x)$ is identically equal to zero.

---

* A function $f(x)$ is said to be periodic if there exists a number $T \neq 0$ such that the identity $f(x+T) \equiv f(x)$ is fulfilled for all the permissible values of x. The number T is then called a period of the function.

SOLUTIONS and ANSWERS

SOLUTIONS AND ANSWERS

ALGEBRA

1. Arithmetic and Geometric Progressions

1. By the hypothesis, we have

$$b-a=c-b=d \text{ and } c-a=2d.$$

Denote
$$A_1=\frac{1}{\sqrt{c}+\sqrt{a}}-\frac{1}{\sqrt{b}+\sqrt{c}}$$

and
$$A_2=\frac{1}{\sqrt{a}+\sqrt{b}}-\frac{1}{\sqrt{c}+\sqrt{a}}.$$

Let us show that $A_1=A_2$. If $d=0$ then $a=b=c$ and $A_1=A_2=0$. Therefore we suppose that $d\neq 0$. Rationalizing the denominators we obtain

$$A_1=\frac{\sqrt{c}-\sqrt{a}}{2d}+\frac{\sqrt{b}-\sqrt{c}}{d}=\frac{2\sqrt{b}-\sqrt{c}-\sqrt{a}}{2d}$$

and

$$A_2-\frac{\sqrt{b}-\sqrt{a}}{d}=\frac{\sqrt{c}-\sqrt{a}}{2d}=\frac{2\sqrt{b}-\sqrt{c}-\sqrt{a}}{2d}.$$

Thus, $A_1=A_2$ which completes the proof.

2. If the common difference d of the given progression is equal to zero the validity of the formula is obvious. Therefore we suppose that $d\neq 0$.

Denote the left-hand side member of the desired equality by S. Rationalizing the denominators we get

$$S=\frac{\sqrt{a_2}-\sqrt{a_1}}{a_2-a_1}+\frac{\sqrt{a_3}-\sqrt{a_2}}{a_3-a_2}+\ldots+\frac{\sqrt{a_n}-\sqrt{a_{n-1}}}{a_n-a_{n-1}}.$$

Since, by the hypothesis, $a_2-a_1=a_3-a_2=\ldots=a_n-a_{n-1}=d$ we obviously obtain

$$S=\frac{\sqrt{a_n}-\sqrt{a_1}}{d}.$$

Now we can write

$$S=\frac{a_n-a_1}{(\sqrt{a_n}+\sqrt{a_1})d}=\frac{(n-1)d}{d(\sqrt{a_n}+\sqrt{a_1})}=\frac{n-1}{\sqrt{a_1}+\sqrt{a_n}},$$

which is what we set out to prove.

3. By the hypothesis we have

$$a_2-a_1=a_3-a_2=\ldots=a_n-a_{n-1}=d.$$

If $d=0$ then the desired equality is obvious. Assuming that $d\neq 0$ we can write

$$\frac{1}{a_1a_2}+\frac{1}{a_2a_3}+\frac{1}{a_3a_4}+\ldots+\frac{1}{a_{n-1}a_n}=$$
$$=\left(\frac{1}{a_1}-\frac{1}{a_2}\right)\frac{1}{d}+\left(\frac{1}{a_2}-\frac{1}{a_3}\right)\frac{1}{d}+\left(\frac{1}{a_3}-\frac{1}{a_4}\right)\frac{1}{d}+\ldots$$
$$+\left(\frac{1}{a_{n-2}}-\frac{1}{a_{n-1}}\right)\frac{1}{d}+\left(\frac{1}{a_{n-1}}-\frac{1}{a_n}\right)\frac{1}{d}=$$
$$=\frac{1}{d}\left(\frac{1}{a_1}-\frac{1}{a_n}\right)=\frac{a_n-a_1}{da_1a_n}=\frac{n-1}{a_1a_n}.$$

which is what we set out to prove.

4. At $n=3$ we have $\frac{1}{a_1a_2}+\frac{1}{a_2a_3}=\frac{2}{a_1a_3}$.

Whence , $\frac{1}{a_1a_2}-\frac{1}{a_1a_3}=\frac{1}{a_1a_3}-\frac{1}{a_3a_2}$

and consequently $a_3-a_2=a_2-a_1$ .Therefore it is sufficient to show that

$$a_n-a_{n-1}=a_{n-1}-a_{n-2}$$

for any $n\geq 4$. Let us write down, in succession, the equality given in the formulation of the problem for the cases $n-2$, $n-1$ and n:

$$\frac{1}{a_1a_2}+\frac{1}{a_2a_3}+\ldots+\frac{1}{a_{n-3}a_{n-2}}=\frac{n-3}{a_1a_{n-2}}, \tag{1}$$
$$\frac{1}{a_1a_2}+\frac{1}{a_2a_3}+\ldots+\frac{1}{a_{n-2}a_{n-1}}=\frac{n-2}{a_1a_{n-1}}, \tag{2}$$
$$\frac{1}{a_1a_2}+\frac{1}{a_2a_3}+\ldots+\frac{1}{a_{n-1}a_n}=\frac{n-1}{a_1a_n}. \tag{3}$$

Subtracting termwise equality (2) from (3) and (1) from (2) we get

$$\frac{1}{a_{n-1}a_n}-\frac{1}{a_1a_n}=(n-2)\frac{a_{n-1}-a_n}{a_1a_{n-1}a_n}$$

and

$$\frac{1}{a_{n-2}a_{n-1}}-\frac{1}{a_1a_{n-2}}=(n-2)\frac{a_{n-2}-a_{n-1}}{a_1a_{n-1}a_{n-2}}.$$

Reducing the fractions to a common denominator and cancelling we find

$$a_1-a_{n-1}=(n-2)(a_{n-1}-a_n),$$
$$a_1-a_{n-1}=(n-2)(a_{n-2}-a_{n-1}).$$

Hence, $a_{n-1}-a_n=a_{n-2}-a_{n-1}$ which is the required result.

5. We shall use the method of induction. Note that the equality holds for $n = 2$ since $a_2 - a_1 = a_3 - a_2$ and, consequently, $a_1 - 2a_2 + a_3 = 0$. Suppose that the desired formula is valid for a certain n or, in other words, for any arithmetic progression $x_1, x_2, \dots, x_{n+1}$ the equality

$$x_1 - C_n^1 x_2 + C_n^2 x_3 + \dots + (-1)^{n-1} C_n^{n-1} x_n + (-1)^n C_n^n x_{n+1} = 0 \qquad (1)$$

holds. Now passing to $n+1$ we use the identity

$$C_n^k = C_{n-1}^k + C_{n-1}^{k-1}$$

which results in

$$a_1 - C_{n+1}^1 a_2 + C_{n+1}^2 a_3 + \dots + (-1)^n C_{n+1}^n a_{n+1} + (-1)^{n+1} C_{n+1}^{n+1} a_{n+2}$$
$$= [a_1 - C_n^1 a_2 + \dots + (-1)^n C_n^n a_{n+1}] -$$
$$-[a_2 - C_n^1 a_3 + \dots + (-1)^{n-1} C_n^{n-1} a_{n+1} + (-1)^n C_n^n a_{n+2}].$$

By the hypothesis, both expressions in square brackets are equal to zero because they are of form (1). Therefore, the desired formula is valid for $n + 1$ as well. Thus, the assertion is proved.

6. We carry out the proof by induction. For $n = 3$ it readily follows that

$$a_1^2 - 3(a_1 + d)^2 + 3(a_1 + 2d)^2 - (a_1 + 3d)^2 = 0.$$

Suppose we have already established that for a certain n and any arithmetic progression $x_1, x_2, \dots, x_{n+1}$ the identity

$$x_1^2 - C_n^1 x_2^2 + \dots + (-1)^n C_n^n x_{n+1}^2 = 0$$

holds. Then passing to $n + 1$ as in the preceding problem we obtain

$$a_1^2 - C_{n+1}^1 a_2^2 + C_{n+1}^2 a_3^2 + \dots + (-1)^n C_{n+1}^n a_{n+1}^2 + (-1)^{n+1} C_{n+1}^{n+1} a_{n+2}^2$$
$$= [a_1^2 - C_n^1 a_2^2 + \dots + (-1)^n C_n^n a_{n+1}^2] -$$
$$-[a_2^2 - C_n^1 a_3^2 + \dots + (-1)^n C_n^n a_{n+2}^2] = 0,$$

and thus the required formula has been proved.

It should be noted that for an arithmetic progression $a_1, a_2, \dots, a_n, a_{n+1}$ the more general formula

$$a_1^k - C_n^1 a_2^k + C_n^2 a_3^k - \dots + (-1)^{n-1} C_n^{n-1} a_n^k + (-1)^n C_n^n a_{n+1}^k = 0$$

holds where $k \geq 1$ is an integer.

7. By the well-known property of the terms of an arithmetic progression we have

$$2 \log_m x = \log_n x + \log_k x.$$

Whence we obtain (see (3), page 17)

$$\frac{2}{\log_x m} = \frac{1}{\log_x n} + \frac{1}{\log_x k}$$

and, consequently,

$$2 = \frac{\log_x m}{\log_x n} + \frac{\log_x m}{\log_x k}.$$

Using formula (2) given on page 24 we deduce

$$2 = \log_n m + \log_k m.$$

Let us rewrite this equality as

$$\log_n n^2 = \log_n m + \log_n (n^{\log_k m}).$$

Now, raising we obtain $n^2 = mn^{\log_k m}$ or

$$n^2 = (kn)^{\log_k m}$$

which is what we set out to prove.

8. Let

$$\frac{a_1 + a_2 + \ldots + a_n}{a_{n+1} + a_{n+2} + \ldots + a_{n+kn}} = c. \qquad (1)$$

Denote the common difference of the progression by d. We are only interested in the case $d \neq 0$ since for $d = 0$ all terms of the progression are equal and the equality (1) is automatically fulfilled. Using the formula for the sum of terms of an arithmetic progression we get from (1) the equality

$$\frac{n}{2}[a_1 + a_1 + d(n-1)] = \frac{kn}{2}[a_1 + nd + a_1 + (n + kn - 1)d]\, c$$

from which, after cancelling $\frac{n}{2}$ and rearranging the terms, we find

$$(2a_1 - 2a_1kc - d + cdk) + n(d - cdk^2 - 2cdk) = 0.$$

Since this equality holds for any n we conclude that

$$2a_1 - 2a_1kc - d + cdk = 0$$

and

$$d - cdk^2 - 2cdk = 0.$$

Cancelling out $d \neq 0$ in the second equality we obtain

$$c = \frac{1}{k(k+2)}. \qquad (2)$$

The first equality can be represented in the form

$$(2a_1 - d)(1 - ck) = 0.$$

By virtue of (2), the second factor is different from zero and hence $d = 2a_1$.

Thus, if $d \neq 0$ equality (1) can be valid for all n only in the case of the progression

$$a,\ 3a,\ 5a,\ \ldots \qquad (a \neq 0). \qquad (3)$$

Now it is easy to verify directly that progression (3) in fact satisfies the condition of the problem. Thus, the sought-for progression is given by (3).

9. Let d be the common difference of the progression. We have

$$b^2 = x_1^2 + (x_1 + d)^2 + \ldots + [x_1 + (n-1)d]^2$$
$$= nx_1^2 + 2x_1 d\,[1 + 2 + \ldots + (n+1)] +$$
$$+ d^2[1^2 + 2^2 + \ldots + (n-1)^2]$$
$$= nx_1^2 + n(n-1)x_1 d + \frac{n(n-1)(2n-1)}{6} d^2$$

and, besides

$$a = nx_1 + \frac{n(n-1)}{2} d.$$

Eliminating x_1 from these equations, after some simple transformations we obtain

$$d^2 \frac{n(n^2-1)}{12} = b^2 - \frac{a^2}{n}.$$

Hence,

$$d = \pm\sqrt{\frac{12(nb^2 - a^2)}{n^2(n^2-1)}};$$

x_1 is then defined in either case by the formula

$$x_1 = \frac{1}{n}\left[a - \frac{n(n-1)}{2} d\right].$$

Thus, there are two progressions satisfying the conditions of the problem for $n^2b^2 - a^2 \neq 0$.

10. Let the sequence $a_1, a_2, \ldots, a_n$ possess the property that

$$a_2 - a_1 = d,\ a_3 - a_2 = 2d,\ \ldots,\ a_n - a_{n-1} = (n-1)d.$$

Adding together the equalities we find that

$$a_n = a_1 + d\frac{n(n-1)}{2}.$$

Using this formula we get

$$S_n = a_1 + a_2 + \ldots + a_n = a_1 n$$
$$+ \left[\frac{1\cdot 2}{2} + \frac{2\cdot 3}{3} + \ldots + \frac{(n-1)n}{2}\right] d.$$

In Problem 266 it is proved that

$$\frac{1\cdot 2}{2} + \frac{2\cdot 3}{2} + \ldots + \frac{(n-1)n}{2} = \frac{n(n^2-1)}{6}.$$

Consequently, $S_n = a_1 n + \frac{n(n^2-1)}{6} d$.

For the problem in question we have $d = 3$, $a_1 = 1$. Therefore,

$$a_n = 1 + \frac{3}{2} n(n-1) \text{ and } S_n = \frac{1}{2} n(n^2+1).$$

11. The nth row contains the numbers n, $n+1, \ldots, 3n-3$, $3n-2$(the total of $2n-1$ numbers). The sum of these numbers is equal to

$$\frac{(n+3n-2)\ (2n-1)}{2} = (2n-1)^2.$$

12. Let q be the common ratio of the progression. Then

$$a_{m+n} = a_1 q^{m+n-1} = A,$$
$$a_{m-n} = a_1 q^{m-n-1} = B.$$

Whence $q^{2n} = \dfrac{A}{B}$ and, hence, $q = \sqrt[2n]{\dfrac{A}{B}}$. Now we have

$$a_m = a_{m-n} q^n = B\left(\sqrt[2n]{\frac{A}{B}}\right)^n = \sqrt{AB},$$

$$a_n = a_{m+n} q^{-m} = A\left(\frac{A}{B}\right)^{-\frac{m}{2n}} = A^{\frac{2n-m}{2n}} B^{\frac{m}{2n}}.$$

13. We have $S_n = a_1 + a_1 q + \ldots + a_1 q^{n-1}$,

$$S_{2n} - S_n = a_1 q^n + a_1 q^{n+1} + \ldots + a_1 q^{2n-1} = q^n S_n$$

and, furthermore,

$$S_{3n} - S_{2n} = a_1 q^{2n} + a_1 q^{2n+1} + \ldots + a_1 q^{3n-1} = q^{2n} S_n.$$

It follows that

$$\frac{1}{q^n} = \frac{S_n}{S_{2n} - S_n} = \frac{S_{2n} - S_n}{S_{3n} - S_{2n}}$$

which is what we set out to prove.

14. We have

$$\Pi_n = a_1 . a_1 q_1 \ldots a_1 q^{n-1} = a_1^n q^{\frac{n(n-1)}{2}} = \left(a_1 q^{\frac{n-1}{2}}\right)^n.$$

Noting that $\quad S_n = a_1 + a_1 q + \ldots + a_1 q^{n-1} = a_1 \dfrac{q^n - 1}{q-1}$

and $\quad \overline{S}_n = \dfrac{1}{a_1} + \dfrac{1}{a_1 q} + \ldots + \dfrac{1}{a_1 q^{n-1}} = \dfrac{1}{a_1} \dfrac{\left(\frac{1}{q}\right)^n - 1}{\frac{1}{q} - 1} = \dfrac{1}{a_1 q^{n-1}} \dfrac{q^n - 1}{q-1}$

we conclude that $\dfrac{S_n}{\overline{S}_n} = a_1^2 q^{n-1} = \left(a_1 q^{\frac{n-1}{2}}\right)^2$,

and thus we obtain $\Pi_n = \left(\dfrac{S_n}{\overline{S}_n}\right)^{\frac{n}{2}}$.

15. Denote the sought-for sum by S_n. Multiplying each item of this sum by x and subtracting the resulting quantity from S_n we obtain

$$S_n - xS_n = 1 + x + x^2 + \ldots + x^n - (n+1)x^{n+1}.$$

Applying the formula for the sum of a geometric progression to the one entering into the right-hand side for $x \neq 1$ we find

$$(1-x)S_n = \frac{1-x^{n+1}}{1-x} - (n+1)x^{n+1}.$$

Hence,

$$S_n = \frac{1-x^{n+1}}{(1-x)^2} - \frac{(n+1)x^{n+1}}{1-x} \qquad (x \neq 1).$$

For $x = 1$ we thus obtain

$$S_n = \frac{(n+1)(n+2)}{2}.$$

16. Let us denote the desired sum by S_n. Transforming the terms of the sum by using the formula for the sum of terms of a geometric progression we can write

$$1 + 10 = \frac{10^2 - 1}{9},$$

$$1 + 10 + 100 = \frac{10^3 - 1}{9},$$

$$\ldots\ldots\ldots\ldots\ldots\ldots$$

$$1 + 10 + 100 + \ldots + 10^{n-1} = \frac{10^n - 1}{9}.$$

Since we have $1 = \frac{10-1}{9}$ the addition of the right-hand sides of the latter equalities yields

$$S_n = \frac{1}{9}(10 + 10^2 + \ldots + 10^n - n) = \frac{1}{9}\left(\frac{10^{n+1} - 10}{9} - n\right).$$

17. By adding together the elements of the columns we can represent the required sum in the form

$$(x + x^2 + x^3 + \ldots + x^{n-2} + x^{n-1} + x^n) +$$
$$+ (x + x^2 + x^3 + \ldots + x^{n-2} + x^{n-1}) +$$
$$+ (x + x^2 + x^3 + \ldots + x^{n-2}) +$$
$$\ldots\ldots\ldots\ldots\ldots\ldots$$
$$+ (x + x^2) +$$
$$+ x.$$

Now summing the terms in the brackets we find that for $x \neq 1$ the sought-for sum is equal to

$$x\frac{x^n-1}{x-1}+x\frac{x^{n-1}-1}{x-1}+x\frac{x^{n-2}-1}{x-1}+\ldots+x\frac{x^2-1}{x-1}+x\frac{x-1}{x-1}$$

$$=\frac{x}{x-1}[x+x^2+\ldots+x^n-n]$$

$$=\frac{x}{x-1}\left[x\frac{x^n-1}{x-1}-n\right]=\frac{x^2(x^n-1)}{(x-1)^2}-\frac{nx}{x-1}.$$

For $x=1$ this sum is equal to $\frac{n(n+1)}{2}$ as the sum of terms of an arithmetic progression.

18. Let S_n denote the required sum. Then

$$2S_n=1+\frac{3}{2}+\frac{5}{2^2}+\frac{7}{2^3}+\ldots+\frac{2n-1}{2^{n-1}}=1+\left(\frac{2}{2}+\frac{1}{2}\right)+\left(\frac{2}{2^2}+\frac{3}{2^2}\right)+$$

$$+\left(\frac{2}{2^3}+\frac{5}{2^3}\right)+\ldots+\left(\frac{2}{2^{n-1}}+\frac{2n-3}{2^{n-1}}\right)$$

$$=1+\frac{1-\frac{1}{2^{n-1}}}{1-\frac{1}{2}}+S_n-\frac{2n-1}{2^n},$$

whence

$$S_n=3-\frac{2n+3}{2^n}.$$

19. The general form of these numbers is

$$\overbrace{44\ldots4}^{n}\,\overbrace{88\ldots89}^{n-1}=4\cdot\overbrace{11\ldots1}^{n}\cdot10^n+8\cdot\overbrace{11\ldots1}^{n}+1.$$

The number $\overbrace{1\,1\ldots1}^{n}$ can be written in the form of the sum of the terms of a geometric progression with the common ratio 10 :

$$\overbrace{11\ldots1}^{n}=1+10+10^2+\ldots+10^{n-1}=\frac{10^n-1}{9}.$$

Thus we have

$$\frac{4}{9}(10^n-1)10^n+\frac{8}{9}(10^n-1)+1=\frac{4}{9}10^{2n}+$$

$$+\frac{4}{9}10^n+\frac{1}{9}=\left(\frac{2\cdot10^n+1}{3}\right)^2.$$

20. By the hypothesis, we have $|q|<1$ and, consequently,

$$q^n = k(q^{n+1}+q^{n+2}+\dots) = kq^{n+1}\frac{1}{1-q}. \quad (1)$$

Hence, $1-q=kq$ and thus, if the problem has a solution, we have

$$q=\frac{1}{k+1}. \quad (2)$$

It is, however, easily seen that if, conversely, equality (2) implies that $|q|<1$, then the equality (2) implies equality (1), and the corresponding progression satisfies the condition of the problem. Thus, the problem is solvable for any k satisfying the inequality $\left|\frac{1}{k+1}\right|<1$.The latter holds for $k>0$ and $k<-2$.

21. The proof is carried out by complete induction. Let us first consider a sequence of three terms x_1, x_2, x_3. Opening brackets in the formula

$$(x_1^2+x_2^2)(x_2^2+x_3^2)=(x_1x_2+x_2x_3)^2,$$

we find that

$$x_2^4+x_1^2x_3^2-2x_1x_2^2x_3=0,$$

whence $(x_2^2-x_1x_3)^2=0$, and, consequently,$x_1x_3=x_2^2$. If $x_1\neq 0$ this implies that the numbers x_1, x_2, $x_3$form a geometric progression. Now assume that the suggested assertion is proved for a sequence consisting of $k(k\geq 3)$ terms

$$x_1, x_2, \dots, x_k. \quad (1)$$

Let q be the common ratio of the progression. Consider a sequence of k+1terms

$$x_1, x_2, \dots, x_k, x_{k+1}. \quad (2)$$

Let us write down the corresponding condition

$$(x_1^2+x_2^2+\dots+x_{k-1}^2+x_k^2)(x_2^2+x_3^2+\dots+x_k^2+x_{k+1}^2)$$
$$=(x_1x_2+x_2x_3+\dots+x_{k-1}x_k+x_kx_{k+1})^2 \quad (3)$$

and put, for brevity, $x_1^2+x_2^2+\dots+x_{k-1}^2=a^2$. Note that $a\neq 0$ since $x_1\neq 0$. By the induction hypothesis we have

$$x_2=qx_1;\ x_3=qx_2;\dots;\ x_k=qx_{k-1}. \quad (4)$$

Therefore equality (3) can be rewritten as

$$(a^2+x_k^2)(q^2a^2+x_{k+1}^2)=(qa^2+x_kx_{k+1})^2.$$

Opening the brackets and grouping the terms we see that

$$(x_kq-x_{k+1})^2a^2=0.$$

Since $a\neq 0$, then alongside with (4) we get $x_{k+1}=qx_k$. Hence, the sequence $x_1, x_2, \dots, x_k, x_{(k+1)}$ is a geometric progression with the same common ratio $q=\frac{x_2}{x_1}$.

It follows that a sequence composed of first n terms of the given sequence is a geometric progression for any natural n. Therefore, the given infinite sequence is also a geometric progression which is what we set out to prove.

22. Let $a_1 = b_1 = a$. Then, by virtue of the condition $a_2 = b_2$, we have

$$a + d = aq, \tag{1}$$

where d and q are the common difference and ratio of the corresponding progressions. Note that the condition $a_n > 0$ for all n implies that the difference d must be non-negative. Since, in addition, $a_1 \neq a_2$ do we conclude that $d > 0$.

Therefore, formula (1) implies

$$q = 1 + \frac{d}{a} > 1.$$

Now we have to prove that

$$a + (n-1)d < aq^{n-1} \tag{2}$$

for $n > 2$. Since, by equality (1), $d = a(q-1)$, relation (2) is equivalent to the inequality

$$a(n-1)(q-1) < a(q^{n-1} - 1).$$

Dividing both sides by the positive quantity $a\ (q-1)$ we obtain

$$n - 1 < 1 + q + \ldots + q^{n-2}.$$

Since $q > 1$, this inequality holds true. The problem has thus been solved.

23. By the hypothesis, we have

$$a_1 > 0, \frac{a_2}{a_1} = q > 0 \text{ and } b_2 - b_1 = d > 0,$$

where q is the common ratio of the geometric progression and d the difference of the arithmetic progression. Taking advantage of the fact that $a_n = a_1 q^{n-1}$ and $b_n = b_1 + (n-1)d$ we obtain

$$\log_\alpha a_n - b_n = (n-1)(\log_\alpha q - d) + \log_\alpha a_1 - b_1.$$

For the difference on the left-hand side to be independent of n it is necessary and sufficient that $\log_\alpha q - d = 0$. Solving this equation we find

$$\alpha = q^{\frac{1}{d}}. \tag{1}$$

Consequently, the number α exists and is defined by formula (1).

2. Algebraic Equations and Systems of Equations

24. Rewrite the system in the form

$$\left.\begin{aligned}(x+y)(x^2-xy+y^2)&=1,\\ y(x+y)^2&=2,\end{aligned}\right\}$$

and divide the first equation by the second. Discarding the denominator and then collecting similar terms we obtain

$$y^2-3xy+2x^2=0. \tag{3}$$

Solving quadratic equation (3) in y we get the two roots $y=x$ and $y=2x$ and thus obtain two new equations. Solving then each of these equations simultaneously with equation (2) we find real solutions of the corresponding systems.

There are only two solutions:

$$x_1=\frac{1}{2}\sqrt[3]{4},\qquad y_1=\frac{1}{2}\sqrt[3]{4}$$

and

$$x_2=\frac{1}{3}\sqrt[3]{3},\qquad y_2=\frac{2}{3}\sqrt[3]{3}.$$

Each of these pairs of numbers satisfies the original system as well. This can be verified either by the direct substitution or by analyzing the method by which the solutions were found.

25. Let us transform the equations of the system to the form

$$\left.\begin{aligned}(x+y)^2-xy&=4,\\ (x+y)+xy&=2.\end{aligned}\right\}$$

Whence we obtain

$$(x+y)^2+(x+y)=6$$

and, hence, either $x+y=2$ or $x+y=-3$ Combining either of the latter equations with the second equation of the original system we arrive at the following two systems of equations:

$$\left.\begin{aligned}x+y&=2,\\ xy&=0,\end{aligned}\right\}\ (1)\qquad \left.\begin{aligned}x+y&=-3,\\ xy&=5.\end{aligned}\right\}\ (2)$$

System (1) has two solutions

$$x_1=2,\ y_1=0$$

and

$$x_2=0,\ y_2=2.$$

System (2) also has two solutions

$$x_3=-\frac{3}{2}+i\frac{\sqrt{11}}{2},\ y_3=-\frac{3}{2}-i\frac{\sqrt{11}}{2}$$

and

$$x_4=-\frac{3}{2}-i\frac{\sqrt{11}}{2},\ y_4=-\frac{3}{2}+i\frac{\sqrt{11}}{2}.$$

It is obvious that each solution of the original system belongs to the set of solutions of the above system. A simple argument shows that the converse is also true. By the way, it is still easier to verify it by, a direct substitution. Thus, the problem has four solutions.

26. Transform the equations of the system to the form

$$\left.\begin{aligned}(x+y)\,[(x+y)^2-3xy]=5a^3,\\ xy(x+y)=a^3,\end{aligned}\right\}$$

and then put $x+y=u$ and $xy=v$. Substituting $xy\,(x+y)=a^3$ into the first equation we find $u^3=8a^3$. Since we are only interested in real solutions, we have $u=2a$. From the second equation we now find

$$v=\frac{a^3}{u}=\frac{1}{2}a^2.$$

Thus, we have arrived at the following system of equations in x and y:

$$x+y=2a,\ xy=\frac{1}{2}a^2.$$

Solving this system we get

$$x_1=a\,\frac{2+\sqrt{2}}{2},\quad y_1=a\,\frac{2-\sqrt{2}}{2}$$

and

$$x_2=a\,\frac{2-\sqrt{2}}{2},\quad y_2=a\,\frac{2+\sqrt{2}}{2}.$$

These numbers also satisfy the original system and consequently the latter has two real solutions.

27. Reducing the equations to a common denominator we then transform the system to the form

$$\left.\begin{aligned}(x+y)\,[(x+y)^2-3xy]=12xy,\\ 3(x+y)=xy.\end{aligned}\right\}$$

Putting $x+y=u$, $xy=v$ and substituting $xy=v=3\,(x+y)=3u$ into the first equation we see that

$$u(u^2-9u)=36u. \qquad (1)$$

Note that $u\neq 0$(if otherwise, the second equation would imply $xy=0$ which contradicts the original equation). Therefore, it follows from equation (1) that either $u=12$ or $u=-3$.

In the first case ($u=12$) we get the system

$$\left.\begin{aligned}x+y=12,\\ xy=36,\end{aligned}\right\}$$

whence $x = y = 6$.

In the second case ($u = -3$) we have

$$\left.\begin{aligned} x + y &= -3, \\ xy &= -9. \end{aligned}\right\}$$

This system has two solutions

$$x = \frac{3}{2}(\pm\sqrt{5} - 1), \quad y = \frac{3}{2}(\mp\sqrt{5} - 1).$$

The three solutions thus found satisfy the original system as well. Thus, the system has three solutions.

28. Squaring the second equation and subtracting it from the first equation we obtain

$$xy(x^2 + y^2 - xy) = 21. \qquad (1)$$

Whence, by virtue of the second equation of the system, we derive $xy = 3$.

Substituting y into the second equation of the system, we arrive at the biquadratic equation

$$x^4 - 10x^2 + 9 = 0.$$

It follows that $x_1 = 3$, $x_2 = -3$, $x_3 = 1$, $x_4 = -1$ and therefore the corresponding values of y are $y_1 = 1$, $y_2 = -1$, $y_3 = 3$, $y_4 = -3$. A direct veritication shows that all the four pairs of numbers are solutions of the original system. Consequently, the system has four solutions:

$$x_1 = 3, \ y_1 = 1; \ x_2 = -3, \ y_2 = -1;$$
$$x_3 = 1, \ y_3 = 3; \ x_4 = -1, \ y_4 = -3.$$

29. Transform the system to the form

$$\left.\begin{aligned} (x - y)(x^2 + y^2 + xy - 19) &= 0, \\ (x + y)(x^2 + y^2 - xy - 7) &= 0. \end{aligned}\right\}$$

The original system is thus reduced to the following four systems of equations:

$$\left.\begin{aligned} x - y &= 0, \\ x + y &= 0, \end{aligned}\right\} \qquad (1)$$

$$\left.\begin{aligned} x - y &= 0, \\ x^2 + y^2 - xy - 7 &= 0, \end{aligned}\right\} \qquad (2)$$

$$\left.\begin{aligned} x^2 + y^2 + xy - 19 &= 0, \\ x - y &= 0, \end{aligned}\right\} \qquad (3)$$

$$\left.\begin{aligned} x^2 + y^2 + xy - 19 &= 0, \\ x^2 + y^2 - xy - 7 &= 0, \end{aligned}\right\} \qquad (4)$$

The first system has a single solution $x=0, y=0$. The second one has two solutions $x=\pm\sqrt{7}$, $y=\pm\sqrt{7}$. The third system also has two solutions $x=\pm\sqrt{19}$, $y=\mp\sqrt{19}$. Now taking the fourth system we note that the addition and subtraction of both equations leads to the equivalent system

$$\left.\begin{aligned} xy &= 6, \\ x^2+y^2 &= 13. \end{aligned}\right\}$$

This system has four solutions: $x=\pm 2$, $y=\pm 3$ and $x=\pm 3$, $y=\pm 2$.
Thus, the system under consideration has nine solutions:

$$(0,0),\ (\sqrt{7},\sqrt{7}),\ (-\sqrt{7},\ -\sqrt{7}),$$
$$(\sqrt{19},\ -\sqrt{19}),\ (-\sqrt{19},\ \sqrt{19}),\ (2,\ 3), (-2,\ -3),\ (3,\ 2),\ (-3,\ -2).$$

30. Transforming the system to the form

$$\left.\begin{aligned} 2(x+y) &= 5xy, \\ 8(x+y)[(x+y)^2-3xy] &= 65, \end{aligned}\right\}$$

substituting $x+y$ found from the first equation into the second one and putting $xy=v$ we get

$$25v^3-12v^2-13=0.$$

This equation is obviously satisfied by $v=1$. Dividing the left-hand side by $v-1$ we arrive at the equation

$$25v^2+13v+13=0.$$

The latter equation has no real roots. Thus, there is only one possibility: $v=1$. Substituting this value into the first equation we obtain the system

$$\left.\begin{aligned} xy &= 1, \\ x+y &= \frac{5}{2}. \end{aligned}\right\}$$

Hence, $x_1=2,\ y_1=\dfrac{1}{2}$ and $x_2=\dfrac{1}{2},\ y_2=2.$

Both pairs of numbers also satisfy the original equation. Thus, the system has two and only two real solutions.

31. Adding together the equations and then subtracting the second equation from the first one we get the equivalent system

$$\left.\begin{aligned} (x-y)(x^2+y^2+xy) &= 7, \\ (x-y)\,xy &= 2. \end{aligned}\right\} \qquad (1)$$

Representing the first equation in the form

$$(x-y)^3+3xy(x-y)=7,$$

we see that, by virtue of the second equation, $(x-y)^3=1$.
Since we are only interested in real solutions we have $x-y=1$. Taking this into account we easily deduce $xy=2$.

Solving then the system

$$\left.\begin{aligned} xy &= 2, \\ x - y &= 1, \end{aligned}\right\}$$

we find its two solutions

$$x_1 = 2, \ y_1 = 1; \qquad x_2 = -1, \ y_2 = -2.$$

It can be readily verified that both pairs of numbers satisfy the original system. Thus, the system has two real solutions.

32. Transforming the second equation to the form

$$(x^2 + y^2)^2 - 2x^2y^2 = 7$$

and putting $x^2 + y^2 = u$, $xy = v$ we rewrite this equation as

$$u^2 - 2v^2 = 7.$$

Squaring the first equation of the system we get another relationship between u and v:

$$u + 2v = 1.$$

Eliminating u from the last two equations we obtain

$$v^2 - 2v - 3 = 0,$$

whence

$$v_1 = 3, \quad v_2 = -1.$$

Then the corresponding values of u are

$$u_1 = -5 \text{ and } u_2 = 3.$$

Since $u = x^2 + y^2$ and we are only interested in real solutions of the original equation, the first pair of the values of u and v should be discarded. The second pair leads to the system

$$\left.\begin{aligned} x^2 + y^2 &= 3, \\ xy &= -1. \end{aligned}\right\}$$

This system has four real solutions

$$\left(\frac{1+\sqrt{5}}{2}, \frac{1-\sqrt{5}}{2}\right), \left(\frac{1-\sqrt{5}}{2}, \frac{1+\sqrt{5}}{2}\right),$$

$$\left(\frac{-1+\sqrt{5}}{2}, \frac{-1-\sqrt{5}}{2}\right), \left(\frac{-1-\sqrt{5}}{2}, \frac{-1+\sqrt{5}}{2}\right).$$

It is easy, however, to verify that the original system is satisfied only by the first two of them. Thus, the problem has two real solutions.

33. Raising the first equation to the fifth power and subtracting the second equation from the result we get, after some simplifications, the equation

$$xy(x^3 + y^3) + 2x^2y^2 + 6 = 0. \qquad (1)$$

From the first equation after it has been cubed it follows that $x^3 + y^3 = 1 - 3xy$ which makes it possible to transform equation (1) to the form $$x^2y^2 - xy - 6 = 0.$$

Solving the latter equation we obtain

$$(xy)_1 = 3, \quad (xy)_2 = -2.$$

Combining these relations with $x + y = 1$ we find the four pairs of numbers

$$(2, -1); \quad (-1, 2); \left(\frac{1+i\sqrt{11}}{2}, \frac{1-i\sqrt{11}}{2}\right) \text{ and } \left(\frac{1-i\sqrt{11}}{2}, \frac{1+i\sqrt{11}}{2}\right).$$

It can be easily checked that they all satisfy the original system of equations.

34. Transform the equations of the given system to the form

$$\left.\begin{aligned} (x^2 - y^2)^2 + x^2y^2 = 13, \\ x^2 - y^2 + 2xy = 1. \end{aligned}\right\}$$

Substituting $x^2 - y^2$ found from the second equation into the first one we get

$$5(xy)^2 - 4xy - 12 = 0.$$

It follows that

$$(xy)_1 = 2, \quad (xy)_2 = -\frac{6}{5}. \qquad (1)$$

Since we are only interested in the solutions for which $xy \geq 0$, there is only one possibility, namely

$$xy = 2.$$

Substituting y expressed from the latter relation into the second equation we get

$$x^4 + 3x^2 - 4 = 0.$$

Among all the roots of this equation there are only two real roots $x_1 = 1$ and $x_2 = -1$. By virtue of (2), the corresponding values of y are $y_1 = 2$ and $y_2 = -2$. Both pairs of the numbers (x, y) satisfy the original system as well. Thus, the problem has two solutions

$$x_1 = 1, \quad y_1 = 2 \text{ and } x_2 = -1. \qquad y_2 = -2.$$

35. Opening the brackets in the equations of the system and putting $x + y = u$, $xy = v$ we rewrite the system in the form

$$\left.\begin{aligned} u^2 + v^2 - 2v = 9, \\ uv - u = 3. \end{aligned}\right\} \qquad (1)$$

If now both sides of the second equation are multiplied by 2 and then the corresponding sides of the first equation are added to and subtracted from the obtained result, then system (1) is replaced by the equivalent system

$$\left.\begin{aligned} (u + v)^2 - 2(u + v) = 15, \\ (u - v)^2 + 2(u - v) = 3. \end{aligned}\right\} \qquad (2)$$

From the first equation of system (2) we find

$$(u+v)_1 = 5; \qquad (u+v)_2 = -3.$$

From the second equation we get

$$(u-v)_1 = -3; \qquad (u-v)_2 = 1.$$

Thus, the determination of all solutions of system (2) is reduced to solving the following four systems:

$$\left.\begin{aligned} u+v&=5,\\ u-v&=-3,\end{aligned}\right\}(3) \qquad \left.\begin{aligned} u+v&=5,\\ u-v&=1,\end{aligned}\right\} \qquad (4)$$

$$\left.\begin{aligned} u+v&=-3,\\ u-v&=-3,\end{aligned}\right\}(5) \qquad \left.\begin{aligned} u+v&=-3,\\ u-v&=1,\end{aligned}\right\} \qquad (6)$$

The solutions of systems (3), (4), (5) and (6) are, respectively,

$$u_1 = 1,\ v_1 = 4;$$
$$u_2 = 3, v_2 = 2;$$
$$u_3 = -3,\ v_3 = 0;$$

and

$$u_4 = -1,\ v_4 = -2;$$

To find all the solutions of the original system we now have to solve the following four systems of two equations which only differ in their right-hand sides:

$$\left.\begin{aligned} x+y&=1,\\ xy&=4,\end{aligned}\right\} \quad (7) \quad \left.\begin{aligned} x+y&=3,\\ xy&=2,\end{aligned}\right\} \qquad (8)$$

$$\left.\begin{aligned} x+y&=-3,\\ xy&=0,\end{aligned}\right\} \quad (9) \quad \left.\begin{aligned} x+y&=-1,\\ xy&=-2.\end{aligned}\right\} \qquad (10)$$

Solving these equations we find all the solutions of the original system. We obviously obtain eight solutions:

$$\left(\frac{1}{2}+i\frac{\sqrt{15}}{2}, \frac{1}{2}-i\frac{\sqrt{15}}{2}\right), \left(\frac{1}{2}-i\frac{\sqrt{15}}{2}, \frac{1}{2}+i\frac{\sqrt{15}}{2}\right),$$

$$(2, 1), (1, 2), (-3, 0), (0, -3), (1, -2), (-2, 1).$$

36. Note first that according to the meaning of the problem we have $x \neq 0$ and $y \neq 0$. Multiplying the left-hand and right-hand sides of the equations we obtain

$$x^4 - y^4 = 6. \qquad (1)$$

Multiplying either equations by xy and adding them together we obtain

$$x^4 - y^4 + 2x^2y^2 = 7xy. \qquad (2)$$

By (1) and (2), we now can write

$$2x^2y^2 - 7xy + 6 = 0.$$

whence

$$(xy)_1 = 2; \quad (xy)_2 = \frac{3}{2}. \qquad (3)$$

Thus, every solution of the original system satisfies equation (1) and one of the equations (3). We can therefore combine each of the equations (3) with equation (1) and solve the corresponding systems. But this leads to an equation of eighth degree and complicates the solution of the problem. Therefore we shall apply another technique. Note that if either equation of the original system is again multiplied by xy and then the second equation is subtracted from the first one this results in the equation

$$x^4 + y^4 = 5xy. \qquad (4)$$

which is also satisfied by every solution of the original system.

Let us consider the two possibilities:

(1) Let
$$xy = 2 \qquad (5)$$
in accordance with (3). Then, by (4), we have $x^4 + y^4 = 10$. Combining this equation with (1) and solving the resulting system we find

$$x^4 = 8,$$

and, hence,

$$x_1 = \sqrt[4]{8},\ x_2 = -\sqrt[4]{8},$$
$$x_3 = i\sqrt[4]{8},\ x_4 = -i\sqrt[4]{8}.$$

By virtue of (5), the corresponding values of y are

$$y_1 = \frac{2}{\sqrt[4]{8}} = \sqrt[4]{2},\ y_2 = -\sqrt[4]{2},$$
$$y_3 = -i\sqrt[4]{2},\ \ y_4 = i\sqrt[4]{2}.$$

(2) In the second case we have

$$xy = \frac{3}{2}. \qquad (6)$$

Equation (4) then results in the relation $x^4 + y^4 = \frac{15}{12}$. Combining it with (1) we obtain $x^4 = \frac{27}{4}$. It follows that

$$x_5 = \sqrt[4]{\frac{27}{4}},\ x_6 = -\sqrt[4]{\frac{27}{4}},$$
$$x_7 = i\sqrt[4]{\frac{27}{4}},\ \ x_8 = -i\sqrt[4]{\frac{27}{4}}$$

and the corresponding values of y are

$$y_5 = \sqrt[4]{\frac{3}{4}},\ y_6 = -\sqrt[4]{\frac{3}{4}}, y_7 = -i\sqrt[4]{\frac{3}{4}},\ y_8 = i\sqrt[4]{\frac{3}{4}}.$$

Thus, every solution of the original system belongs to the set of the eight pairs of numbers thus found. It is, however, readily seen that all the eight pairs of numbers satisfy the original system. Consequently, all the solutions of the system have been found.

37. Let us rewrite the second equation in the form

$$(x^2+y^2)^2-2x^2y^2=bx^2y^2.$$

Substituting the expression $x^2+y^2=axy$ found from the first equation we obtain

$$(a^2-2-b)x^2y^2=0$$

There are two possible cases here:

(1) $a^2-2-b\neq 0$. It is easily seen that in this case the system has only one solution $x=0, y=0$.

(2) $a^2-2-b=0$. If this condition is satisfied, the second equation is obtained by squaring both sides of the first equation. Therefore, if any x and y form a pair of numbers satisfying the first equation, the same pair satisfies the second equation as well. Consequently, the system has an infinitude of solutions.

38. Let us transform the left-hand side to the form

$$\frac{x+a}{x+b}\left(\frac{x+a}{x+b}-\frac{a}{b}\frac{x-a}{x-b}\right)+\frac{x-a}{x-b}\left(\frac{x-a}{x-b}-\frac{b}{a}\frac{x+a}{x+b}\right)=0.$$

Noting that the expressions in the brackets differ by the factor $-\dfrac{a}{b}$ we obtain

$$\left(\frac{x+a}{x+b}-\frac{a}{b}\frac{x-a}{x-b}\right)\left(\frac{x+a}{x+b}-\frac{b}{a}\frac{x-a}{x-b}\right)=0.$$

For $a\neq b$ the latter equality implies

$$[x^2-(a+b)x-ab]\ [x^2+(a+b)x-ab]=0,$$

and thus we find the four roots of the original equation:

$$x_{1,2}=\frac{(a+b)\pm\sqrt{(a+b)^2+4ab}}{2},$$

$$x_{3,4}=\frac{-(a+b)\pm\sqrt{(a+b)^2+4ab}}{2}.$$

If $a=b$ the equation is satisfied by any x.

39. Putting $\frac{x}{3}-\frac{4}{x}=t$ we transform the equation to the form

$$3t^2-10t+8=0.$$

Whence we obtain

$$t_1=2 \text{ and } t_2=\frac{4}{3}.$$

Solving then the two quadratic equations for x we find the four roots of the original equation:

$$x_1=3+\sqrt{21},\ x_2=3-\sqrt{21},$$
$$x_3=6,\ x_4=-2.$$

40. Let us put

$$\frac{x+y}{xy} = u \quad \text{and} \quad \frac{x-y}{xy} = v. \tag{1}$$

Then the equations of the system can be written as

$$\left.\begin{aligned} u + \frac{1}{u} &= a + \frac{1}{a}, \\ v + \frac{1}{v} &= b + \frac{1}{b}. \end{aligned}\right\}$$

Solving either equation we find

$$v_1 = a, \ u_2 = \frac{1}{a} \tag{2}$$

and

$$v_1 = b, \ v_2 = \frac{1}{b}. \tag{3}$$

Now we have to solve the four systems of form (1) whose right-hand sides contain all the possible combinations of the values of u and v determined by formulas (2) and (3). Write system (1) in the form

$$\left.\begin{aligned} \frac{1}{y} + \frac{1}{x} &= u, \\ \frac{1}{y} - \frac{1}{x} &= v. \end{aligned}\right\} \tag{4}$$

This yields

$$\left.\begin{aligned} \frac{1}{x} &= \frac{1}{2}(u-v), \\ \frac{1}{y} &= \frac{1}{2}(u+v). \end{aligned}\right\} \tag{5}$$

It follows from formula (5) that for system (4), and, hence, for the original system to be solvable, the numbers a and b must satisfy, besides $ab \neq 0$, some additional conditions implied by the form of the equations of the original system.

Let

$$|a| \neq |b|. \tag{6}$$

Then, substituting the values $u = a$, $v = b$ and then $u = \frac{1}{a}$, $v = \frac{1}{b}$ into the right-hand sides of formulas (5) we find two solutions, namely

$$x_1 = \frac{2}{a-b}, \quad y_1 = \frac{2}{a+b}$$

and

$$x_2 = \frac{2ab}{b-a}, \quad y_2 = \frac{2ab}{a+b}.$$

Furthermore, let

$$|ab| \neq 1. \qquad (7)$$

Then substituting the values $u = a$, $u = \frac{1}{b}$ and then $u = \frac{1}{a}$, $v = b$ into the right-hand sides of formulas (5) we find two more solutions:

$$x_3 = \frac{2b}{ab-1}, \quad y_3 = \frac{2b}{ab+1} \text{ and}$$

$$x_4 = \frac{2a}{1-ab}, \quad y_4 = \frac{2a}{1+ab}$$

Thus, if both conditions (6) and (7) are fulfilled the system has four solutions; if one of the conditions is violated then the system has only two solutions and finally, if both conditions are violated (which may happen only in the case $|a| = |b| = 1$) then the system has no solutions at all.

41. As is easily seen, the numbers

$$x_1 = 4.5 \text{ and } x_2 = 5.5$$

satisfy the equation. Therefore, the polynomial $(x-4.5)^4 + (x-5.5)^4 - 1$ is divisible by the product $(x-4.5) + (x-5.5)$. To perform the division and reduce the problem to a quadratic equation it is convenient to represent the above polynomial in the form

$$[(x-4.5)^4 - 1] + (x-5.5)^4.$$

Factoring the expression in the square brackets by the formula

$$\alpha^4 - 1 = (\alpha - 1)(\alpha + 1)(\alpha^2 + 1)$$

$$= (\alpha - 1)(\alpha^3 + \alpha^2 + \alpha + 1),$$

we come to the equation

$$(x-5.5)\{(x-4.5)^3 + (x-4.5)^2 + (x-4.5) + 1\} + (x-5.5)^4 = 0.$$

Now taking the common factor outside the brackets we obtain

$$(x-5.5)\{(x-4.5)^3 + (x-4.5)^2 + (x-4.5) + 1 + [(x-4.5) - 1]^3\} = 0$$

$$(x-5.5)(x-4.5)\{2(x-4.5)^2 - 2(x-4.5) + 4\} = 0.$$

Hence, we have

$$x_1 = 5.5, \; x_2 = 4.5, \; x_{3,4} = \frac{10 \pm i\sqrt{7}}{2}.$$

42. From the second equation of the system we conclude that $y - 5 = |x-1| \geq 0$, and, consequently, $y \geq 5$. Therefore, the first equation can be rewritten in the form

$$y - 5 = 1 - |x-1|.$$

Adding this equation to the second one we get

$$2(y-5) = 1.$$

Whence we find $\qquad y = \frac{11}{2}.$

From the second equation we now obtain $|x-1|=\frac{1}{2}$ and, hence, $x-1=\pm\frac{1}{2}$. Therefore $x_1=\frac{1}{2}$ and $x_2=\frac{3}{2}$. The system thus has two solutions

$$x_1=\frac{1}{2},\quad y_1=\frac{11}{2} \text{ and } x_2=\frac{3}{2},\quad y_2=\frac{11}{2}.$$

43. Grouping the terms we reduce the left-hand side to the form

$$(2x+y-1)^2+(x+2y+1)^2=0.$$

Thus we obtain

$$2x+y-1=0,\quad x+2y+1=0,$$

whence it follows that

$$x=1,\quad y=-1.$$

Let us demonstrate another method of solution. Arranging the summands in the lelt-hand side in the ascending powers of x we get the following quadratic equation in x:

$$5x^2+(8y-2)x+(5y^2+2y+2)=0. \tag{1}$$

For real values of y this equation has real roots if and only if its discriminant is non-negative, i.e.

$$(8y-2)^2-4.5\,(5y^2+2y+2)\geq 0. \tag{2}$$

Removing the brackets we transform this inequality to the form

$$-36\,(y+1)^2\geq 0.$$

The latter is fulfilled only for $y=-1$, and then equation (1) implies that $x=1$.

44. We transform the equation to the form

$$[x+2\cos(xy)]^2+4[1-\cos^2(xy)]=0.$$

Both summands being non-negative, we have

$$x+2\cos(xy)=0,\quad \cos^2(xy)=1,$$

It follows that $\cos(xy)=\pm 1$. In the case of the plus sign we have the system

$$\cos(xy)=1,\quad x+2\cos(xy)=0.$$

Whence we find $x=-2$ and $y=k\pi$ where $k=0,\pm 1,\pm 2,\ldots$.

In the case of the minus sign we have

$$\cos(xy)=-1,\qquad x+2\cos(xy)=0.$$

This implies $x=2$ and $y=\frac{\pi}{2}(2m+1)$ where $m=0,\pm 1,\pm 2\ldots$. Thus, the equation has two infinite sequences of different real solutions, the value of x in either sequence being the same.

45. Eliminating z from the system we obtain

$$2xy-(2-x-y)^2=4$$

or

$$x^2-4x+4+y^2-4y+4=0,$$

i.e.

$$(x-2)^2+(y-2)^2=0.$$

For real numbers x and y the latter equality holds only for $x=2$ and $y=2$.

From the first equation of the system we find $z=-2$. The system thus has only one real solution:

$$x=2, \qquad y=2, \qquad z=-2.$$

46. *First method.* Note that from the given x and y the value of z is uniquely determined by the first equation in the form

$$z=x^2+y^2. \tag{1}$$

Substituting this value of z into the second equation we get

$$x^2+x+y^2+y=a.$$

The latter equation is equivalent to the equation

$$\left(x+\frac{1}{2}\right)^2+\left(y+\frac{1}{2}\right)^2=a+\frac{1}{2}. \tag{2}$$

If now $a+\frac{1}{2}<0$, then equation (2) has no real solutions because real x and y result in a non-negative number on the left-hand side. But if $a+\frac{1}{2}>0$, equation (2) and, consequently, the whole system, has obviously more than one solution.

Consequently, a unique real solution exists only if $a+\frac{1}{2}=0$. In this case equation (2) takes the form

$$\left(x+\frac{1}{2}\right)^2+\left(y+\frac{1}{2}\right)^2=0$$

and has the only real solution $x=-\frac{1}{2}$, $y=-\frac{1}{2}$. Finding then z from equation (1) we conclude that the given system has a unique real solution only for $a=-\frac{1}{2}$, namely:

$$x=-\frac{1}{2}, \qquad y=-\frac{1}{2}, \qquad z=\frac{1}{2}.$$

Second method. It is easily seen that if the given system has a solution $x=x_0$, $y=y_0$, $z=z_0$, then it also has another solution $x=y_0$, $y=x_0$, $z=z_0$.

Therefore, for the solution to be unique it is necessary that $x = y$. Under this condition the system takes the form

$$\left.\begin{aligned} 2x^2 &= z, \\ 2x + z &= a. \end{aligned}\right\}$$

Eliminating z we obtain the quadratic equation for x:

$$2x^2 + 2x - a = 0.$$

For this equation also to have a unique real root it is necessary and sufficient that the discriminant of the equation be equal to zero:

$$D = 2^2 - 4 \times 2(-a) = 4(1 + 2a) = 0.$$

Hence $a = -\frac{1}{2}$, and the corresponding value of x is equal to $-\frac{1}{2}$. Thus, we arrive at the former result.

47. Let x_0, y_0 be a solution of the system. By virtue of the first equation we have

$$[(x_0^2 + y_0^2) - a]^2 = x_0^2 y_0^2 + \frac{1}{x_0^2 y_0^2} + 2, \tag{1}$$

and, according to the second equation,

$$(x_0^2 + y_0^2)^2 = x_0^2 y_0^2 + \frac{1}{x_0^2 y_0^2} + 2 + b^2. \tag{2}$$

Removing the square brackets on the left-hand side of equality (1) and subtracting equality (2) from it we get

$$-2a(x_0^2 + y_0^2) + a^2 = -b^2.$$

Hence, we obtain

$$x_0^2 + y_0^2 = \frac{a^2 + b^2}{2a}.$$

Since a and b are real, the assertion has been proved.

48. It is readily seen that the system always has the solution

$$x = 1, \quad y = 1, \quad z = 1. \tag{1}$$

It is also obvious that in the case

$$a = b = c \tag{2}$$

all the three equations take the form $x + y + z = 3$, and the system has an infinitude of solutions.

Let us show that if condition (2) is inot fulfilled, i.e. if among a, b, c there are unequal numbers, then solution (1) is unique.

First adding together all the three equations of the given system we obtain

$$(a + b + c)(x + y + z) = 3\,(a + b + c).$$

Cancelling out $a + b + c$ we receive

$$x + y + z = 3. \tag{3}$$

Whence, we find $z = 3 - x - y$. Substituting this expression into the first two equations of the system we obtain

$$\left.\begin{aligned}(a-c)x + (b-c)y &= a + b - 2c,\\ (b-a)x + (c-a)y &= -2a + b + c.\end{aligned}\right\} \tag{4}$$

Multiplying the first of these equations by $c - a$, the second by $c - b$ and adding them together we get

$$[-(a-c)^2 + (b-a)(c-b)]\,x = (a+b-2c)(c-a) + (c-b)(-2a+b+c). \tag{5}$$

Equation (5) being satisfied by $x = 1$, the coefficient in x must identically coincide with the right-hand side of the equation for all a, b and c. Opening the brackets in both expressions we see that they actually coincide and are equal:

$$-\frac{1}{2}[2a^2 - 4ac + 2c^2 - 2bc + 2b^2 + 2ac - 2ab]$$

$$= -\frac{1}{2}[(a-c)^2 + (b-c)^2 + (a-b)^2].$$

Thus, if there are unequal numbers among a,b and c, the equation (5) is satisfied only by $x = 1$. From equations (4) it then readily follows that $y = 1$, and from relation (3) we see that $z = 1$. Thus, if the condition

$$(a-c)^2 + (b-c)^2 + (a-b)^2 \neq 0,$$

holds, the system has the unique solution

$$x = 1, \quad y = 1, \quad z = 1.$$

49. Adding together all the equations we get

$$(a+2)(x+y+z) = 1 + a + a^2. \tag{1}$$

If $a \neq -2$, we have $x + y + z = \dfrac{1 + a + a^2}{a + 2}$.

Combining this equation with each equation of the original system and solving the systems thus obtained we find, for $a \neq 1$, the values

$$x = -\frac{1+a}{a+2}, \quad y = \frac{1}{a+2}, \quad z = \frac{(a+1)^2}{a+2}.$$

For $a = -2$ the system is inconsistent because equality (1) is not fulfilled for any x,y and z. For $a = 1$ the system is indefinite and any three numbers satisfying the condition $x + y + z = 1$ form its solution.

50. It is easily seen that if among the numbers a_1, a_2, a_3 two numbers are equal to zero, the system has an infinite number of solutions. Indeed, let, for instance, $a_2 = 0$ and $a_3 = 0$. Putting then $x = 0$ and choosing y and z so that the equation $y + z = 1$ is satisfied we thus satisfy all the three equations of the system.

Therefore, when establishing the condition for uniqueness we may suppose that at least two numbers are different from zero, Let, for example,

$$a_2 \neq 0 \text{ and } a_3 \neq 0. \tag{1}$$

Subtracting the first equation from the second and the second equation from the third one we find $a_1x = a_2y = a_3z$. It follows, by virtue of (1), that

$$y = \frac{a_1}{a_2}x, \; z = \frac{a_1}{a_3}x. \tag{2}$$

Substituting these expressions into the first equation we get

$$x\left(1 + a_1 + \frac{a_1}{a_2} + \frac{a_1}{a_3}\right) = 1. \tag{3}$$

This equation is solvable only if the expression in the brackets is different from zero.

Taking into account (1) we arrive at the condition

$$D = a_1a_2 + a_2a_3 + a_1a_3 + a_1a_2a_3 \neq 0. \tag{4}$$

If this condition is fulfilled, we find from (3) and (2) the values

$$x = \frac{a_2a_3}{D}, \quad y = \frac{a_1a_3}{D}, \quad z = \frac{a_1a_2}{D}. \tag{5}$$

These three numbers yield a solution of the system, and this solution is unique according to the method by which it is obtained.

Thus, (4) is a necessary condition for the system to be solvable and have a unique solution.

It can be readily veryfied that if we assumed another pair of numbers a_1, a_3 or a_1, a_2 to be different from zero, an analogous argument would again lead us to condition (4) and to the same solution (5). Furthermore, since from condition (4) it follows that at least one of the three pairs of the numbers is non-zero, the above condition is not only necessary but also sufficient.

51. Let us multiply the equations by $a, -b, -c$ and $-d$, respectively, and then add them together. We get $(a^2 + b^2 + c^2 + d^2)\, x = ap - bq - cr - ds$ which implies

$$x = \frac{ap - bq - cr - ds}{a^2 + b^2 + c^2 + d^2}.$$

Analogously, we find

$$y = \frac{bp + aq - dr + cs}{a^2 + b^2 + c^2 + d^2}; \qquad z = \frac{cp + dq + ar - bs}{a^2 + b^2 + c^2 + d^2};$$

$$t = \frac{dp - cq + br + as}{a^2 + b^2 + c^2 + d^2}.$$

52. Adding together all equations of the system we find

$$x_1 + x_2 + \ldots + x_n = \frac{2(a_1 + a_2 + \ldots + a_n)}{n(n+1)}. \quad (1)$$

Let us denote the right-hand side of this equation by A. Now subtracting the second equation from the first one we get

$$(x_1 + x_2 + \ldots + x_n) - nx_1 = a_1 - a_2.$$

By virtue of (1), we can write

$$x_1 = \frac{A - (a_1 - a_2)}{n}.$$

Generally, $x_k (1 \le k \le n-1)$ is obtained by subtracting the $(k+1)$th equation from the kth equation. Similarly, we obtain

$$x_k = \frac{A - (a_k - a_{k+1})}{n}.$$

Finally, subtracting the first equation from the last one we get

$$x_n = \frac{A - (a_n - a_1)}{n}.$$

The values thus found can be expressed by the general formula

$$x_i = \frac{A - (a_i - a_{i+1})}{n} \qquad (1 \le i \le n), \quad (2)$$

where a_{n+1} is understood as being equal to a_1. The direct substitutions. shows that the set of numbers (2) in fact satisfies all the equations of the system. Thus, the given system has a unique solution.

53. Adding up all the equalities and dividing the result by 3 we obtain

$$x_1 + x_2 + x_3 + \ldots + x_{100} = 0 \quad (1)$$

The left-hand side of the new equality contains a hundred of summands, and it can be represented in the form

$$(x_1 + x_2 + x_3) + (x_4 + x_5 + x_6) + \ldots + (x_{97} + x_{98} + x_{99}) + x_{100} = 0.$$

But each of the sums in the brackets is equal to zero by virtue of the original equalities. Therefore, $x_{100} = 0$. Similarly, transposing x_{100} to the first place and representing equality (1) in the form

$$(x_{100} + x_1 + x_2) + (x_3 + x_4 + x_5) + \ldots + (x_{96} + x_{97} + x_{98}) + x_{99} = 0$$

we find that $x_{99} = 0$. Transferring the x_{99} to the first place and regrouping the summands in triads we conclude that $x_{98} = 0$ and so on. Thus,

$$x_1 = x_2 = \ldots = x_{100} = 0,$$

which is what we set out to prove.

54. Adding together the equalities we get

$$(x+y+z)^2-(x+y+z)-12=0. \qquad (1)$$

Putting $x+y+z=t$ we find from equation (1) that

$$t_1=-3, \quad t_2=4. \qquad (2)$$

Substituting the sum $y+z=t-x$ into the first equation of the original system we get

$$x^2+x(t-x)-x=2,$$

whence we obtain

$$x=\frac{2}{t-1}. \qquad (3)$$

Analogously, substituting $x+z=t-y$ into the second equation and $x+y=t-z$ into the third equation we receive

$$y=\frac{4}{t-1} \qquad (4)$$

and

$$z=\frac{6}{t-1}. \qquad (5)$$

Substituting the two values of t [see (2)] into formulas (3), (4) and (5) we find the two solutions of the original system:

$$\left(-\frac{1}{2}, -1, -\frac{3}{2}\right); \quad \left(\frac{2}{3}, \frac{4}{3}, 2\right).$$

55. We rewrite the system in the form

$$\left.\begin{aligned} x+y&=7+z,\\ x^2+y^2&=37+z^2,\\ x^3+y^3&=1+z^3.\end{aligned}\right\} \qquad (1)$$

Squaring the first equation and eliminating x^2+y^2 by means of the second equation we find

$$(7+z)^2=37+z^2+2xy,$$

which implies

$$xy=6+7z$$

Further, we obtain

$$(7+z)^3=x^3+y^3+3xy(x+y),$$

that is

$$x^3+y^3=(7+z)^3-3(6-7z)(7+z)=z^3-18z+217. \qquad (2)$$

Comparing (2) with the last equation of system (1) we find that $z=12$. But then we have

$$\left.\begin{aligned} x+y&=19,\\ xy&=90.\end{aligned}\right\}$$

Solving this system of two equations we receive

$$x_1 = 9, y_1 = 10, z_1 = 12, \text{ and } x_2 = 10, y_2 = 9, z_2 = 12.$$

It is readily verified by substitution that these two sets of numbers satisfy the original system as well. Thus, the original system has two solutions.

56. Dividing the first equation by the second one and by the third we obtain

$$\frac{y+z}{x+y} = \frac{5}{3}, \frac{z+x}{x+y} = \frac{4}{3}.$$

Multiplying both equations by $x + y$ we find

$$\left.\begin{aligned} 5x + 2y - 3z = 0, \\ x + 4y - 3z = 0. \end{aligned}\right\}$$

These equations imply that $y = 2x$ and $z = 3x$. Substituting the latter expressions into the first equation of the original system we see that $x^2 = 1$. Finally, we get

$$x_1 = 1, y_1 = 2, z_1 = 3 \text{ and } x_2 = -1, y_2 = -2, z_2 = -3.$$

The direct verification shows that both solutions satisfy the original system as well.

57. Noting that the difference of every two equations of the system can be factorized, we form the differences between the first and second equations and between the first and third ones. Combining the two equations thus obtained with the third equations of the original system we arrive at the following system:

$$\left.\begin{aligned} (u - w)(u + w - 1) = 0, \\ (v - w)(v + w - 1) = 0, \\ w^2 + u^2 + v = 2. \end{aligned}\right\} \quad (1)$$

It is obvious that any solution of the original system satisfies system (1). Since conversely, all equations of the original system can be obtained by addition and subtraction of the equations of system (1), any solution of system (1) is a solution of the original system, and, hence, these two systems are equivalent.

System (1) can be decomposed into the following four systems:

$$\left.\begin{aligned} u - w = 0, \\ v - w = 0, \\ w^2 + u^2 + v = 2, \end{aligned}\right\} (2) \quad \left.\begin{aligned} u - w = 0, \\ v + w - 1 = 0, \\ w^2 + u^2 + v = 2, \end{aligned}\right\} \quad (3)$$

$$\left.\begin{aligned} u + w - 1 = 0, \\ v - w = 0, \\ w^2 + u^2 + v = 2, \end{aligned}\right\} (4) \quad \left.\begin{aligned} u + w - 1 = 0, \\ v + w - 1 = 0, \\ w^2 + u^2 + v = 2, \end{aligned}\right\} \quad (5)$$

It apparently follows that all the solutions of the above four systems and only they are the solutions of the original system.

Each of the four systems is readily reduced to a quadratic equation and has two solutions. Below, omitting the calculations, we give the corresponding solutions (u, v, w). The solutions of system (2):

$$\left(\frac{-1+\sqrt{17}}{4}, \frac{-1+\sqrt{17}}{4}, \frac{-1+\sqrt{17}}{4}\right),$$

$$\left(\frac{-1-\sqrt{17}}{4}, \frac{-1-\sqrt{17}}{4}, \frac{-1-\sqrt{17}}{4}\right).$$

The solutions of system (3):

$$(1, 0, 1); \left(-\frac{1}{2}, \frac{3}{2}, -\frac{1}{2}\right).$$

The solutions of system (4):

$$(0, 1, 1); \left(\frac{3}{2}, -\frac{1}{2}, -\frac{1}{2}\right).$$

The solutions of system (5):

$$(1, 1, 0); \left(-\frac{1}{2}, -\frac{1}{2}, \frac{3}{2}\right).$$

Thus the original system has the total of eight solutions.

58. Subtracting the first equation from the second we get $z^2 - y^2 + x(z - y) = 3$ whence we find $(z - y)(x + y + z) = 3$. Subtracting the second equation from the third we similarly find

$$(y - x)(x + y + z) = 3.$$

From the two latter equations is follows that

$$z - y = y - x. \qquad (1)$$

Now we rewrite the original system in the form

$$\left.\begin{aligned} (x-y)^2 &= 1 - 3xy, \\ (x-z)^2 &= 4 - 3xz, \\ (y-z)^2 &= 7 - 3yz. \end{aligned}\right\} \qquad (2)$$

From (1) we conclude that the right-hand sides of the first and third equations of system (2) are equal, i.e. $1 - 3xy = 7 - 3yz$, whence it follows that

$$z - x = \frac{2}{y}. \qquad (3)$$

According to (1) we have

$$z + x = 2y, \qquad (4)$$

and therefore, solving (3) and (4) as simultaneous equations we find

$$x = y - \frac{1}{y}, \; z = y + \frac{1}{y}.$$

Substituting the expression of x thus obtained into the first equation of the original system we obtain

$$3y^4-4y^2+1=0,$$

which implies

$$y_{1,2}=\pm 1,\ y_{3,4}=\pm\frac{1}{\sqrt{3}}.$$

As a result, we find the following four sets of numbers:

$$(0,1,2),\ (0,-1,-2);$$
$$\left(\frac{2}{\sqrt{3}},-\frac{1}{\sqrt{3}},-\frac{4}{\sqrt{3}}\right);$$
$$\left(-\frac{2}{\sqrt{3}},\frac{1}{\sqrt{3}},\frac{4}{\sqrt{3}}\right).$$

The corresponding verification shows that they all satisfy the original system.

59. Multiplying the left-hand and right-hand sides of the equations we get

$$(x_1x_2\dots x_n)^{n-2}=a_1a_2\dots a_n,$$

whence $\quad x_1x_2\dots x_n=\sqrt[n-2]{a_1a_2\dots a_n}.\qquad(1)$

Let us rewrite the kth equation of the system in the form

$$a_kx_k^2=x_1x_2\dots x_n.$$

It follows, by virtue of (1), that

$$x_k=\sqrt{\frac{\sqrt[n-2]{a_1a_2\dots a_n}}{a_k}}\quad(k=1,2,\dots,n).$$

The substitution into the original system indicates that this set of numbers satisfies it. Thus, the problem has a unique solution.

60. First note that for $a=1$ the system takes the form

$$\left.\begin{aligned}(x+y+z)^2&=k^2,\\(x+y+z)^2&=l^2,\\(x+y+z)^2&=m^2.\end{aligned}\right\}$$

The latter system is solvable only if the additional condition

$$k^2=l^2=m^2\qquad(1)$$

holds. In this case we obviously obtain an infinite number of solutions. In what follows we may thus suppose that

$$a\neq 1.\qquad(2)$$

Adding together all equations of the system and putting, for brevity,

$$x+y+z=t$$

we get $\quad t^2(a+2)=k^2+l^2+m^2.$

By the hypothesis, the right-hand side is positive and therefore for $a = -2$ the system has no solutions at all. For

$$a \neq -2 \tag{3}$$

we find

$$t = \pm \sqrt{\frac{k^2 + l^2 + m^2}{a + 2}}. \tag{4}$$

Now, transforming the equations of the system to the form

$$\left.\begin{aligned} t^2 + t(a-1)\,x &= k^2, \\ t^2 + t(a-1)y &= l^2, \\ t^2 + t(a-1)\,z &= m^2, \end{aligned}\right\}$$

and solving them we determine, according to (4), two sets of values of x and y:

$$x = \pm \sqrt{\frac{a+2}{k^2 + l^2 + m^2}} \frac{k^2(a+1) - l^2 - m^2}{(a+2)(a-1)},$$

$$y = \pm \sqrt{\frac{a+2}{k^2 + l^2 + m^2}} \frac{l^2(a+1) - k^2 - m^2}{(a+2)(a-1)},$$

$$z = \pm \sqrt{\frac{a+2}{k^2 + l^2 + m^2}} \frac{m^2(a+1) - k^2 - l^2}{(a+2)(a-1)}.$$

Finally, we check by substitution that both triplets of numbers satisfy the original system. Thus, in the general case when $a \neq 1$ and $a \neq -2$ the system has two different solutions.

61. Squaring the first equation and subtracting the second equation from the resulting relation we find

$$xy + yz + zx = 11. \tag{1}$$

The third equation then implies that

$$(xy)^2 + 3xy - 10 = 0.$$

Solving this equation we get

$$(xy)_1 = 2, (xy)_2 = -5. \tag{2}$$

Now there can be two possibilities here:

(1) Let

$$xy = 2 \tag{3}$$

Eliminating $x + y$ from the first and third equations of the original system we arrive at the following equation in z:

$$z^2 - 6z + 9 = 0.$$

Hence, $$z^{(1)} = 3.$$

The first equation of the original system then gives

$$x + y = 3.$$

Combining this equation with equation (3) and solving them we get

$$x_1^{(1)} = 1,\ y_1^{(1)} = 2,$$
$$x_2^{(1)} = 2,\ y_1^{(1)} = 1.$$

(2) Now, in conformity with (2), we suppose that

$$xy = -5 \tag{4}$$

From the first and third equations we then obtain

$$z^2 - 6z + 16 = 0.$$

This equation has no real roots and, consequently, we may not consider the case (4).

Thus, the set of possible solutions (x, y, z) consists of $(1, 2, 3)$ and $(2, 1, 3)$.

Substituting these values into the original system we check the both triplets satisfy it. Thus, all real solutions of the system have been found.

62. One can easily note that the left-hand sides of the equations can be factorized which brings that system to the form

$$\left.\begin{aligned} (x+y)(x+z) &= a,\\ (x+y)(y+z) &= b,\\ (x+z)(y+z) &= c. \end{aligned}\right\} \tag{1}$$

Let us put, for brevity,

$$x + y = u,\ x + z = v,\ y + z = w.$$

Then we can write

$$\left.\begin{aligned} uv &= a,\\ uw &= b,\\ vw &= c. \end{aligned}\right\} \tag{2}$$

Multiplying all the equations we find

$$(uvw)^2 = abc,$$

whence

$$uvw = \pm\sqrt{abc}. \tag{3}$$

Now all the solutions of system (2) are found without difficulty. First taking the plus sign in formula (3) and then minus sign we conclude that system (2) has two solutions, namely

$$u_1 = \frac{\sqrt{abc}}{c},\ v_1 = \frac{\sqrt{abc}}{b},\ w_1 = \frac{\sqrt{abc}}{a} \tag{4}$$

and

$$u_2 = \frac{-\sqrt{abc}}{c},\ v_2 = \frac{-\sqrt{abc}}{b},\ w_2 = \frac{-\sqrt{abc}}{a}. \tag{5}$$

Now we have only to solve the two systems of equations obtained after the values (4) and (5) have been substituted into the right-hand sides of the equations

$$\left.\begin{aligned} x+y&=u,\\ x+z&=v,\\ y+z&=w.\end{aligned}\right\}\qquad (6)$$

Adding together equations (6) we get $x+y+z=\dfrac{u+v+w}{2}$. Whence, by virtue of (6), it readily follows that

$$x=\frac{u+v-w}{2},\ y=\frac{u-v+w}{2},\ z=\frac{-u+v+w}{2}. \qquad (7)$$

Thus, the original system has only two solutions which are determined by formulas (7) after the values (4) and (5) have been substituted into them.

63. Adding together all the equations we find

$$xy+xz+yz=\frac{a^2+b^2+c^2}{2}. \qquad (1)$$

By virtue of the equations of the system we now easily obtain

$$\left.\begin{aligned} xy&=\frac{a^2+b^2-c^2}{2}=\alpha,\\ xz&=\frac{a^2-b^2+c^2}{2}=\beta,\\ yz&=\frac{-a^2+b^2+c^2}{2}=\gamma.\end{aligned}\right\}\qquad (2)$$

For brevity we have denoted the obtained fractions by α, β and γ. It should also be noted that if the original system is solvable, all the three numbers α, β and γ are different from zero. Indeed, let, for instance, $\alpha=0$. Then $\beta\gamma=xyz^2=0$. Adding the first equation of system (2) to the second and third ones we get

$$a^2=\beta,\ b^2=\gamma$$

which implies $a^2b^2=0$ and thus, according to the conditions of the problem, we arrive at a contradiction. Hence, $\alpha\beta\gamma\neq 0$. System (2) therefore coincides with system (2) of the preceding problem. Consequently, it has two solutions

$$x_1=\frac{\sqrt{\alpha\beta\gamma}}{\gamma},\ y_1=\frac{\sqrt{\alpha\beta\gamma}}{\beta},\ z_1=\frac{\sqrt{\alpha\beta\gamma}}{\alpha} \qquad (3)$$

and

$$x_2=\frac{-\sqrt{\alpha\beta\gamma}}{\gamma},\ y_2=\frac{-\sqrt{\alpha\beta\gamma}}{\beta},\ z_2=\frac{-\sqrt{\alpha\beta\gamma}}{\alpha}. \qquad (4)$$

It can be readily verified that the same two sets of numbers satisfy the original system as well. Thus, all the solutions of the system are given by formulas (3) and (4).

64. Let us put

$$xy + xz + yz = t^3. \qquad (1)$$

Then the system is written in the form

$$\left.\begin{aligned} y^3 + z^3 &= 2at^3, \\ z^3 + x^3 &= 2bt^3, \\ x^3 + y^3 &= 2ct^3. \end{aligned}\right\} \qquad (2)$$

Adding together all equations of this system we find that

$$x^3 + y^3 + z^3 = (a + b + c)\, t^3. \qquad (3)$$

Subtracting in succession the equations of system (2) from the latter equation we obtain

$$x^3 = (b + c - a)t^3,\ y^3 = (c + a - b)t^3,\ z^3 = (a + b - c)t^3,$$

whence we find

$$x = \sqrt[3]{b + c - a}.t,\ y = \sqrt[3]{c + a - b}.t,\ z = \sqrt[3]{a + b - c}\cdot t \qquad (4)$$

Substituting these expressions into equation (1) we conclude that either $t_1 = 0$ or

$$t_2 = \sqrt[3]{(b + c - a)(c + a - b)} + \sqrt[3]{(b + c - a)(a + b - c)} + \sqrt[3]{(c + a - b)(a + b - c)}.$$

Substituting these values of t into formulas (4) we find two solutions of the original system.

65. Put

$$x + y = u,\ x + z = v, y + z = w.$$

Then the system is rewritten in the form

$$\left.\begin{aligned} u + v &= auv, \\ u + w &= buw, \\ v + w &= cvw. \end{aligned}\right\} \qquad (1)$$

Obviously, system (1) has the following solution:

$$u = 0,\ v = 0,\ w = 0. \qquad (2)$$

Note furthermore, that if $u = 0$ then the first equation (1) implies $v = 0$ and the third equation implies $w = 0$. Therefore we shall only limit ourselves to the cases when $uvw \neq 0$.

From system (1) we find

$$\left.\begin{aligned} \frac{1}{v} + \frac{1}{u} &= a, \\ \frac{1}{w} + \frac{1}{u} &= b, \\ \frac{1}{w} + \frac{1}{v} &= c. \end{aligned}\right\}$$

This system has the same form as system (6) in Problem 62. Applying the same method we obtain

$$\left.\begin{aligned}\frac{1}{u}&=\frac{a+b-c}{2},\\ \frac{1}{v}&=\frac{a-b+c}{2},\\ \frac{1}{w}&=\frac{-a+b+c}{2}.\end{aligned}\right\} \quad (3)$$

Hence, system (1) can have a solution other than solution (2) only if the additional condition

$$\left.\begin{aligned}a+b-c=\alpha\neq 0, a-b+c=\beta\neq 0,\\ -a+b+c=\gamma\neq 0\end{aligned}\right\} \quad (4)$$

holds. If conditions (4) is fulfilled, we obtain from formulas (3) the expressions

$$u=\frac{2}{\alpha}, v=\frac{2}{\beta}, w=\frac{2}{\gamma}. \quad (5)$$

To complete the solution we have to solve the following two systems:

$$\left.\begin{aligned}x+y&=0,\\ x+z&=0,\\ y+z&=0,\end{aligned}\right\} (6) \qquad \left.\begin{aligned}x+y&=\frac{2}{\alpha},\\ x+z&=\frac{2}{\beta},\\ y+z&=\frac{2}{\gamma}.\end{aligned}\right\} \quad (7)$$

System (7) appears only if condition (4) is fulfilled. Either system has exactly one solution. Namely, the solution of system (6) is

$$x=0, y=0, z=0,$$

and system (7) has the solution

$$\left.\begin{aligned}x=\frac{1}{\alpha}+\frac{1}{\beta}-\frac{1}{\gamma}, y=\frac{1}{\alpha}-\frac{1}{\beta}+\frac{1}{\gamma},\\ z=-\frac{1}{\alpha}+\frac{1}{\beta}+\frac{1}{\gamma}.\end{aligned}\right\} \quad (8)$$

Thus, the original system has only a zero solution $x=y=z=0$, and if the additional condition (4) is fulfilled, there appears one more solution determined by formulas (8) and (4).

66. The form of the seconds equation of the system indicates that $x\neq 0$, $y\neq 0$ and $z\neq 0$. Reducing the fractions on the left-hand side of the second equation to a common denominator we get, by virtue of the third equation, the relation

$$xyz=27. \quad (1)$$

Multiplying then the third equation by z and taking into account (1), we can write

$$27+(x+y)z^2=27z.$$

Substituting the expression $x+y=9-z$ found from the first equation of the system into the latter equation we obtain

$$z^3-9z^2+27z-27=0$$

i.e. $(z-3)^3=0$. Therefore $z=3$. Substituting this value both in the first equation and in (1) we find that $x=3$ and $y=3$. This result is, by way, quite obvious since all the unknowns are involved symmetrically into the equations of the system. Thus, if the system is solvable, the only solution is the triplet of number $x=3, y=3, z=3$. The direct substitution into the original system confirms that this set of numbers is in fact a solution. Thus, the system is solvable and has the unique solution $x=3, y=3, z=3$.

67. Substituting the quantity $x+y$ found from the first equation into the second one we get

$$xy+z(a-z)=a^2.$$

Expressing xy from this equation and substituting it into the third equation we obtain

$$z^3-az^2+a^2z-a^3=0.$$

The left-hand side of the latter equation is readily factorized.

$$(z-a)(z-ai)(z+ai)=0.$$

If follows that $\qquad z_1=a, z_2=ai, z_3=-ai$

Substituting $z=a$ into the first and second equations we arrive at the system

$$x+y=0, xy=a^2$$

whose solution is $x=\pm ia$, $y=\mp ia$. It is readily verified that both triplets of numbers (x, y, z) of the form

$$(ia, -ia, a) \text{ and } (-ia, ia, a)$$

satisfy the original system. Analogously, we find more pairs of solutions corresponding to the values z_2 and z_3:

$$(a, -ia, ia), (-ia, a, ia) \text{ and } (ia, a-ia), (a, ia, -ia).$$

Thus, the system is satisfied by the above six solutions, and there are no other solutions. The result can be achieved in a shorter way if we use a relationship between the system under consideration and the roots of the cubic equation

$$t^3-at^2+a^2t-a^3=0. \qquad (1)$$

Namely, according to Vieta's formulas [see (2), page 10] the three roots

$$t_1=a, t_2=ia, t_3=-ia$$

of equation (1) (taken in any order) form a solution of the system in question. Thus, we have already obtained six (i.e. 3!) solutions. Let us show that the system has no other solutions. Indeed, let (x_1, y_1, z_1) be a solution of the system. Consider the cubic equation

$$(t-x_1)(t-y_1)(t-z_1)=0 \qquad (2)$$

whose roots are the numbers x_1, y_1 and z_1. Removing the brackets in equation (2) and using the equalities

$$x_1 + y_1 + z_1 = a,$$
$$x_1y_1 + y_1z_1 + x_1z_1 = a^2,$$
$$x_1y_1z_1 = a^3,$$

we reveal that equations (2) and (1) coincide. Consequently, x_1, y_1 and z_1 are the roots of equation (1) which is what we set out to prove. The same argument can be used in solving the preceding problem.

68. Substituting x found from the first equation into the second one we get

$$3y^2 + z^2 = 0. \qquad (1)$$

By virtue of the third equation, if follows that

$$3y^2 - xy = 0. \qquad (2)$$

Therefore, we have either $y = 0$ or $x = 3y$.

In the case $y = 0$ we see that according to (1) we have $z = 0$. By virtue of the first equation of the given system we also conclude that $x = 0$.

In the case $y = -2z$ we substitute x expressed by the equality $x = 3y$ into the second equation of the system and thus obtain

$$2y^2 + 4yz = 0. \qquad (3)$$

If now $y = 0$, we arrive at the former case, and if $y = -2z$, then condition (1) implies that $z = 0$, and consequently, $y = 0$ and $x = 0$. The assertion has thus been proved.

69. From the identity

$$(x + y + z)^2 = x^2 + y^2 + z^2 + 2(xy + xz + yz), \qquad (1)$$

by virtue of the first and second equations of the system, we get

$$xy + xz + yz = 0. \qquad (2)$$

Now let us consider the identity obtained by cubing the trinomial $x + y + z$:

$$(x + y + z)^3 = x^3 + y^3 + z^3 + 3x^2y + 3x^2z + 3xy^2 + 6xyz + 3xz^2 + 3y^2z + 3yz^2. \qquad (3)$$

Its right-hand side can be represented in the form

$$x^3 + y^3 + z^3 + 3x(xy + xz + yz) + 3y(xy + yz + xz) + 3z^2(x + y).$$

Consequently identity (3),by virtue of the equations of the system and equality (2) implies that

$$3z^2(x + y) = 0. \qquad (4)$$

There can be the following two cases here:

(1) If $z = 0$, then, according to (2), we have $xy = 0$. Taking into account the first equation of the system, we get the two sets of values

$$x_1 = a,\ y_1 = 0,\ z_1 = 0 \qquad (5)$$

$$\text{and } x_2 = 0,\ y_2 = a,\ z_2 = 0 \qquad (6)$$

It can be easily seen that formulas (5) and (6) determine two solutions of the original system.

(2) If $x + y = 0$, then from the condition (2) we again get $xy = 0$, and hence, $x = 0$ and $y = 0$. From the first equation of the system it then follows, that $z = a$, and we thus arrive at another solution of the original system

$$x_3 = 0, y_3 = 0, z_3 = a. \tag{7}$$

Thus, if $a \neq 0$ the system has three different solutions, and if $a = 0$ it possesses only a zero solution.

70. Let us consider the identity

$$(x + y + z)^3 = x^3 + y^3 + z^3 + 3x^2y + 3x^2z + 3xy^2 + 6xy^2 + 3xz^2 + 3y^2z + 3yz^2. \tag{1}$$

Transforms its right-hand member as follows:

$$x^3 + y^3 + z^3 + 3x(xy + xz + yz) + 3y (xy + xz + yz) + 3z (xy + xz + yz) - 3xyz.$$

It follows that identity (1) can be rewritten as

$$(x + y + z)^3 = x^3 + y^3 + z^3 + 3(x + y + z)(xy + xz + yz) - 3xyz. \tag{2}$$

From relation (2) it is seen that for determining the sum $x^3 + y^3 + z^3$ it is sufficient to express $xy + xz + yz$ and xyz from the original system.

Squaring the first equation and subtracting the second one from the result we get

$$xy + xz + yz = \frac{1}{2}(a^2 - b^2). \tag{3}$$

Let us rewrite the third equation in the form

$$xyz = c(xy + xz + yz) \tag{4}$$

Now taking into consideration (3) and (4) we finally find from (2) the expression

$$x^3 + y^3 + z^3 = a^3 - \frac{3}{2}a(a^2 - b^2) + \frac{3}{2}c(a^2 - b^2) = a^3 + \frac{3}{2}(a^2 - b^2)(c - a).$$

71. Removing the brackets we rewrite the second equation in the form

$$x^2 + y^2 + z^2 + 3xy + 3xz + 3yz = 1,$$

which implies

$$(x + y + z)^2 + xy + xz + yz = 1.$$

Now using the first equation of the system we derive

$$xy + xz + yz = -3. \tag{1}$$

The third equation of the system can be represented in the form

$$x(xy + xz) + y(yz + xy) + z(xz + yz) = -6$$

and therefore, taking into account (1), we obtain

$$x(3 + yz) + y(3 + xz) + z(3 + xy) = 6,$$

which implies

$$x + y + z + xyz = 2,$$

i.e. $$xyz = 0.$$

We thus arrive at the following system:

$$\left.\begin{aligned} x + y + z &= 2, \\ xy + xz + yz &= -3, \\ xyz &= 0. \end{aligned}\right\} \qquad (2)$$

From the last equation of this system it follows that at least one of the unknowns is equal to zero. Let $x = 0$, then

$$y + z = 2,\ yz = -3,$$

whence either $y = 3, z = -1$ or $y = -1, z = 3$. The cases $y = 0$ and $z = 0$ are treated analogously. Thus, we get the following six solutions (x, y, z) of system (2):

$$(0, 3, -1); \quad (-1, 0, 3); \quad (0, -1, 3);$$
$$(3, -1, 0); \quad (3, 0, -1); \quad (-1, 3, 0).$$

It is readily checked that all these solutions satisfy the original system as well. Thus, the problem has six solutions.

72. Removing the brackets in all the equations we note that if the third equation is subtracted from the sum of the first two, then the following equation is obtained:

$$(x - y + z)^2 = a - b + c. \qquad (1)$$

Similarly, we deduce

$$(x + y - z)^2 = a + b - c \qquad (2)$$

and

$$(y + z - x)^2 = b + c - a. \qquad (3)$$

It can be easily shown that, conversely, the original system is a consequence of the system of equations (1), (2) and (3). Indeed, adding, for example equations (2) and (3), we obtain the second equation of the original system and so on. Thus, the original system is equivalent to that obtained. Therefore, it is sufficient to find all solutions of the system of equations (1), (2) and (3).

Let us put, for brevity,

$$\sqrt{b + c - a} = a_1,\ \sqrt{a - b + c} = b_1,\ \sqrt{a + b - c} = c_1.$$

Then the system of equations (1), (2), (3) is equivalent to the following eight linear systems

$$\left.\begin{aligned} x-y+z&=\pm b_1,\\ x+y-z&=\pm c_1,\\ -x+y+z&=\pm a_1.\end{aligned}\right\} \qquad (4)$$

Taking the plus sign on the right-hand side of all equations we easily find the following unique solution of the corresponding system:

$$x=\frac{b_1+c_1}{2},\ y=\frac{a_1+c_1}{2},\ z=\frac{b_1+a_1}{2}.$$

Considering all the possible combinations of signs of the right-hand members, we find another seven solutions:

$$\left(\frac{-b_1+c_1}{2},\quad \frac{a_1+c_1}{2},\quad \frac{-b_1+a_1}{2}\right);$$

$$\left(\frac{b_1-c_1}{2},\quad \frac{a_1-c_1}{2},\quad \frac{b_1+a_1}{2}\right);$$

$$\left(\frac{b_1+c_1}{2},\quad \frac{-a_1+c_1}{2},\quad \frac{b_1-a_1}{2}\right);$$

$$\left(\frac{-b_1-c_1}{2},\quad \frac{a_1-c_1}{2},\quad \frac{-b_1+a_1}{2}\right);$$

$$\left(\frac{-b_1+c_1}{2},\quad \frac{-a_1+c_1}{2},\quad \frac{-b_1-a_1}{2}\right);$$

$$\left(\frac{b_1-c_1}{2},\quad \frac{-a_1-c_1}{2},\quad \frac{b_1-a_1}{2}\right);$$

$$\left(\frac{-b_1-c_1}{2},\quad \frac{-a_1-c_1}{2},\quad \frac{-b_1-a_1}{2}\right).$$

The eight solutions thus found obviously represent all the possible solutions of the system.

73. Rewrite the third equation of the system in the form

$$z^2+xy-z(x+y)=2. \qquad (1)$$

Substituting z^2 found from the second equation and $z(x+y)$ expressed from the first one into (1) we get

$$x^2+y^2+xy-47+xy=2, \text{ or } (x+y)^2=49.$$

Whence we derive

$$x+y=\pm 7. \qquad (2)$$

Multiplying both sides of the first equation by 2 and adding the second equation to it we obtain

$$(x+y)^2+2z(x+y)=94+z^2. \qquad (3)$$

There are two possible cases here:

(1) If in formula (2) the plus sign is chosen, then substituting $x+y$ expressed from the equation $x+y=7$ into (3) we get $z^2-14z+45=0$. Denoting the roots of the latter equation by $z_1^{(1)}$ and $z_2^{(1)}$ we find

$z_1^{(1)} = 9$ and $z_2^{(1)} = 5$. For $z = 9$ it follows from equation (1) that $xy = -16$. Combining this equation with $x + y = 7$ and solving them we find

$$x_1^{(1)} = \frac{7+\sqrt{113}}{2}, \; y_1^{(1)} = \frac{7-\sqrt{113}}{2}$$

and

$$x_2^{(1)} = \frac{7-\sqrt{113}}{2}, y_2^{(1)} = \frac{7+\sqrt{113}}{2}.$$

Finally, if $z = 5$, then from (1) we determine $xy = 12$. Solving the system

$$\left.\begin{aligned} xy &= 12, \\ x + y &= 7, \end{aligned}\right\}$$

we obtain $x_3^{(1)} = 4, \; y_3^{(1)} = 3$ and $x_4^{(1)} = 3, \; y_4^{(1)} = 4$.

(2) In the case $x + y = -7$ we similarly obtain the equation $z^2 + 14z + 45 = 0$. Its roots are $z_1^{(2)} = -9$ and $z_2^{(2)} = -5$. Solving then in succession the two systems of equations of form

$$\left.\begin{aligned} xy &= -16, \\ x + y &= -7. \end{aligned}\right\} \qquad (4)$$

and

$$\left.\begin{aligned} xy &= 12, \\ x + y &= -7, \end{aligned}\right\} \qquad (5)$$

we find from system (4) the roots

$$x_1^{(2)} = \frac{-7-\sqrt{113}}{2}, \quad y_1^{(2)} = \frac{-7+\sqrt{113}}{2}$$

and

$$x_2^{(2)} = \frac{-7+\sqrt{113}}{2}, y_2^{(2)} = \frac{-7-\sqrt{113}}{2},$$

and from system (5) the roots

$$x_3^{(2)} = -4, y_3^{(2)} = -3$$

and

$$x_4^{(2)} = -3, y_4^{(2)} = -4.$$

Our argument implies that only the following eight triplets of numbers $x, y, z)$ can represent the solutions of the original system:

$$\left(\frac{7+\sqrt{113}}{2}, \frac{7-\sqrt{113}}{2}, 9\right);$$

$$\left(\frac{7-\sqrt{113}}{2}, \frac{7+\sqrt{113}}{2}, 9\right);$$

$$(4, \;\; 3, \;\; 5); (3, \;\; 4, \;\; 5);$$

$$\left(\frac{-7-\sqrt{113}}{2}, \frac{-7+\sqrt{113}}{2}, -9\right);$$

$$\left(\frac{-7+\sqrt{113}}{2}, \frac{-7-\sqrt{113}}{2}, -9\right);$$

$$(-4,\ -3,\ -5); (-3,\ -4,\ -5).$$

Substituting these values into the system we check that they all are in fact solutions.

74. Let (x, y, z) be a real solution of the system. Consider the first equation of the system. By equality (1) on page 13, we have

$$\frac{2z}{1+z^2} \le 1.$$

The first equation then implies that

$$x \le z. \tag{1}$$

Similarly, from the second and third equations of the system we obtain

$$y \le x \tag{2}$$

and

$$z \le y. \tag{3}$$

The system of inequalities (1)-(3) is satisfied only if

$$x = y = z. \tag{4}$$

Substituting $z = x$ into the first equation we find

$$x_1 = 0, \quad x_2 = 1.$$

From (4) we finally conclude that the system has two real solutions, namely (0, 0, 0) and (1, 1, 1).

75. Let $x_1, x_2, \ldots, x_n$ be a real solution of the system. The numbers $x_k (k = 1, \ldots, n)$ are obviously of the same sign. For definiteness, let us suppose that they all are positive: $x_k > 0$ (if otherwise, we can change the signs in all equations of the system). Let us show that

$$x_k \ge \sqrt{2} \ (k = 1, 2, \ldots, n). \tag{1}$$

Indeed, by inequality (1) on page 13, we have

$$x_k + \frac{2}{x_k} \ge 2\sqrt{x_k \cdot \frac{2}{x_k}} = 2\sqrt{2},$$

whence it follows, by virtue of the equation of the system, that inequality (1) is fulfilled.

Now adding together all the equations of the system we obtain

$$x_1 + x_2 + \ldots + x_n = \frac{2}{x_1} + \frac{2}{x_2} + \ldots + \frac{2}{x_n}. \tag{2}$$

According to condition (1) equality (2) is only possible if all the unknowns are equal to $\sqrt{2}$. It can be easily verified that the numbers $x_1 = x_2 = \ldots = x_n = \sqrt{2}$ satisfy the original system and therefore it has a positive solution which is unique. Changing the signs of the values of the unknowns we get another real solution

$$x_1 = x_2 = \ldots = x_n = -\sqrt{2}.$$

Thus, the system has only two real solutions.

76. Let x, y, z be a solution of the system. Expressing x from the first equality and substituting it into the second and third ones we obtain

$$\left.\begin{aligned}(a-b)+(c-b)y+(d-b)z=0,\\ (a^2-b^2)+(c^2-b^2)y+(d^2-b^2)z=0.\end{aligned}\right\}$$

Whence we find, after some simple transformations, the expressions

$$y=-\frac{(a-b)\,(a-d)}{(c-b)(c-d)},\quad z=-\frac{(a-b)\,(a-c)}{(d-b)\,(d-c)}.$$

Substituting these values of y and z into the first equality we obtain

$$x=-\frac{(a-c)\,(a-d)}{(b-c)\,(b-d)}.$$

Consequently, we can write the inequality

$$xyz=\frac{(a-b)^2\,(a-c)^2\,(a-d)^2}{(b-c)^2\,(c-d)^2(d-b)^2}>0.$$

77. If $a \neq 0$, then $x = a$ is not a root of the equation. Dividing both sides of the equation by $\sqrt[3]{(a-x)^2}$ we replace it by the equivalent equation

$$\sqrt[3]{\left(\frac{a+x}{a-x}\right)^2}+4=5\sqrt[3]{\frac{a+x}{a-x}}.$$

Putting $t=\sqrt[3]{\dfrac{a+x}{a-x}}$ we find $t_1 = 4$, $t_2 = 1$. It follows that $x_1 = \dfrac{63}{65}a$ and $x_2 = 0$. If $a = 0$, the original equation has only one root $x = 0$.

78. By substitution we verify that $x = 1$ is not a root. Therefore, after both sides have been divided by $\sqrt[m]{(1-x)^2}$ the equation turns into the equivalent equation

$$\sqrt[m]{\left(\frac{1+x}{1-x}\right)^2}-1=\sqrt[m]{\frac{1+x}{1-x}}.$$

Denoting $\sqrt[m]{\dfrac{1+x}{1-x}}$ by t we get the equation $t^2-1=t$, i.e. $t^2-t-1=0$.

Whence we find $t_1=\dfrac{1+\sqrt{5}}{2}$ and $t_2=\dfrac{1-\sqrt{5}}{2}$. Since the second value is negative, then if m is even, the value t_2 should be discarded according to our convention concerning the roots of equations. Thus, for even m we have

$$\sqrt[m]{\frac{1+x}{1-x}}=\frac{1+\sqrt{5}}{2},\ \frac{1+x}{1-x}=\left(\frac{1+\sqrt{5}}{2}\right)^m$$

and, consequently,

$$x=\frac{\left(\frac{1+\sqrt{5}}{2}\right)^m-1}{\left(\frac{1+\sqrt{5}}{2}\right)^m+1}.$$

If m is odd, the equation has the roots, namely

$$x_{1,\,2}=\frac{\left(\frac{1\pm\sqrt{5}}{2}\right)^m-1}{\left(\frac{1\pm\sqrt{5}}{2}\right)^m+1}.$$

79. Making the substitution $\sqrt{2y-5}=t\geq 0$ we obtain

$$\sqrt{t^2+2t+1}+\sqrt{t^2+6t+9}=14.$$

This implies $t+1+t+3=14$ and $t=5$. Solving the equation

$$\sqrt{2y-5}=5,$$

we find

$$y=15.$$

80. Multiplying both sides of the equation by $\sqrt{x+\sqrt{x}}$ we get

$$x-\sqrt{x^2-x}=\frac{1}{2}\sqrt{x} \tag{1}$$

Since $x>0$ (for $x=0$ the right-hand side of the original equation makes no sense), equation (1) is equivalent to the equation

$$2\sqrt{x}-1=2\sqrt{x-1}.$$

Squaring both sides of the latter equation we see that it has the unique root $x=\frac{25}{16}$ which also satisfies the original equation.

81. Multiplying both sides of the equation by $\sqrt{x+1}$ and putting $x^2+8x=t$ we arrive at the equation

$$\sqrt{t}+\sqrt{t+7}=7.$$

This equation has a unique root: $t=9$. Solving then the equation $x^2+8x-9=0$ we find $x_1=-9$ and $x_2=1$. The original equation, by virtue of the convention concerning the values of roots, is only satisfied by $x=1$.

82. Cubing both sides of the equation we obtain

$$x-1+3\sqrt[3]{(x-1)^2}\ \sqrt[3]{x+1}+3\sqrt[3]{x-1}\ \sqrt[3]{(x+1)^2}+x+1=2x^3.$$

Whence we find

$$2x+3\sqrt[3]{x^2-1}\,(\sqrt[3]{x-1})+\sqrt[3]{x+1})=2x^3. \qquad (1)$$

On the basis of the original equation we thus can write

$$2x+3\sqrt[3]{x^2-1}\;x\;\sqrt[3]{2}=2x^3. \qquad (2)$$

After some simple transformations we deduce

$$x\sqrt[3]{x^2-1}\,[3\sqrt[3]{2}-2\sqrt[3]{(x^2-1)^2}]=0.$$

Thus we find all the numbers which can serve as the roots of the original equation. Indeed, we obviously have

$$x_1=0, \quad x_2=1, \quad x_3=-1.$$

Solving then the equation

$$3\sqrt[3]{2}=2\sqrt[3]{(x^2-1)^2},$$

we find

$$27=4\,(x^2-1)^2,\ (x^2-1)^2=\frac{27}{4},\ x^2=1\pm\frac{3\sqrt{3}}{2}.$$

Since we are only interested in real roots, it follows that

$$x^2=1+\frac{3\sqrt{3}}{2}.$$

Consequently, $x_4=\sqrt{1+\dfrac{3\sqrt{3}}{2}}$, $x_5=-\sqrt{1+\dfrac{3\sqrt{2}}{2}}$.

It is readily checked by substitution that x_1, x_2, and x_3 are roots of the original equation. But the direct substitution of the values x_4 and x_5 involves some difficulties. We proceed therefore as follows. Let us put

$$a=\sqrt[3]{x_4-1},\quad b=\sqrt[3]{x_4+1}$$

and $\qquad c=\sqrt[3]{2}\,x_4,$

and show that

$$a+b=c. \qquad (3)$$

Since x_4 satisfies equation (2), we have

$$a^3+3abc+b^3=c^3, \qquad (4)$$

and thus we must show that (4) implies (3). Note that if $a+b$ is substituted for c into (4) this results in an identity. Consequently, according to Bezout's theorem, the expression $c^3-3abc-a^3-b^3$ regarded as a polynomial in c is divisible by the binomial $c-(a+b)$. Performing the division we get

$$c^3-3abc-a^3-b^3=[c-(a+b)]\{c^2+c\,(a+b)+a^2-ab+b^2\}. \qquad (5)$$

By (4), the left-hand side of (5) is equal to zero. It is however readily seen that $a>0,\ b>0,\ c>0$, which implies that the expression in the braces is positive. Thus, equality (3) has been proved. We then similarly prove that x_5 is also a root of the original equation.

83. Transposing $\sqrt{x}$ to the left-hand side and squaring both members of the equation we get

$$\sqrt{x}.\sqrt{x-4a+16}=x-2a.$$

Squaring then both sides of the resulting equation we find that $x=\frac{a^2}{4}$ is the only root of the equation. Substituting it into the equation we obtain

$$\sqrt{a^2-16a+64}=2\sqrt{a^2-8a+16}-\sqrt{a^2},$$

which implies, since the radicals are positive, the relation

$$|a-8|=2|a-4|-|a|. \qquad (1)$$

For $a\geq 8$ equality (1) is fulfilled. Consequently, for $a\geq 8$ the original equation has a root $x=\frac{a^2}{4}$. For $4\leq a<8$ condition (1) is not fulfilled because

$$8-a\neq 2(a-4)-a.$$

For $0\leq a<4$ condition (1) takes the form

$$8-a=2(4-a)-a$$

and is only fulfilled for $a=0$. Finally, for $a<0$ condition (1) turns into the identity $8-a=2(4-a)+a$. Hence, for $a\geq 8$ and $a\leq 0$ the equation has the only root

$$x=\frac{a^2}{4}.$$

For $0<a<8$ there are no roots at all.

84. Squaring both members of the first equation and substituting the expression of x^2+y^2 found from the second equation into the resulting equation we obtain

$$36xy-1=\sqrt{-\frac{11}{5}+64xy+256(xy)^2}.$$

Again squaring both members of the equation we arrive at a quadratic equation with respect to $t=xy$:

$$650t^2-85t+2=0.$$

Solving this equation we find $t_1=\frac{1}{10}$ and $t_2=\frac{2}{65}$. Now consider the following two systems of equations:

$$\left.\begin{aligned} x^2+y^2+4xy&=\frac{1}{5},\\ xy&=\frac{1}{10},\end{aligned}\right\}(1) \qquad \left.\begin{aligned} x^2+y^2+4xy&=\frac{1}{5},\\ xy&=\frac{2}{65}.\end{aligned}\right\} \qquad (2)$$

Obviously, all the solutions of the original system are solutions of these systems.

Solving system (1) we find

$$(x+y)^2=\frac{1}{5}-2xy=\frac{1}{5}-\frac{1}{5}=0.$$

Consequently, $x+y=0$, and thus we get two solutions of system (1):

$$x_1=\frac{t}{\sqrt{10}},\ y_1=-\frac{t}{\sqrt{10}};\ x_2=-\frac{t}{\sqrt{10}},\ y_2=\frac{t}{\sqrt{10}}.$$

Transforming the first equation of system (2) to the form $(x+y)^2=\dfrac{9}{65}$ we reduce the system to the following two systems:

$$\left.\begin{aligned} x+y&=\frac{3}{\sqrt{65}},\\ xy&=\frac{2}{65},\end{aligned}\right\}(2') \qquad \left.\begin{aligned} x+y&=-\frac{3}{\sqrt{65}},\\ xy&=\frac{2}{65}.\end{aligned}\right\} \qquad (2'')$$

System (2') has two solutions, namely

$$x_3=\frac{2}{\sqrt{65}},\ y_3=\frac{1}{\sqrt{65}} \text{ and } x_4=\frac{1}{\sqrt{65}},\ y_4=\frac{2}{\sqrt{65}}.$$

System (2″) also has two solutions:

$$x_5=-\frac{2}{\sqrt{65}},\ y_5=-\frac{1}{\sqrt{65}} \text{ and } x_6=-\frac{1}{\sqrt{65}},\ y_6=-\frac{2}{\sqrt{65}}.$$

As is readily verified, the original system is only satisfied by the first, second, third and sixth sets of numbers. Thus, the system has exactly four solutions.

85. Putting

$$\sqrt[3]{x}=u,\ \sqrt[3]{y}=v$$

we can rewrite the given system in the form

$$\left.\begin{aligned} u^3-v^3&=\frac{7}{2}(u^2v-uv^2),\\ u-v&=3.\end{aligned}\right\}$$

The first equation is transformed to the form

$$(u-v)^2+3uv=\frac{7}{2}uv,$$

whence we find

$$uv=18.$$

Combining the latter equation with the second equation of the system and solving them we find $u_1 = 6$, $v_1 = 3$ and $u_2 = -3$, $v_2 = -6$. Returning to the original system we get its two solutions:

$$x_1 = 216, y_1 = 27 \text{ and } x_2 = -27, \ y_2 = -216.$$

86. Making the substitution $\sqrt{\frac{x}{y}} = t \geq 0$ we transform the first equation to the form

$$2t^2 - 3t - 2 = 0.$$

It follows that $t = 2$ (the second root $-\frac{1}{2}$ is discarded). Solving the system

$$\left.\begin{array}{r} \sqrt{\frac{x}{y}} = 2, \\ x + xy + y = 9, \end{array}\right\}$$

we find its two solutions

$$x_1 = 4, \ \ y_1 = 1 \text{ and } x_2 = -9, \quad y_2 = -\frac{9}{4},$$

which are also solutions of the original system. Thus, the original system has two solutions.

87. Let us put

$$\sqrt{\frac{y+1}{x-y}} = t > 0.$$

Then the first equation takes the form

$$t^2 - 3t + 2 = 0,$$

whence we find $t_1 = 1$ and $t_2 = 2$.

Consider now the following two systems of equations:

$$\left.\begin{array}{r} \sqrt{\frac{y+1}{x-y}} = 1, \\ x + xy + y = 7, \end{array}\right\} (1) \qquad \left.\begin{array}{r} \sqrt{\frac{y+1}{x-y}} = 2, \\ x + xy + y = 7, \end{array}\right\} \qquad (2)$$

System (1) possesses two solutions:

$$(-5, -3); \ (3, 1).$$

System (2) also has two solutions:

$$\left(\sqrt{10} - 1, \ \frac{\sqrt{160} - 5}{5}\right); \left(-\sqrt{10} - 1, \ \frac{-\sqrt{160} - 5}{5}\right).$$

Hence, the original system has four solutions.

88. Taking into account that

$$\sqrt{\frac{x+y}{x-y}} = \frac{1}{|x-y|}\sqrt{x^2 - y^2},$$

and multiplying the first equation by $x - y$ we obtain

$$x^2-y^2-\sqrt{x^2-y^2}-12=0 \quad \text{for} \quad x-y>0$$

and

$$x^2-y^2+\sqrt{x^2-y^2}-12=0 \quad \text{for} \quad x-y<0.$$

Whence

$$(\pm\sqrt{x^2-y^2})_1=4, (\pm\sqrt{x^2-y^2})_2=-3.$$

Thus, we now must consider the two systems of equations

$$\left.\begin{array}{r} x^2-y^2=16, \\ xy=15, \end{array}\right\} (1) \quad \left.\begin{array}{r} x^2-y^2=9, \\ xy=15. \end{array}\right\} \quad (2)$$

System (1) has two real solutions:

$$x_1=5, \quad y_1=3 \text{ and } x_2=-5, \quad y_2=-3.$$

System (2) also has two real solutions:

$$x_3=\sqrt{\frac{9+\sqrt{981}}{2}}, y_3=\sqrt{\frac{\sqrt{981}-9}{2}}$$

and

$$x_4=-\sqrt{\frac{9+\sqrt{981}}{2}}, y_4=-\sqrt{\frac{\sqrt{981}-9}{2}}.$$

It can be, however, easily checked that the original system is satisfied only by two of these pairs of numbers, namely by

$$(5, 3); \left(-\sqrt{\frac{\sqrt{981}+9}{2}}, -\sqrt{\frac{\sqrt{981}-9}{2}}\right).$$

Thus, the original system has two real solutions.

89. Put

$$\sqrt{x^2-12y+1}=t.$$

Then, the first equation can be written in the form

$$t^2-8t+16=0.$$

It follows that $t_{1,\,2}=4$, and thus we obtain

$$x^2-12y=15. \quad (1)$$

Noting that $y \neq 0$, we multiply the second equation by $\frac{2x}{y}$ which transform it to the form

$$\left(\frac{x}{2y}\right)^2-2\left(\frac{x}{2y}\right)\sqrt{1+\frac{4x}{3y}}+\left(1+\frac{4x}{3y}\right)=0.$$

This implies

$$\frac{x}{2y}-\sqrt{1+\frac{4x}{3y}}=0. \quad (2)$$

Raising to the second power we arrive at the equation

$$3\left(\frac{x}{y}\right)^2 - 16\left(\frac{x}{y}\right) - 12 = 0,$$

wherefrom we find

$$\left(\frac{x}{y}\right)_1 = 6, \quad \left(\frac{x}{y}\right)_2 = -\frac{2}{3}.$$

It is obvious that the second value does not satisfy equation (2) and therefore we confine ourselves to the system

$$\left.\begin{aligned} x^2 - 12y = 15, \\ \frac{x}{y} = 6. \end{aligned}\right\}$$

This system has two solutions $\left(5, \frac{5}{6}\right)$ and $\left(-3, -\frac{1}{2}\right)$ which, as is readily seen, satisfy the original system as well.

90. Rationalizing the denominators of the first equation, we obtain

$$\frac{4x^2 - 2y^2}{y^2} = \frac{17}{4}.$$

Whence we find

$$\left(\frac{x}{y}\right)_1 = \frac{5}{4} \quad \text{and} \quad \left(\frac{x}{y}\right)_2 = -\frac{5}{4}.$$

In the second equation we put

$$\sqrt{x^2 + xy + 4} = t, \tag{1}$$

and rewrite it in the form

$$t^2 + t - 56 = 0.$$

Hence, we obtain $t_1 = 7$ and $t_2 = -8$. Since in (1) we have $t \geq 0$, the second root must be discarded. As a result, we arrive at the following two systems of equations:

$$\left.\begin{aligned} x = \frac{5}{4}y, \\ x^2 + xy - 45 = 0 \end{aligned}\right\} \tag{2}$$

and

$$\left.\begin{aligned} x = -\frac{5}{4}y, \\ x^2 + xy - 45 = 0. \end{aligned}\right\} \tag{3}$$

The solutions of system (2) are (5, 4) and (–5, –4). The solutions of (3) are (15, –12) and (–15, 12). These four solutions satisfy the original system as well.

91. Expressing x from the second equation and substituting it into the first one we obtain

$$y^2+\sqrt{3y^2-\frac{4}{3}y-\frac{1}{3}}=\frac{2}{3}\frac{2y+5}{3}+5.$$

Putting here $\sqrt{\dfrac{9y^2-4y-1}{3}}=t\geq 0$ we arrive at the equation

$$t^2+3t-18=0.$$

Whence we find

$$t_1=3,\quad t_2=-6.$$

Since, by the hypothesis, t is non-negative, we have only one equation

$$9y^2-4y-28=0.$$

Combining this equation with the second equation of the original system, we find their two solutions

$$x_1=3,\quad y_1=2\quad \text{and}\quad x_2=\frac{17}{27},\quad y_2=-\frac{14}{9}.$$

92. Let us put

$$\sqrt{x^2-6y+1}=t\geq 0.$$

Then the first equation is written in the form

$$t^2-8t+16=0.$$

Whence we obtain $t=4$, and thus

$$x^2-6y-15=0. \qquad (1)$$

If now we put $x^2y=u$ in the second equation and take into account (1), we get the equation

$$9u^2-241u-13230=0,$$

from which we obtain $u_1=54$ and $u_2=-\dfrac{245}{9}$.

We thus arrive at the two systems of equations

$$\left.\begin{aligned}x^2-6y-15=0,\\ x^2y=54,\end{aligned}\right\}\ (2) \qquad \left.\begin{aligned}x^2-6y-15=0,\\ x^2y=-\frac{245}{9}.\end{aligned}\right\}\ (3)$$

Eliminating x^2 from system (2), we obtain the equation

$$2y^2+5y-18=0,$$

whose roots are $y_1=2$ and $y_2=-4\dfrac{1}{2}$. The second root must be discarded because, by virtue of the equation $x^2y=54$, it leads to nonreal values of x. Hence, system (2) has two real solutions:

$$x_1=\sqrt{27},\,y_1=2;\,x_2=-\sqrt{27},\,y_2=2.$$

System (3) is reduced to the equation

$$54y^2+135y+245=0,$$

which has no real solutions. Thus, the original system has two real solutions.

93. Put

$$\sqrt{x}=u\geq 0,\quad \sqrt{y}=v\geq 0. \qquad (1)$$

Then the system is rewritten in the following way:

$$\left.\begin{aligned}(u^2-v^2)v&=\frac{u}{2},\\ (u^2+v^2)u&=3v.\end{aligned}\right\} \qquad (2)$$

System (2) has an obvious solution, namely

$$u=0,\quad v=0. \qquad (3)$$

Therefore, in what follows we suppose that $u\neq 0$, and hence (by virtue of the equations) we also have $v\neq 0$. Multiplying the right-hand and left-hand sides of equations (2) we obtain

$$u^4-v^4=\frac{3}{2}. \qquad (4)$$

Multiply then the first equation of system (2) by v, the second by u and adding them together we obtain the following equation:

$$u^4-v^4+2u^2v^2=\frac{7}{2}uv.$$

By virtue of (4), we have

$$4(uv)^2-7uv+3=0. \qquad (5)$$

Whence we find

$$(uv)_1=1,\qquad (uv)_2=\frac{3}{4}.$$

Now consider the two systems of equations

$$\left.\begin{aligned}uv&=1,\\ (u^2+v^2)u&=3v,\end{aligned}\right\}(6)\qquad \left.\begin{aligned}uv&=\frac{3}{4},\\ (u^2+v^2)u&=3v,\end{aligned}\right\} \qquad (7)$$

It is obvious that any solution of system (2) other than (3) is among the solutions of these systems.

Multiplying the second equation of system (6) by u we find, by virtue of the first equation, that $u^4=2$. Whence, taking into account (1). we get

$$u=\sqrt[4]{2},\quad v=\frac{\sqrt[4]{8}}{2}.$$

Analogously, we also find the solution of system (7) satisfying the condition (1):

$$u=\frac{\sqrt[4]{27}}{2},\quad v=\frac{\sqrt[4]{3}}{2}.$$

It is easy to check that both solutions also satisfy system (2). Thus, the original system has three solutions:

$$(0,\ 0);\left(\sqrt{2},\ \frac{\sqrt{2}}{2}\right);\left(\frac{3\sqrt{3}}{4},\ \frac{\sqrt{3}}{4}\right).$$

94. Squaring both members of the first equation we obtain

$$\sqrt{x^2-y^2}=x-\frac{a^2}{2}. \qquad (1)$$

By virtue of the second equation, we have

$$\sqrt{x^2+y^2}=\frac{3a^2}{2}-x. \qquad (2)$$

Now squaring both sides of the second equation of the original system we receive

$$\sqrt{x^2+y^2}\sqrt{x^2-y^2}=\frac{a^4}{2}-x^2.$$

Whence, by virtue of (1) and (2), we find

$$\frac{a^4}{2}-x^2=\left(x-\frac{a^2}{2}\right)\left(\frac{3a^2}{2}-x\right).$$

Removing the brackets we obtain $x=\frac{5}{8}a^2$. After this we easily get from equation (1) the two values of y

$$y_1=a^2\sqrt{\frac{3}{8}}\ y_2=-a^2\sqrt{\frac{3}{8}}.$$

The verification by substitution shows however that the original system has only one solution $\left(\frac{5}{8}a^2,\ a^2\sqrt{\frac{3}{8}}\right)$.

95. Let us put

$$\sqrt{x}=u\geq 0 \text{ and } \sqrt{y}=v\geq 0. \qquad (1)$$

This reduces the system to the form

$$\left.\begin{aligned}u^3-v^3&=a(u-v),\\ u^4+u^2v^2+v^4&=b^2.\end{aligned}\right\} \qquad (2)$$

It appears obvious that the latter system falls into two systems of the form

$$\left.\begin{aligned}u-v=0,\\ u^4+u^2v^2+v^4=b^2,\end{aligned}\right\}(2') \text{ and } \left.\begin{aligned}u^2+uv+v^2=a,\\ u^4+u^2v^2+v^4=b^2,\end{aligned}\right\} \qquad (2'')$$

Solving system (2′) we find $3u^4=b^2$, whence, taking into consideration (1), we get

$$u=\frac{\sqrt{b}\sqrt[4]{27}}{3},v=\frac{\sqrt{b}\sqrt[4]{27}}{3}. \qquad (3)$$

Passing to system (2″), we transform both equations in the following way:

$$u^2+v^2=a-uv,\ (u^2+v^2)^2=b^2+u^2v^2.$$

This yields the values of uv and 4^2+v^2:

$$\left.\begin{aligned} uv&=\frac{a^2-b^2}{2a},\\ u^2+v^2&=\frac{a^2+b^2}{2a}.\end{aligned}\right\} \qquad (4)$$

It can easily be shown that the system of equations (4) is equivalent to system (2″).

From equations (4) we receive

$$\left.\begin{aligned} (u+v)^2&=\frac{3a^2-b^2}{2a}.\\ (u-v)^2&=\frac{3b^2-a^2}{2a}.\end{aligned}\right\} \qquad (5)$$

It should be noted that, by virtue of (1), the right-hand member of the first equation of system (4) must be non-negative; the right-hand member of the second equation of system (5) must also be non-negative. Thus, we must impose the condition

$$3b^2\ge a^2\ge b^2 \qquad (6)$$

because, if otherwise, system (5), and, hence, system (2″) have no solutions satisfying condition (1).

Solving system (5) we get

$$u+v=\sqrt{\frac{3a^2-b^2}{2a}},\quad u-v=\pm\sqrt{\frac{3b^2-a^2}{2a}}.$$

Finally we obtain

$$u=\frac{1}{2}\left(\sqrt{\frac{3a^2-b^2}{2a}}\pm\sqrt{\frac{3b^2-a^2}{2a}}\right),$$

$$v=\frac{1}{2}\left(\sqrt{\frac{3a^2-b^2}{2a}}\mp\sqrt{\frac{3b^2-a^2}{2a}}\right),$$

As is easily seen, by virtue of condition (6), both pairs of values (u, v) are non-negative. Indeed, we have $a^2\ge b^2$ and therefore $3a^2-b^2\ge 3b^2-a^2$.

Thus, if the additional condition (6) is fulfilled, the original system has three solutions, namely

$$x_1=\frac{b}{\sqrt{3}},\quad y_1=\frac{b}{\sqrt{3}};$$

$$x_2 = \frac{1}{4}\left(\sqrt{\frac{3a^2-b^2}{2a}}+\sqrt{\frac{3b^2-a^2}{2a}}\right)^2,$$

$$y_2 = \frac{1}{4}\left(\sqrt{\frac{3a^2-b^2}{2a}}-\sqrt{\frac{3b^2-c^2}{2a}}\right)^2;$$

$$x_3 = \frac{1}{4}\left(\sqrt{\frac{3a^2-b^2}{2a}}-\sqrt{\frac{3b^2-a^2}{2a}}\right)^2,$$

$$y_3 = \frac{1}{4}\left(\sqrt{\frac{3a^2-b^2}{2a}}+\sqrt{\frac{3b^2-a^2}{2a}}\right)^2.$$

If condition (6) is violated, then only the first solution remains valid.

3. Algebraic Inequalities

96. For the quadratic trinomial

$$ax^2+bx+c \qquad (a \neq 0)$$

to be positive for all x it is necessary and sufficient that $a>0$ and the discriminant D of the trinomial be negative. In our case we have

$$a = r^2-1>0 \tag{1}$$

and

$$D = 4(r-1)^2-4(r^2-1) = -8(r-1)<0. \tag{2}$$

Inequalities (1) and (2) are fulfilled simultaneously for $r>1$. It should also be noted that for $r=1$ the polynomial under consideration is identically equal to 1. Thus, all the sought-for values of r are determined by the inequality

$$r \geq 1.$$

97. If we put

$$\frac{x}{y}+\frac{y}{x}=u$$

and take into account that $\frac{x^2}{y^2}+\frac{y^2}{x^2}=u^2-2$, the given expression is readily transformed to the form

$$3u^2-8u+4. \tag{1}$$

If x and y are of opposite signs, then $u<0$ and trinomial (1) is positive. If x and y are of the same sign, it is easily seen that $u \geq 2$. The roots of quadratic trinomial (1) being equal to $\frac{2}{3}$ and 2, the trinomial is non-negative for $u \geq 2$. Thus, the trinomial is non-negative both for $u<0$ and $u \geq 2$, and, consequently, the original expression is non-negative for all real nonzero values of x and y.

98. Note that $x^2 - x + 1 > 0$ for all values of x because the discriminant of the quadratic trinomial is equal to $-3 < 0$ and the coefficient in x^2 is positive. Therefore it is permissible to multiply both inequalities by the denominator. This results in

$$-3x^2 + 3x - 3 < x^2 + ax - 2,$$
$$x^2 + ax - 2 < 2x^2 - 2x + 2,$$

that is

$$4x^2 + (a-3)x + 1 > 0,$$
$$x^2 - (a+2)x + 4 > 0.$$

The first inequality is fulfilled for all x if and only if the discriminant of the quadratic trinomial is negative, i.e. if $(a-3)^2 - 16 < 0$. Similarly, the second inequality is fulfilled if and only if

$$(a+2)^2 - 16 < 0.$$

Now combining the two inequalities $(a-3)^2 - 16 < 0$ and $(a+2)^2 - 16 < 0$ and solving them as a system with respect to a we get

$$-4 < a - 3 < 4, \quad -1 < a < 7$$

and

$$-4 < a + 2 < 4, \quad -6 < a < 2.$$

Hence, we finally obtain $-1 < a < 2$.

99. By virtue of inequality (1) on page 20, we have

$$a^4 + b^4 \ge 2a^2b^2,$$
$$c^4 + d^4 \ge 2c^2d^2.$$

Adding together these inequalities, we obtain

$$a^4 + b^4 + c^4 + d^4 \ge 2(a^2b^2 + c^2d^2). \tag{1}$$

According to inequality (3) on page 20, after putting $u = a^2b^2$ and $v = c^2d^2$, we receive

$$a^2b^2 + c^2d^2 \ge 2\sqrt{a^2b^2c^2d^2}. \tag{2}$$

We always have $\sqrt{a^2b^2c^2d^2} \ge abcd$ (the sign > appears if $abcd < 0$), and therefore comparing (1) and (2) we arrive at the required proof.

100. The given system is equivalent to the system

$$x^2 + (x+a)^2 + 2x \le 1, \quad y = x + a.$$

The inequality $2x^2 + 2(a+1)x + a^2 - 1 \le 0$

has a unique solution with respect to x if and only if the discriminant of the trinomial is equal to zero:

$$(a+1)^2 - 2(a^2 - 1) = 0,$$

i.e.

$$a^2 - 2a - 3 = 0.$$

Solving the latter equation we find

$$a_1 = 3, \quad a_2 = -1.$$

Finally, we consider the two possible cases:

(1) If $a = 3$, then $x^2 + 4x + 4 = 0$ and $x = -2$, $y = 1$.

(2) If $a = -1$, then $x^2 = 0$ and $x = 0$, $y = -1$.

101. Rewrite the given system of inequalities in the following way:

$$y+\frac{1}{2}>\left|x^2-2x\right|,\ y<2-|x-1|.$$

Since we always have $\left|x^2-2x\right|\geq 0$ and $|x-1|\geq 0$, we can write

$$-\frac{1}{2}<y<2.$$

The only integers y satisfying this inequality are 0 and 1. Consequently, the given system of inequalities considered for integral x and y can be consistent only for the values $y=0$ and $y=1$. Let us consider both cases.

Case 1. If $y=0$, the system of inequalities takes the form

$$\left|x^2-2x\right|<\frac{1}{2},\ \ |x-1|<2.$$

The second of these inequalities is satisfied only by the integral numbers 0, 1 and 2. It can easily be checked by substitution that 0 and 2 satisfy the first inequality as well, but it is not satisfied by 1. Thus, for the case $y=0$ two solutions are found, namely

$$x_1=0,\ y_1=0 \text{ and } x_2=2,\ y_2=0.$$

Case 2. If $y=1$, the original system of inequalities reduces to

$$\left|x^2-2x\right|<\frac{3}{2},\ \ |x-1|<1.$$

The second inequality is satisfied by the only integral number $x=1$ which also satisfies the first inequality. Hence, in this case we have one more solution of the problem: $x_3=1$, $y_3=1$. Thus, the system of inequalities is satisfied by three pairs of integers.

102. There are n summands on the left-hand side of the inequality, the first $n-1$ summands being greater than the last one. Therefore,

$$\frac{1}{n+1}+\frac{1}{n+2}+\ldots+\frac{1}{2n}>n\frac{1}{2n}=\frac{1}{2}.$$

103. Let S_m denote the left member of the inequality to be proved. Then, as is easily seen,

$$S_{m+1}-S_m=\frac{1}{3m+4}+\frac{1}{3m+3}+\frac{1}{3m+2}-\frac{1}{m+1}.$$

Reducing the fractions to a common denominator we find

$$S_{m+1}-S_m=\frac{2}{(3m+2)(3m+3)(3m+4)}>0.$$

Thus, $S_{m+1}>S_m$. We have $S_1=\frac{1}{2}+\frac{1}{3}+\frac{1}{4}>1$,

and, consequently, $S_m>S_{m-1}>\ldots.>S_2>S_1>1$, i.e. $S_m>1$ which is what we set out to prove.

104. Write the following obvious inequalities:

$$\frac{1}{2^2} < \frac{1}{1\cdot 2} = \frac{1}{1} - \frac{1}{2},$$

$$\frac{1}{3^2} < \frac{1}{2\cdot 3} = \frac{1}{2} - \frac{1}{3},$$

. . . .

$$\frac{1}{n^2} < \frac{1}{(n-1)n} = \frac{1}{n-1} - \frac{1}{n}.$$

Adding them termwise, we get

$$\frac{1}{2^2} + \frac{1}{3^2} + \ldots + \frac{1}{n^2} < 1 - \frac{1}{n} = \frac{n-1}{n},$$

which is the required result.

105. Rewrite both sides of the given inequality in the following way:

$$(n!)^2 = \underbrace{(1\cdot n)[2(n-1)]\ldots[k(n-k+1)]\ldots(n\cdot 1)}_{n\,\text{factors}}$$

and $$n^n = \underbrace{n\cdot n\ldots.n.}_{n\,\text{factors}}$$

Let us prove that

$$(n-k+1)k \ge n \qquad (1)$$

for $n \ge k \ge 1$. Indeed, we have

$$nk - k^2 + k - n = k(n-k) - (n-k) = (n-k)(k-1) \ge 0. \qquad (2)$$

Thus, we have proved that

$$(n!)^2 \ge n^n. \qquad (3)$$

Let us note that if a number k is greater than unity and less than n, formula (1), as it follows from (2), assumes the form of a strict inequality which obviously leads to a strict inequality in formula (3) as well. For $n > 2$ there exists such k. Hence, in this case we have the strict inequality $(n!)^2 > n^n$.

106. It can easily be checked that for constructing a triangle with sides a,b and c it is necessary and sufficient that the numbers a, b *and* c satisfy the three inequalities

$$\left.\begin{aligned} a+b-c&>0,\\ a+c-b&>0,\\ b+c-a&>0.\end{aligned}\right\} \qquad (1)$$

Let us prove that this system of simultaneous inequalities is equivalent to the condition set in the problem. Let us put

$$K = pa^2 + qb^2 - pqc^2.$$

Since, $q = 1 - p$, this expression can be rewritten in the form

$$\begin{aligned} K &= pa^2 + (1-p)b^2 - p\,(1-p)c^2 \\ &= c^2p^2 + (a^2 - b^2 - c^2)p + b^2, \end{aligned}$$

where a, b and c are constants, and p may assume arbitrary values. Thus, K is a quadratic trinomial in p. In the general case the trinomial K can take on values of different sign depending on p. The inequality indicated in the problem is equivalent to the condition that $K > 0$ for all p. As is known,for this to be so, it is necessary and sufficient that the discriminant

$$D = (a^2 - b^2 - c^2)^2 - 4b^2c^2$$

of the trinomial be negative (here we take into consideration that the coefficient in p^2 is equal to $c^2 > 0$).

The discriminant can be represented in the following form:

$$D = (a^2 - b^2 - c^2)^2 - 4b^2c^2 = (a^2 - b^2 - c^2 - 2bc)(a^2 - b^2 - c^2 + 2bc) =$$
$$= [a^2 - (b + c)^2][a^2 - (b - c)^2] = (a + b + c)$$
$$(a - b - c)(a + b - c)(a - b + c) =$$
$$= -(a + b + c)(a + b - c)(b + c - a)(c + a - b).$$

If a triangle can be constructed, inequalities (1) are fulfilled, and, hence, $D < 0$. Thus, we have proved that the existence of such a triangle implies the inequality $D < 0$.

Conversely, if $D < 0$, then

$$(a + b - c)(b + c - a)(c + a - b) > 0. \qquad (2)$$

Let us show that (2) implies inequalities (1). Indeed, suppose that only one expression in the brackets on the left-hand side of (2) is positive and the other two are negative. For instance, let $a + b - c < 0$ and $b + c - a < 0$. Adding together these inequalities we get $2b < 0$ which is impossible. Thus, we have also proved that the condition $D < 0$ implies the existence of a triangle with given sides a, b and c.

107. Transform the left member of the inequality in the following way:

$$4(x + y)(x + z)x(x + y + z) + y^2z^2 = 4$$
$$(x^2 + xy + xz + yz)(x^2 + xy + xz) + y^2z^2 =$$
$$= 4(x^2 + xy + xz)^2 + 4yz(x^2 + xy + xz) + y^2z^2 = [2(x^2 + xy + xz) + yz]^2.$$

The obtained expression is non-negative for any real x, y and z which is what we set out to prove.

108. Denoting the left member of the inequality by z we transform z in the following way:

$$z = x^2 + 2xy + 3y^2 + 2x + 6y + 4 = (x + y + 1)^2 + 2(y + 1)^2 + 1.$$

For real x and y the first two summands are non-negative, and, consequently, $z \geq 1$.

109. Since $x = \frac{1-4y}{2}$, the inequality to be proved is equivalent to the inequality

$$\left(\frac{1-4y}{2}\right)^2 + y^2 \ge \frac{1}{20},$$

which is readily transformed to the equivalent form

$$100y^2 - 40y + 4 = (10y-2)^2 \ge 0,$$

the latter inequality being automatically fulfilled.

110. Since $d > 0$ and $R \ge r > 0$, we have

$$d^2 + R^2 - r^2 > 0 \quad \text{and} \quad 2dR > 0.$$

Consequently, the given inequality is equivalent to the inequality

$$d^2 + R^2 - r^2 \le 2dR.$$

Reducing it to the form $(d-R)^2 \le r^2$, we get

$|d - R| \le r$, i.e. $-r \le d - R \le r$.

Hence,

$$R - r \le d \le R + r.$$

111. Multiplying both members of the desired inequality by $a+b+c$, we get an equivalent inequality whose left member is equal to

$$(a+b+c)\left(\frac{1}{a}+\frac{1}{b}+\frac{1}{c}\right) = 3 + \left(\frac{a}{b}+\frac{b}{a}\right) + \left(\frac{b}{c}+\frac{c}{b}\right) + \left(\frac{c}{a}+\frac{a}{c}\right)$$

$$= 9 + \left(\sqrt{\frac{a}{b}} - \sqrt{\frac{b}{a}}\right)^2 + \left(\sqrt{\frac{b}{c}} - \sqrt{\frac{c}{b}}\right)^2 + \left(\sqrt{\frac{c}{a}} - \sqrt{\frac{a}{c}}\right)^2 \ge 9.$$

112. Note that the given expression turns into zero for $b = c, c = a$ and $a = b$. Therefore, according to Bezout's theorem, it is divisible by the differences $a - b$, $a - c$ and $b - c$. Arranging the summands in descending powers of the letter a and performing the division by $a - b$, we receive

$$a^3(b^2 - c^2) + a^2(c^3 - b^3) + b^3c^2 - c^3b^2$$
$$= (a-b)[a^2(b^2-c^2) + ac^2(c-b) + bc^2(c-b)].$$

Taking the factor $(b - c)$ outside the square brackets and dividing the remaining polynomial by $a - c$, we obtain

$$a^3(b^2 - c^2) + b^3(c^2 - a^2) + c^3(a^2 - b^2)$$
$$= -(b-a)(c-b)(c-a)[ac + bc + ab].$$

Since, by the hypothesis, $a < b < c$

and a, b and c are of the same sign, the expression on the right-hand side is negative.

113. We have $1-2\sqrt{a_k}+a_k=(1-\sqrt{a_k})^2\geq 0$,

whence $1+a_k\geq 2\sqrt{a_k}$.

Writing these inequalities for $k=1, 2, ..$ and multiplying them termwise we receive

$$(1+a_1)(1+a_2)\dots(1+a_n)\geq 2^n\sqrt{a_1a_2\dots a_n}=2^n.$$

114. It is sufficient to consider the case when a and b are of the same sign (i.e. positive), since otherwise one of the numbers is greater than unity and the inequality becomes obvious. We have

$$a^2+b^2=(a+b)^2-2ab=1-2ab,$$
$$a^4+b^4=(1-2ab)^2-2a^2b^2.$$

But if $a+b=1$, then $0\leq ab\leq\frac{1}{4}$, since

(see formula (3) on page 20).

Consequently,

$$a^4+b^4\geq\left(1-2\cdot\frac{1}{4}\right)^2-2\cdot\frac{1}{16}=\frac{1}{8}.$$

115. Consider the following three cases:

(1) $x\leq 0$; then $x^8-x^5+x^2-x+1>0$ because the first four summands are non-negative.

(2) $0<x<1$; transform the polynomial to the form

$$x^8+(x^2-x^5)+(1-x)=x^8+x^2(1-x^3)+(1-x).$$

Here all the summands are obviously positive and, consequently, the polynomial is greater than zero.

(3) $x\geq 1$; write the polynomial in the form

$$x^5(x^3-1)+x(1-x)+1.$$

The first two summands being non-negative, we also have in this case

$$x^8-x^5+x^2-x+1>0.$$

116. We have

$$(1+x)^n+(1-x)^n=2\,(1+C_n^2x^2+C_n^4x^4+\dots),\qquad(1)$$

the last term of the sum in the brackets being equal to x^n for even n and to nx^{n-1} for odd n. By the hypothesis, we have $-1<x<1$, whence it follows that $C_n^{2k}x^{2k}<C_n^{2k}$ for all integral k. Therefore,

$$(1+x)^n+(1-x)^n<A_n,$$

where A_n is the value of polynomial (1) for $x=\pm 1$, i.e. $A_n=2^n$.

117. The inequality to be proved is equivalent to the inequality

$$\varepsilon^2(a_1^2+a_2^2+\dots+a_n^2)+4\,(x_1^2+x_2^2+\dots+x_n^2)\pm\,4\varepsilon(x_1a_1+x_2a_2+\dots+x_na_n)\geq 0,$$

which holds true because the left-hand side is equal to

$$(\varepsilon a_1\pm 2x_1)^2+(\varepsilon a_2\pm 2x_2)^2+\dots+(\varepsilon a_n\pm 2x_n)^2$$

118. The radicand must be ≥ 0, and therefore

$$-\frac{1}{2} \leq x \leq \frac{1}{2}. \tag{1}$$

For nonzero values of x satisfying condition (1) we have $\sqrt{1-4x^2} < 1$. Therefore, if $-\frac{1}{2} \leq x < 0$, the inequality indicated in the problem is fulfilled, because its left-hand side is negative.

But if $0 < x \leq \frac{1}{2}$, then rationalizing the numerator of the left-hand side we obtain

$\frac{1-\sqrt{1-4x^2}}{x} = -\frac{4x^2}{(1+\sqrt{1-4x^2})x} = \frac{4x}{1+\sqrt{1-4x^2}}$. It is readily seen that the numerator of the fraction on the right-hand side does not exceed 2 for $0 < x \leq \frac{1}{2}$, and the denominator is not less than unity. Therefore,

$$\frac{1-\sqrt{1-4x^2}}{x} \leq 2 < 3.$$

Thus, the inequality in question is true for the values $x \neq 0$ satisfying condition (1). For $x = 0$ and $|x| > \frac{1}{2}$ the left member of the inequality makes no sense.

119. For definiteness, let $x \geq y$. Then putting $\frac{y}{x} = \alpha \leq 1$ we get an equivalent inequality:

$$\sqrt[m]{1+\alpha^m} \geq \sqrt[n]{1+\alpha^n}. \tag{1}$$

Raising both members of (1) to the power mn we obtain the inequality

$$(1+\alpha^m)^n \geq (1+\alpha^n)^m.$$

It is easily seen that this inequality holds true because $0 \leq \alpha \leq 1$ and $n \geq m$.

120. Put

$$x_n = \underbrace{\sqrt{a+\sqrt{a+\ldots+\sqrt{a}}}}_{n\,\text{radicals}} \tag{1}$$

It is obvious that $x_n = \sqrt{a + x_{n-1}}$ $(n = 2, 3, \ldots)$, and, consequently, $x_n^2 = a + x_{n-1}$. Furthermore, let us note that $x_n > x_{n-1}$ because when passing from $n-1$ to n the radical $\sqrt{a}$ is replaced by a greater number $\sqrt{a+\sqrt{a}}$. For this reason we have $x_n^2 < a + x_n$ and, consequently, the quantities we are interested in satisfy the inequality

$$x^2 - x - a < 0. \tag{2}$$

The roots of the trinomial on the left- hand side are equal to

$$x^{(1)} = \frac{1-\sqrt{1+4a}}{2}, \quad x^{(2)} = \frac{1+\sqrt{1+4a}}{2}.$$

The numbers x_n satisfying inequality (2), the relation $x^{(1)} < x_n < x^{(2)}$ is fulfilled (see page 21). Hence,

$$x_n < \frac{1+\sqrt{1+4a}}{2} \quad (n = 2, 3, \ldots), \tag{3}$$

which completes the proof.

For $n = 1$ we have $x_1 = \sqrt{a}$ and the inequality (3) becomes obvious.

121. Let us denote the expression containing k radical signs by x_k:

$$\sqrt{2+\sqrt{2+\ldots+\sqrt{2+\sqrt{2}}}} = x_k.$$

Note that $x_k < 2$. Indeed, let us replace 2 in the radical $\sqrt{2}$ by 4. Then all the roots are extracted and the left member becomes equal to 2. This means that $x_k < 2$. Hence, in particular, it follows that both the numerator and denominator on the left-hand side of the original inequality are different from zero. Using then the fact that

$$x_n = \sqrt{2 + x_{n-1}}$$

we transform the left-hand side of the original inequality in the following way:

$$\frac{2-\sqrt{x_{n-1}+2}}{2-x_{n-1}} = \frac{\sqrt{x_{n-1}+2}-2}{(x_{n-1}+2)-4} = \frac{1}{\sqrt{x_{n-1}+2}+2} = \frac{1}{x_n+2}.$$

Since $x_n < 2$, we have $\dfrac{1}{x_n+2} > \dfrac{1}{4}$ which is what we set out to prove.

122. As is known, for any real numbers a and b the following inequality holds true:

$$|a \cdot b| \le \frac{a^2+b^2}{2}$$ (see formula (1), page 20).

Taking advantage of the fact that the absolute value of a sum does not exceed the sum of the absolute values of the summands we get

$$|a_1 b_1 + a_2 b_2 + \ldots + a_n b_n| \le |a_1 b_1| + |a_2 b_2| + \ldots + |a_n b_n|$$
$$\le \frac{a_1^2+b_1^2}{2} + \frac{a_2^2+b_2^2}{2} + \ldots + \frac{a_n^2+b_n^2}{2} =$$
$$= \frac{a_1^2+b_1^2+\ldots+a_n^2+b_1^2+b_2^2+\ldots+b_n^2}{2} \le \frac{1+1}{2} = 1,$$

which completes the proof.

123. If $n = 1$, then $x_1 = 1$ and, hence, $x_1 \geq 1$, the assertion being therefore true. Suppose it is true for all m such that $1 \leq m \leq n-1$; let us prove that then it holds for $m = n$. If all the numbers $x_1, x_2, \dots, x_n$ are equal to unity, the assertion is obviously true. If atleast one of these numbers is greater than unity, then, by virtue of the equality $x_1 x_2 \dots x_n = 1$, there must be a number among $x_1, x_2, \dots, x_n$ which is less than unity. Let the numeration of $x_1, x_2, \dots, x_n$ be such that $x_n > 1$, $x_{n-1} < 1$. The induction hypothesis and the condition

$$x_1 x_2 \dots x_{n-2}(x_{n-1} x_n) = 1$$

imply

$$x_1 + x_2 + \dots + x_{n-2} + x_{n-1} x_n \geq n-1,$$

i.e.

$$x_1 + x_2 + \dots + x_{n-2} + x_{n-1} x_n + 1 \geq n.$$

We have $(x_n - 1)(1 - x_{n-1}) > 0$ and therefore

$$x_n + x_{n-1} - x_n x_{n-1} - 1 > 0.$$

Consequently

$$x_{n-1} + x_n > x_{n-1} x_n + 1.$$

Thus, $x_1 + x_2 + \dots + x_{n-1} + x_n > x_1 + x_2 + \dots + x_{n-2} + x_{n-1} x_n + 1 \geq n$,

and the assertion has been proved.

4. Logarithmic and Exponential Equations, Identities and Inequalities

124. As is seen from the equation, it only makes sense for $a > 0$, $a \neq 1$ and $b > 0$, $b \neq 1$. For solving the equation let us make use of the formula for change of base of logarithms

$$\log_b a = \frac{\log_c a}{\log_c b}$$

(see formula (2) on page 24). Here c is an arbitrary base ($c > 0$, $c \neq 1$). The choice of the base c is inessential here because we only want to reduce all logarithms to one base. We may, for instance, take a as a common base, since $a > 0$ and $a \neq 1$. Then the equation takes the form

$$\frac{\log_a x}{\log_a 2} \log_a^2 2 - 2 \log_a x \log_a \frac{1}{b} = \frac{\log_a x}{\log_a \sqrt[3]{a}} \log_a x,$$

which yields after some simplifications the new equation

$$(\log_a 2 + 2 \log_a b) \log_a x = 3 \log_a^2 x.$$

Hence, there are two solutions, one being $\log_a x = 0$, i. e. $x = 1$, and the other being

$$\log_a x = \frac{1}{3}(\log_a 2 + 2 \log_a b) = \frac{1}{3} \log_a 2b^2 = \log_a \sqrt[3]{2b^2},$$

i.e.

$$x = \sqrt[3]{2b^2}.$$

125. Let us pass to logarithms to the base 2; using formula (2) on page 24 we get

$$\frac{1}{\log_2 x}\cdot\frac{1}{\log_2 x-4}=\frac{1}{\log_2 x-6}.$$

The latter equation is equivalent to the equation

$$\log_2^2 x-5\log_2 x+6=0.$$

Hence we have

$$(\log_2 x)_1=2,\ x_1=4$$

and

$$(\log_2 x)_2=3,\ x_2=8.$$

126. Raising we obtain

$$9^{x-1}+7=4\,(3^{x-1}+1).$$

Whence we find

$$(3^{x-1})^2-4\,(3^{x-1})+3=0.$$

Consequently,

$$(3^{x-1})_1=3,\ x_1=2 \text{ and } (3^{x-1})_2=1,\ \ x_2=1.$$

127. Let us pass to logarithms to the base 3. By formula (2) on page 24 We have

$$\frac{1-\log_3 x}{1+\log_3 x}+\log_3^2 x=1.$$

This results in

$$(1-\log_3 x)\,[1-(1+\log_3 x)^2]=0$$

and, hence,

$$(\log_3 x)_1=1,\qquad x_1=3;$$
$$(\log_3 x)_2=0,\qquad x_2=1;$$
$$(\log_3 x)_3=-2,\qquad x_3=\frac{1}{9}.$$

128. Let us pass in the given equation to logarithms to the base 2. By formula (2) on page 24, we obtain

$$\frac{1-\log_2 x}{1+\log_2 x}\log_2^2 x+\log_2^4 x=1.$$

Multiplying both members of the equation by the denominator, transposing all the terms to the left-hand side and factorizing we get

$$(\log_2 x-1)\,(\log_2^4 x+2\log_2^3 x+\log_2^2 x+2\log_2 x+1)=0.$$

For $x>1$ the second factor is obviously positive and does not vanish. Equating the first factor to zero we find that for $x>1$ the original equation is solvable and has only one root $x=2$.

129. Let us change the logarithms to bring them to the base a (here $a>0$ and $a \neq 1$ because if otherwise the expression $\log_{\frac{1}{a}} 2x$ makes no sense). By virtue of formula (2) on page 24, we get

$$\frac{\log_a 2x}{\log_a a^2\sqrt{x}} + \frac{\log_a 2x}{\log_a \frac{1}{a} \log_a ax} = 0.$$

This enables us to consider the following possible cases:

(1) $\log_a 2x = 0$ and we obtain $x = \frac{1}{2}$ which does not satisfy the original equation (the logarithm of a number $a \neq 0$ to the base 1 does not exist);

(2) $\log_a ax = \log_a(a^2\sqrt{x})$ which yields $x = a^2$.

Answer: $x = a^2$.

130. Applying the equality $\log_x b = \frac{1}{\log_b x}$ we transform the original equation to the equivalent equation

$$\log_b [x(2\log a - x)] = 2.$$

Whence, after raising, we obtain

$$x^2 - 2\log a \cdot x + b^2 = 0.$$

Solving this equation we find

$$x_{1,2} = \log a \pm \sqrt{\log^2 a - b^2}.$$

For $a \geq 10^b$ and $\log a \neq \frac{1}{2}(b^2+1)$ both roots are positive and unequal to unity and, as is readily verified, satisfy the original equation. For $\log a = \frac{1}{2}(b^2+1)$ we must only take the root $x_1 = b^2$. For $a < 10^b$ the equation has no roots.

131. Passing in the equation to logarithms to the base a we transform it to the form

$$\sqrt{\log_a \sqrt[4]{ax}\left(1+\frac{1}{\log_a x}\right)} + \sqrt{\log_a \sqrt[4]{\frac{x}{a}}\left(1-\frac{1}{\log_a x}\right)} = a.$$

After some transformations we get

$$\sqrt{\frac{(\log_a x+1)^2}{4\log_a x}} + \sqrt{\frac{(\log_a x-1)^2}{4\log_a x}} = a.$$

Taking into consideration that the square roots are understood here in the arithmetic sense we see that the given equation can be rewritten in the following way:

$$|\log_a x + 1| + |\log_a x - 1| = 2a\sqrt{\log_a x}. \quad (1)$$

Now consider the following two cases:

(1) Suppose that

$$\log_a x > 1. \qquad (2)$$

Then equation (1) takes the form

$$\log_a x = a\sqrt{\log_a x},$$

whence we obtain $x_1 = a^{a^2}$.

It can be easily seen that condition (2) is then satisfied only if $a > 1$.

(2) Suppose that

$$0 < \log_a x \le 1. \qquad (3)$$

Then equation (1) turns into $2 = 2a\sqrt{\log_a x}$,

Hence, $$x_2 = a^{\frac{1}{a^2}}.$$

It should be noted that condition (3) is only fulfilled if $a \ge 1$. Since we a priori have $a \ne 1$ (otherwise the original equation makes no sense), the second root x_2 exists only if $a > 1$.

We have considered all the possibilities because it is obvious that the values of x for which $\log_a x \le 0$ cannot satisfy equation (1). Thus, for $a > 1$ the equation under consideration has two roots, namely $x_1 = a^{a^2}$ and $x_2 = a^{\frac{1}{a^2}}$. For $0 < a < 1$ the equation has no roots.

132. We have $\log(\sqrt{x+1} + 1) = \log(x - 40)$.

Putting $\sqrt{x+1} = t$ and raising we get the equation

$$t^2 - t - 42 = 0,$$

whose roots are $t_1 = 7$ and $t_2 = -6$. Since $t = \sqrt{x+1} \ge 0$, the root t_2 is discarded. The value of x corresponding to the root t_1 is equal to 48. By substitution we check that it satisfies the original equation. Thus, the equation has the unique root $x = 48$.

133. Passing over in the equation to logarithms to the base a we get

$$1 + \frac{\log_a(p-x)}{\log_a(x+q)} = \frac{2\log_a(p-q) - \log_a 4}{\log_a(x+q)}.$$

After performing some simplifications and taking antilogarithms we arrive at the quadratic equation

$$(x+q)(p-x) = \frac{1}{4}(p-q)^2.$$

The roots of this equation are

$$x_1 = \frac{1}{2}(p-q) + \sqrt{pq},\ x_2 = \frac{1}{2}(p-q) - \sqrt{pq}.$$

It is easy to verify that both roots satisfy the inequality

$$p > x_{1,2} > -q,$$

and, consequently, the original equation as well.

134. After some simple transformations based on the formula for change of base of logarithms we reduce the given equation to the form

$$\log_{\sqrt{5}} x\sqrt{\frac{3}{\log_{\sqrt{5}} x}+3}=-\sqrt{6}.$$

Putting $\log_{\sqrt{5}} x=t$ we obtain, after performing some simplifications and squaring both sides of the equation, the new equation

$$t^2+t-2=0.$$

Its roots are $t_1=-2$ and $t_2=1$. The first root yields the value $x=\frac{1}{5}$ which, as is readily seen, satisfies the original equation. The second root gives the value $x=\sqrt{5}$ which does not satisfy the original equation.

135. Using the fact that $0.4=\frac{2}{5}$ and $6.25=\left(\frac{5}{2}\right)^2$ we reduce the original equation to the form

$$\left(\frac{2}{5}\right)^{\log^2 x+1}=\left(\frac{2}{5}\right)^{2(\log x^3-2)}$$

Equating the exponents we pass to the equation

$$\log^2 x-6\log x+5=0.$$

After solving it we find

$$(\log x)_1=1,\ x_1=10 \text{ and } (\log x)_2=5, x_2=10^5.$$

136. Passing over to logarithms to the base 10 we obtain

$$1+\frac{\log\left(\frac{4-x}{10}\right)}{\log x}=(\log\log n-1)\frac{1}{\log x}.$$

After simple transformations this leads to the equation

$$\log\left(x\cdot\frac{4-x}{10}\right)=\log\frac{\log n}{10}.$$

Taking antilogarithms we obtain

$$x^2-4x+\log n=0,$$

whence

$$x_{1,2}=2\pm\sqrt{4-\log n}.$$

A simple argument now leads to the following final results:

(a) If $0<n<10^4$ and $n\neq 10^9$, the equation has two different roots, namely

$$x_1=2+\sqrt{4-\log n} \quad\text{and}\quad x_2=2-\sqrt{4-\log n}.$$

(b) If $n=10^9$, there is only one root $x=3$ ($x=1$ should be discarded); for $n=10^4$ we also get one root $x=2$.

(c) If $n>10^4$ there are no roots.

137. Passing to logarithms to the base 2 we obtain the equation

$$\frac{1}{\log_2 \sin x}\frac{\log_2 a}{2\log_2 \sin x}+1=0.$$

Hence,

$$\log_2^2 \sin x=-\frac{\log_2 a}{2}.$$

The quantity on the left-hand side being strictly positive ($\sin x \neq 1$ because otherwise the symbol $\log_{\sin x} 2$ makes no sense), we have $\log_2 a<0$ and, consequently, for $a>1$ the equation has no solutions at all. Supposing that $0<a<1$ we obtain

$$\log_2 \sin x=\pm\sqrt{-\frac{\log_2 a}{2}}.$$

The plus sign in front of the radical must be discarded because $\log_2 \sin x<0$.

Thus we have

$$\sin x=2^{-\sqrt{-\frac{\log_2 a}{2}}}$$

and

$$x=(-1)^k \arcsin 2^{-\sqrt{-\frac{\log_2 a}{2}}}+\pi k$$
$$(k=0, \pm 1, \ldots).$$

It can easily be seen that all this infinite sequence of values of x satisfies the original equation.

138. From the second equation we find

$$x+y=\frac{2}{x-y}. \qquad (1)$$

Substituting this expression for $x+y$ into the first equation we obtain

$$1-\log_2(x-y)-\log_3(x-y)=1,$$

that is

$$\log_2(x-y)+\log_3(x-y)=0.$$

Passing to logarithms to the base 3 we transform the last equation to the form

$$(\log_2 3+1)\log_3(x-y)=0.$$

Since $\log_2 3+1 \neq 0$, it follows that $\log_3(x-y)=0$ and $x-y=1$. Combining this with equation (1) we obtain the system

$$\left.\begin{aligned} x+y&=2,\\ x-y&=1.\end{aligned}\right\}$$

Solving it we get $x=\frac{3}{2},\ y=\frac{1}{2}.$

Finally, we verify by substitution that the above pair of numbers is the solution of the original system.

139. Taking logarithms of the both sides of the first equation to the base c we obtain

$$a\log_c x = b\log_c y. \qquad (1)$$

From the second equation we find

$$\log_c x - \log_c y = \frac{\log_c x}{\log_c y}.$$

Substituting $\log_c y$ expressed from equation (1) into the latter equation we get

$$\log_c x - \frac{a}{b}\log_c x = \frac{b}{a}, \text{ or } \log_c x^{1-\frac{a}{b}} = \frac{b}{a}.$$

Now, raising, we obtain

$$x^{\frac{b-a}{b}} = c^{\frac{b}{a}}, \text{or } x = c^{\frac{b^2}{a(b-a)}}.$$

From the first equation of the system we now find

$$y = x^{\frac{a}{b}} = c^{\frac{b}{b-a}}.$$

140. Using the logarithmic identity $a^{\log_a b} = b$ we write the system in the form

$$\left.\begin{aligned} \log_5 x + y &= 7, \\ x^y &= 5^{12}. \end{aligned}\right\} \qquad (1)$$

Taking antilogarithms in the first equation we get $x \cdot 5^y = 5^7$ whence

$$x = 5^{7-y}. \qquad (2)$$

Substituting x found from equation (2) into the second equation of system (1) we get the equation $5^{12+y^2-7y} = 1$ whose roots are $y_1 = 4$ and $y_2 = 3$.

Finally, we arrive at the two solutions

$$x_1 = 125, y_1 = 4 \text{ and } x_2 = 625, y_2 = 3.$$

141. Taking logarithms of both sides of the first equation to the base y we get a quadratic equation with respect to $\log_y x$ of the form

$$2\log_y^2 x - 5\log_y x + 2 = 0,$$

whose roots are

$$\log_y x = 2, \log_y x = \frac{1}{2}.$$

If $\log_y x = 2$, we have

$$x = y^2. \qquad (1)$$

By virtue of the identity $\log_a b \frac{1}{\log_b a}$, we get from the second equation the relation $\log_y (y-3x) = \log_y 4$, whence we find

$$y - 3x = 4. \qquad (2)$$

Equations (2) and (1) imply a quadratic equation for y of the form

$$3y^2 - y + 4 = 0.$$

This equation has no real solutions. If $\log_y x = \frac{1}{2}$. we have $x = \sqrt{y}$ and $y = x^2$. In this case, by virtue of (2). we get the equation

$$x^2 - 3x - 4 = 0.$$

Answer: $x = 4$, $y = 16$.

142. Taking logarithms to the base a in the first equation we find

$$x + y\log_a b = 1 + \log_a b. \quad (1)$$

In the second equation we pass over to logarithms to the base a. Then we obtain

$$2\log_a x = -\frac{\log_a y}{\log_a b}\frac{\log_a b}{\log_a \sqrt{a}} = -2\log_a y,$$

which yields $x = \frac{1}{y}$.

Substituting $y = \frac{1}{x}$ into (1) we get the equation

$$x^2 - x(1 + \log_a b) + \log_a b = 0,$$

having the roots

$$x_1 = \log_a b \text{ and } x_2 = 1.$$

The final answer is

$$x_1 = \log_a b,\ y_1 = \log_b a;\ x_2 = 1,\ y_2 = 1.$$

143. In the first equation we pass over to logarithms to the base x. Then the equation takes the form

$$3\left(\log_x y + \frac{1}{\log_x y}\right) = 10.$$

Putting here $\log_x y = t$ we get the equation

$$3t^2 - 10t + 3 = 0,$$

having the roots $t_1 = 3$ and $t_2 = \frac{1}{3}$. In the first case $\log_x y = 3$, $y = x^3$ and, by virtue of the second equation of the original system, we obtain $x^4 = 81$. Since $x > 0$ and $y > 0$, here we have only one solution:

$$x_1 = 3,\ y_1 = 27.$$

Putting then $\log_x y = \frac{1}{3}$ we find one more solution

$$x_2 = 27,\ y_2 = 3.$$

144. Let us pass in both equations of the system to logarithms to the base 2. This results in the following system:

$$\left.\begin{aligned} \frac{\log_2 x}{\log_2 12}(\log_2 x+\log_2 y)=\log_2 x,\\ \log_2 x\cdot\frac{\log_2(x+y)}{\log_2 3}=3\frac{\log_2 x}{\log_2 3}\end{aligned}\right\} \qquad (1)$$

Since $x \neq 1$ (if otherwise, the left member of the first equation of the original system makes no sense). we have $\log_2 x \neq 0$, and system (1) can thus be rewritten in the following way:

$$\left.\begin{aligned} \log_2 x+\log_2 y=\log_2 12,\\ \log_2(x+y)=3.\end{aligned}\right\}$$

Taking antilogarithms we get $xy=12$, $x+y=8$,
whence it follows that

$$x_1=6, y_1=2 \text{ and } x_2=2,\ y_2=6.$$

145. Converting the logarithms in each of the given equations to the base 2 we get

$$\left.\begin{aligned} x\log_2 y=y\sqrt{y}(1-\log_2 x),\\ 2\log_2 x=3\log_2 y.\end{aligned}\right\} \qquad (1)$$

From the second equation of system (1) we find $x^2=y^3$, whence

$$x=y^{\frac{3}{2}}. \qquad (2)$$

Using (2), we find from the first equation $y=\sqrt[5]{4}$. Hence,

$$x=2^{\frac{3}{5}},\ y=2^{\frac{2}{5}}.$$

146. Let us transform the system by passing to logarithms to the base 2 in the first equation, to the base 3 in the second and to the base 4 in the third. We obtain

$$\left.\begin{aligned} \log_2 x+\frac{1}{2}\log_2 y+\frac{1}{2}\log_2 z=\log_2 4,\\ \log_3 y+\frac{1}{2}\log_3 z+\frac{1}{2}\log_3 x=\log_3 9,\\ \log_4 z+\frac{1}{2}\log_4 x+\frac{1}{2}\log_4 y=\log_4 16.\end{aligned}\right\}$$

Taking antilogarithms we come to the system

$$\left.\begin{aligned} x\sqrt{yz}=4,\\ y\sqrt{xz}=9,\\ z\sqrt{xy}=16.\end{aligned}\right\} \qquad (1)$$

Multiplying the equations of system (1) termwise we find

$$(xyz)^2=24^2.$$

Since $x > 0$, $y > 0$, $z > 0$, we thus have

$$xyz = 24. \qquad (2)$$

Squaring the first equation of system (1) and using (2) we get

$$x = \frac{16}{24} = \frac{2}{3}.$$

Analogously, we find $y = \frac{27}{8}$ and $z = \frac{32}{3}$. The verification by substitution confirms that the three numbers thus found form a solution.

147. Passing over to logarithms to the base 2 in the first equation and then raising we get

$$y^2 - xy = 4. \qquad (1)$$

Equation (1) and the second equation of the original system form the system

$$\left.\begin{aligned} x^2 + y^2 &= 25, \\ y^2 - xy &= 4, \end{aligned}\right\} \qquad (2)$$

This system has two solutions satisfying the conditions $y > x$, $y > 0$, namely:

$$x_1 = -\frac{7}{\sqrt{2}},\ y_1 = \frac{1}{\sqrt{2}} \text{ and } x_2 = 3,\ y_2 = 4.$$

148. Dividing both members of the equation by 4^x we find

$$1 - \left(\frac{3}{4}\right)^x \cdot \frac{1}{\sqrt{3}} = \left(\frac{3}{4}\right)^x \sqrt{3} - \frac{1}{2}.$$

This yields

$$\left(\frac{3}{4}\right)^x = \frac{3\sqrt{3}}{8} = \left(\frac{3}{4}\right)^{\frac{3}{2}}$$

and, hence,

$$x = \frac{3}{2}.$$

149. Substituting y expressed from the second equation into the first we obtain

$$x^{x + \frac{1}{x^2}} = x^{-2x + \frac{2}{x^2}}.$$

It follows that either $x = 1$ or $x + \frac{1}{x^2} = -2x + \frac{2}{x^2}$

and, consequently, $x = \frac{1}{\sqrt[3]{3}}$.

Answer: $x_1 = y_1 = 1$, $x_2 = \frac{1}{\sqrt[3]{3}}$, $y_2 = \sqrt[3]{9}$.

150. Putting $a^x = u$ and $a^y = v$ we represent the system in the form

$$\left.\begin{aligned} u^2+v^2&=2b,\\ uv&=c.\end{aligned}\right\}$$

These two equations imply

$$(u+v)^2 = 2(b+c) \text{ and } (u-v)^2 = 2(b-c).$$

Since the sought-for values of u and v must be positive, the first equation is reduced to the equation

$$u+v=\sqrt{2(b+c)} \tag{1}$$

The second equation indicates that for the system to be solvable, it is necessary to require, besides the positivity of the numbers b and c, that the inequality

$$b \geq c \tag{2}$$

should be fulfilled. We also have

$$u-v=\pm\sqrt{2(b-c)} \tag{3}$$

and therefore, solving the system of equations (1) and (3), we find, taking the plus sign, the values

$$u_1=\frac{\sqrt{2}}{2}(\sqrt{b+c}+\sqrt{b-c}),\ v_1=\frac{\sqrt{2}}{2}(\sqrt{b+c}-\sqrt{b-c}).$$

In the second case we get

$$u_2=\frac{\sqrt{2}}{2}(\sqrt{b+c}-\sqrt{b-c}),\ v_2=\frac{\sqrt{2}}{2}(\sqrt{b+c}+\sqrt{b-c}).$$

We have found two solutions of system (1), and if condition (2) is fulfilled all the values of the unknowns are obviously positive. The two corresponding solutions of the original system have the form

$$x_1=\log_a u_1,\ y_1=\log_a v_1;\quad x_2=\log_a u_2,\ y_2=\log_a v_2.$$

We now can assert that for the system to be solvable it is necessary and sufficient that $b>0$, $c>0$ and $b \geq c$. If these conditions hold the system has two solutions.

151. Multiplying the equations we get

$$(xy)^{x+y}=(xy)^{2n}.$$

Since x and y are positive, it follows that either $xy=1$ or $xy \neq 1$, and then

$$x+y=2n. \tag{1}$$

Let us first consider the second case. The first equation of the original system then takes the form $x^{2n}=y^n$, whence we obtain

$$y=x^2. \tag{2}$$

Substituting $y=x^2$ into equation (1) we receive

$$x^2+x-2n=0.$$

This equation has only one positive root

$$x_1 = \frac{\sqrt{8n+1}-1}{2}. \quad (3)$$

Using (2) we find the corresponding value of y:

$$y_1 = \frac{1}{4}(\sqrt{8n+1}-1)^2. \quad (4)$$

In the second case when $xy = 1$ we have $y = \frac{1}{x}$, and the first equation of the original system takes the form

$$x^{\frac{1}{x}+x} = x^{-n}.$$

Since x and n are positive this equality is only possible if $x = 1$. Thus, we have found one more solution: $x_2 = 1$, $y_2 = 1$.

152. We transform the system in the form

$$\left.\begin{aligned} (3x+y)^{x-y} &= 9, \\ \sqrt[x-y]{324} &= 2(3x+y)^2. \end{aligned}\right\}$$

From the second equation we find $324 = 2^{x-y}(3x+y)^{2(x-y)}$ and, consequently, by virtue of the first equation, we have $324 = 2^{x-y}\cdot 81$ which results in $2^2 = 2^{x-y}$, i.e.

$$x - y = 2.$$

Combining equation (1) with the first equation of the original system we arrive at the two systems

$$\left.\begin{aligned} x - y &= 2, \\ 3x + y &= 3, \end{aligned}\right\} \quad (2)$$

$$\left.\begin{aligned} x - y &= 2, \\ 3x + y &= -3. \end{aligned}\right\} \quad (3)$$

The solution of system (2) is $x_1 = \frac{5}{4}, y_1 = -\frac{3}{4}$ The solution of system (3) is $x_2 = -\frac{1}{4}$, $y_2 = -\frac{9}{4}$. The substitution in the original system confirms that both pairs of numbers satisfy it.

153. Put $\frac{q}{p} = \alpha$. If $\alpha = 1$, i.e. $p = q$, the system is satisfied by any pair of equal positive numbers. Let us, therefore, suppose that $\alpha \neq 1$. From the second equation we get $x = y^{\alpha}$. Taking logarithms of both sides of the first equation and using the above equality we obtain $y \log y(\alpha - y^{\alpha-1}) = 0$. We have $y > 0$ and therefore either $\log y = 0$ or $\alpha = y^{\alpha-1}$. In the first case we obtain $x_1 = 1$, $y_1 = 1$ and in the second case $x_2 = \alpha^{\frac{\alpha}{\alpha-1}}$, $y_2 = \alpha^{\frac{1}{\alpha-1}}$. Both pairs of numbers satisfy the original system as well.

154. Taking logarithms of both equations we get the system

$$\left.\begin{aligned} y\log x &= x\log y,\\ x\log p &= y\log q,\end{aligned}\right\} \quad (1)$$

which determines the ratio $\frac{x}{y}=\frac{\log q}{\log p}=\alpha$. Consequently,

$$x=\alpha y. \quad (2)$$

If $p=q$, the system has an infinite number of solutions of the form $x=y=a$ where $a > 0$ is an arbitrary number. If $p\neq q$, then, substituting x determined from formula (2) into the first equation of system (1) we find

$$x=\alpha^{\frac{\alpha}{\alpha-1}},\quad y=\alpha^{\frac{1}{\alpha-1}}.$$

Consequently, if $p\neq q$ the system has a unique solution.

155. Taking logarithms of both members of the equality $a^2=c^2-b^2$ we get

$$2=\log_a(c-b)+\log_a(c+b).$$

Whence we obtain $2=\dfrac{1}{\log_{c-b}a}+\dfrac{1}{\log_{c+b}a}$

and, hence, at $\log_{c+b}a+\log_{c-b}a=2\log_{c+b}a\cdot\log_{c-b}a$.

156. Using the formula $\log_n m=\dfrac{1}{\log_m n}$ we obtain

$$\log_{b^{2^{-k}}}a=2^k\log_b a \text{ and } \log_{a^{2^k}}b=\frac{1}{2^k}\log_a b,$$

$$\sum_{k=0}^{n}\log_{b^{2^{-k}}}a-\log_{a^{2^k}}b)^{2^*}=\sum_{k=0}^{n}\left(2^k\log_b a-\frac{1}{2^k}\log_a b\right)^2=$$

$$=\log_b^2 a\sum_{k=0}^{n}4^k+\log_a^2 b\sum_{k=0}^{n}\frac{1}{4^k}-\sum_{k=0}^{n}2=$$

$$=\frac{4^{n+1}-1}{4-1}\log_b^2 a+\frac{\left(\frac{1}{4}\right)^{n+1}-1}{\frac{1}{4}-1}\log_a^2 b-2\,(n+1)=$$

$$=\frac{1}{3}(4^{n+1}-1)\log_b^2 a+\frac{1}{3}(4^{n+1}-1)$$

$$\frac{1}{4^n}\log_a^2 b-2(n+1)=$$

$$=\frac{1}{3}(4^{n+1}-1)\left(\log_b^2 a+\frac{1}{4^n\log_b^2 a}\right)-2(n+1).$$

157. $a^{\frac{\log_b \log_b a}{\log_b a}} = (a^{\log_a b})^{\log b \log_b a} = b^{\log_b \log_b a} = \log_b a.$

158. We have

$$c = a_1 a_2 \dots a_n = a.aq.\dots(aq^{n-1}) = a^n q^{\frac{n(n-1)}{2}}.$$

Using the formula for changing the base of logarithms we obtain

$$\log_c b = \frac{\log_a b}{\log_a c} = \frac{A}{n + \frac{n(n-1)}{2}\log_a q}.$$

But we have $$\log_a q = \frac{\log_b q}{\log_b a} = \frac{\log_a b}{\log_q b} = \frac{A}{B},$$

and therefore

$$\log_c b = \frac{2AB}{2nB + n(n-1)A}.$$

159. Taking advantage of the equality $\log_a b = \frac{1}{\log_b a}$ we transform the given formula as follows:

$$\frac{\log_N c}{\log_N a} = \frac{\frac{1}{\log_N a} - \frac{1}{\log_N b}}{\frac{1}{\log_N b} - \frac{1}{\log_N c}}$$

$$= \frac{\log_N \frac{b}{a}}{\log_N \frac{c}{b}} \cdot \frac{\log_N c}{\log_N a}.$$

This implies

$$\log_N \frac{b}{a} = \log_N \frac{c}{b}, \tag{1}$$

because the factor $\frac{\log_N c}{\log_N a}$ is different from zero. Taking antilogarithms in equality (1) we get

$$\frac{b}{a} = \frac{c}{b}. \tag{2}$$

Thus, b is the mean proportional between a and c. Taking then logarithms of both sides of equality (2) to an arbitrary base N and carrying out the transformations in reverse order we complete the proof of the assertion.

---

* The symbol $\sum_{k=0}^{n} a_k$ denotes the sum $a_0 + a_1 + a_2 + \dots + a_n$.

160. It should be supposed that $N \neq 1$ because, if otherwise, the fraction on the right-hand side becomes indeterminate. Dividing the identity to be proved by $\log_a N \log_b N \ \log_c N$ we replace it by the equivalent relation

$$\frac{1}{\log_a N} + \frac{1}{\log_b N} + \frac{1}{\log_c N} = \frac{1}{\log_{abc} N}.$$

Passing here to logarithms to the base N we get

$$\log_N a + \log_N b + \log_N c = \log_N abc.$$

The last identity being obviously valid, the problem has thus been solved.

161. We have

$$\frac{\log_a x}{\log_{ab} x} = \frac{\log_x ab}{\log_x a} = 1 + \frac{\log_x b}{\log_x a} = 1 + \log_a b,$$

which is what we set out to prove.

162. Using the logarithmic identity $\log_b a = \dfrac{\log_c a}{\log_c b}$ we transform the left member of the given inequality in the following way:

$$\log_{\frac{1}{2}} x + \log_3 x = \frac{\log_3 x}{\log_3 \frac{1}{2}} + \log_3 x$$

$$= \log_3 x \left(\log_{\frac{1}{2}} 3 + 1 \right)$$

$$= \log_3 x \cdot \log_{\frac{1}{2}} \frac{3}{2} = \frac{\log_3 x}{\log_{\frac{3}{2}} \frac{1}{2}} = -\frac{\log_3 x}{\log_{\frac{3}{2}} 2}.$$

Then the given inequality takes the form $-\dfrac{\log_3 x}{\log_{\frac{3}{2}} 2} > 1.$

We have $2 > 1$ and $\dfrac{3}{2} > 1$, and, by property of logarithms, $\log_{\frac{3}{2}} 2 > 0.$

Consequently, the foregoing inequality is equivalent to the inequality

$$\log_3 x < -\log_{\frac{3}{2}} 2.$$

Hence, noting that $x > 0$ according to the meaning of the problem, we finally obtain

$$0 < x < 3^{-\log_{\frac{3}{2}} 2}.$$

163. Since $x > 0$, the given inequality is equivalent to the inequality

$$x^{\log_a x} > a^2.$$

But $a > 1$, and therefore taking logarithms of both sides of the last inequality to the base a we get the equivalent inequality

$$\log_a^2 x > 2.$$

From this we deduce the final result:

either $\log_a x > \sqrt{2}$, and, consequently, $x > a^{\sqrt{2}}$

or $\log_a x < -\sqrt{2}$, and then $0 < x < a^{-\sqrt{2}}$.

164. By the meaning of the problem we have $x > 0$ and therefore the given inequality is equivalent to the inequality

$$\log_a x(x+1) < \log_a(2x+6).$$

Since $a > 1$, it follows that $x(x+1) < 2x + 6$, that is $x^2 - x - 6 < 0$. Solving this quadratic inequality for $x > 0$ we get

$$0 < x < 3.$$

165. The inequality to be established is equivalent to

$$0 < x^2 - 5x + 6 < 1.$$

Since $x^2 - 5x + 6 = (x-2)(x-3)$, the inequality $0 < x^2 - 5x + 6$ holds true for $x < 2$ and for $x > 3$.

Solving then the inequality $x^2 - 5x + 6 < 1$, we find that it is satisfied for

$$\frac{5-\sqrt{5}}{2} < x < \frac{5+\sqrt{5}}{2}.$$

Since $\sqrt{5} > 2$, we have $\frac{5-\sqrt{5}}{2} < 2$ and, consequently, $\frac{5+\sqrt{5}}{2} > 3$.

Therefore, the original inequality holds true for

$$\frac{5-\sqrt{5}}{2} < x < 2 \quad \text{and} \quad 3 < x < \frac{5+\sqrt{5}}{2}.$$

166. Reducing the fractions on the left-hand side to a common denominator, we find

$$\frac{-1}{\log_2 x(\log_2 x - 1)} < 1$$

and, hence,

$$\frac{1 + \log_2 x(\log_2 x - 1)}{\log_2 x(\log_2 x - 1)} > 0.$$

The numerator of the last expression is positive [indeed, we have

$$1 + \log_2^2 x - \log_2 x = \left(\log_2 x - \frac{1}{2}\right)^2 + \frac{3}{4}\Bigg],$$

the inequality is reduced to the relation

$$\log_2 x(\log_2 x - 1) > 0,$$

which is fulfilled for $x > 2$ and $0 < x < 1$.

167. According to the meaning of the problem, we have $x > 0$ and, hence, the given inequality is equivalent to the inequality

$$x^{3-\log_2^2 x-2\log_2 x} > 1.$$

Taking logarithms of both sides of this inequality to the base 2 and putting $y = \log_2 x$, we get an equivalent inequality of the form

$$y(3 - y^2 - 2y) > 0,$$

which, after the quadratic trinomial has been factorized, can be written in the form

$$y(1 - y)(3 + y) > 0.$$

The latter inequality is fulfilled if and only if either all the three factors are positive or one of them is positive and the other two are negative. Accordingly, in the first case,
i. e. when

$$y > 0,\ 1 - y > 0,\ 3 + y > 0,$$

we obtain $0 < y < 1$ and, hence,

$$1 < x < 2. \qquad (1)$$

The second case reduces to three subcases among which only one leads to a consistent system of inequalities. Namely, when

$$y < 0,\ 1 - y > 0,\ 3 + y < 0.$$

We receive $y < -3$ and, hence,

$$0 < x < \frac{1}{8}. \qquad (2)$$

Thus, the original inequality holds if and only if either

$$0 < x < \frac{1}{8},\text{ or } 1 < x < 2.$$

168. Putting $\log_2 x = y$ and noting that $\log_x 2 = \dfrac{1}{\log_x 2} = \dfrac{1}{y}$ we rewrite the given inequality in the form

$$y + \frac{1}{y} + 2\cos\alpha \le 0. \qquad (1)$$

The numbers $z = y + \dfrac{1}{y}$ and y have the same sign, and $|z| \ge 2$ for all y (see (2), page 20). Therefore, if $z > 0$, then the inequality $z \le -2\cos\alpha$ is fulfilled only if $z = 2$ (i.e., $y = 1$) and $\cos\alpha = -1$ or, in other words, if in the original inequality $x = 2$ and $\alpha = (2k + 1)\pi\ (k = 0, \pm 1, \pm 2, \dots)$. For these values the sign of equality appears.

But if $z < 0$, i. e. $y < 0$, then $z \le -2$, and inequality (1) is fulfilled for all α, whence it follows that the original inequality holds for $0 < x < 1$ and all real values of α besides the values found above.

169. The original inequality is equivalent to the relation

$0 < \log_4(x^2-5) < 1,$

whence we find that $1 < x^2-5 < 4$ or $6 < x^2 < 9$ or $\sqrt{6} < |x| < 3$.

Answer : $\sqrt{6} < x < 3$ and $-3 < x < -\sqrt{6}$.

5. Combinatorial Analysis and Newton's Binomial Theorem

170. Taking the ratios of the first term of the proportion to the second and of the second to the third and reducing the fractions to their lowest terms we obtain

$$\frac{(n+1)!}{(m+1)!(n-m)!} : \frac{(n+1)!}{m!(n-m+1)!} = \frac{n-m+1}{m+1}$$

and

$$\frac{(n+1)!}{m!(n-m+1)!} : \frac{(n+1)!}{(m-1)!(n-m+2)!} = \frac{n-m+2}{m}.$$

The conditions of the problem thus lead to the two equations

$$\frac{n-m+1}{m+1} = 1 \text{ and } \frac{n-m+2}{m} = \frac{5}{3}.$$

Solving them as system of simultaneous equations we find $m=3$ and $n=6$.

171. We have

$$(1+x^2-x^3)^9 = 1 + C_9^1(x^2-x^3)$$
$$+ C_9^2(x^2-x^3)^2 + C_9^3(x^2-x^3)^3 + C_9^4(x^2-x^3)^4$$
$$+ C_9^5(x^2-x^3)^5 + \ldots + (x^2-x^3)^9.$$

It is readily seen that x^8 enters only into the fourth and fifth terms on the right-hand side. Using this fact we easily find the coefficient in x^8 which is equal to $3C_9^3 + C_9^4$.

172. The summands of the given sum form a progression with common ratio $1+x$. Therefore,

$$(1+x)^k + (1+x)^{k+1} + \ldots$$
$$+(1+x)^n = \frac{(1+x)^{n+1}-(1+x)^k}{x}. \qquad (1)$$

Writing the sum in the form of a polynomial

$$a_0 + a_1x + \ldots + a_mx^m + \ldots + a_nx^n,$$

and removing brackets in the right-hand member of equality (1) we see that if $m < k$, then

$$a_m = C_{n+1}^{m+1} - C_k^{m+1},$$

and if $m \geq k$, then $a_m = C_{n+1}^{m+1}$.

173. From the conditions of the problem it follows that

$$C_n^2 = C_n^1 + 44, \text{ or } \frac{n(n-1)}{2} = n + 44.$$

Solving this equation for n we find $n = 11$.

The general term of the expansion of the expression

$$\left(x\sqrt{x} + \frac{1}{x^4}\right)^{11}$$

by the binomial formula can be written in the form

$$C_{11}^m x^{\frac{3}{2}(11-m)-4m}.$$

By the hypothesis we have $\frac{3}{2}(11-m)-4m=0$ which yields $m=3$. Hence, the sought-for term is equal to C_{11}^3.

174. Putting $x+\frac{6}{x}=u$ we can write

$$\left(1+x+\frac{6}{x}\right)^{10} = (1+u)^{10} = 1 + C_{10}^1 u + C_{10}^2 u^2 + \ldots + C_{10}^{10} u^{10},$$

where

$$u^k = \left(x+\frac{6}{x}\right)^k = x^k + C_k^1 x^{k-2} 6 + \ldots + C_k^s x^{k-2s} 6^s + \ldots + \frac{6^k}{x^k}. \tag{1}$$

For every summand in expression (1) which does not contain x we have the condition $k-2s=0$. Consequently, this summand is equal to $C_{2s}^s.6^s$. Collecting all these terms we conclude that a summand not containing x in the original expression is equal to

$1 + C_{10}^2 \cdot C_2^1 \cdot 6 + C_{10}^4 \cdot C_4^2 \cdot 6^2 + C_{10}^6 C_6^3 \cdot 6^3 + C_{10}^8 \cdot C_8^4 \cdot 6^4 + C_{10}^{10} \cdot C_{10}^5 \cdot 6^5$.

175. After simplifications the inequalities $T_{k+1} > T_k$ and $T_{k+1} > T_{k+2}$ take the form

$$\frac{\sqrt{3}}{k} > \frac{1}{101-k}, \quad \frac{1}{100-k} > \frac{\sqrt{3}}{k+1}.$$

Solving each of them with respect to k, we get

$$\frac{101\sqrt{3}}{\sqrt{3}+1} > k > \frac{100\sqrt{3}-1}{\sqrt{3}+1}. \tag{1}$$

Both the left and right members of inequality (1) are not integers, the difference between them being equal to unity. Therefore there exists only one integer k satisfying inequality (1). Noting that $1.72 < \sqrt{3} < 1.73$ we establish, by direct computation, that

$$64.64 > k > 63.135.$$

Hence, $k = 64$.

176. The general term T_{k+1} of the expansion is equal to $C_n^k a^k$. If $T_k = T_k + 1$, then $C_n^{k-1} a^{k-1} = C_n^k a^k$, that is

$$\frac{n!a^{k-1}}{(k-1)!(n-k+1)!} = \frac{n!a^k}{k!(n-k)!},$$

whence we obtain $k = \dfrac{n+1}{1+\dfrac{1}{a}}$. We have thus established the required condition:

the number $1+\dfrac{1}{a}$ must be the divisor for the number $n+1$.

Furthermore, the relation $T_k = T_{k+1} = T_{k+2}$ is equivalent to the equalities

$$\frac{1}{(n-k+1)(n-k)} = \frac{a}{k(n-k)} = \frac{a^2}{k(k+1)},$$

that is

$$\frac{k}{n-k+1} = a, \quad \frac{k+1}{n-k} = a.$$

From the latter relations we obtain the equality $n+1=0$ which is impossible.

177. The expansion will contain n terms of the form x_i^3 $(i = 1, 2, \ldots, n)$, $n(n-1)$ terms of the form $x_i^2 x_j$ $(i, j = 1, 2, \ldots, n, i \neq j)$ and, finally, C_n^3 terms of the form $x_i x_j x_k$ where i, j and k are different numbers. Thus, the number of different dissimilar terms is equal to

$$n + n(n-1) + \frac{n(n-1)(n-2)}{6} = \frac{n(n+1)(n+2)}{6}.$$

178. The divisors of the number q are obviously the numbers $p_1, p_2, \ldots, p_k$ and all their possible products. The number of these divisors is equal to $C_k^0 + C_k^1 + \ldots + C_k^k = 2^k$.

The fact that all the divisors are different and that there are no other divisors is implied by the uniqueness of the representation of an integer as a product of prime numbers.

179. The equality to be proved has the form

$$1 + \frac{C_n^1}{2} + \frac{C_n^2}{3} + \ldots + \frac{C_n^k}{k+1} + \ldots + \frac{C_n^{n-1}}{n} + \frac{1}{n+1} = \frac{2^{n+1}-1}{n+1}$$

and is equivalent to the equality

$$1 + (n+1) + \frac{n+1}{2} C_n^1 + \frac{n+1}{3} C_n^2 + \ldots + \frac{n+1}{k+1} C_n^k$$
$$+ \ldots + \frac{n+1}{n} C_n^{n-1} + 1 = 2^{n+1}.$$

Since

$$\frac{n+1}{k+1}C_n^k = \frac{n+1}{k+1}\frac{n!}{k!(n-k)!} = \frac{(n+1)!}{(k+1)!(n-k)!} = C_{n+1}^{k+1},$$

the left-hand side of the last equality is equal to

$$1 + C_{n+1}^1 + C_{n+1}^2 + \ldots + C_{n+1}^{k+1} + \ldots + C_{n+1}^n + 1 = (1+1)^{n+1} = 2^{n+1},$$

which is what we set out to prove.

180. The general term on the left-hand side of the equality can be transformed in the following way:

$$kC_n^k x^k (1-x)^{n-k} = k\frac{n!}{k!(n-k)!}x^k(1-x)^{n-k}$$

$$= nx\frac{(n-1)!}{(k-1)!(n-k)!}x^{k-1}(1-x)^{n-1-(k-1)}$$

$$= nxC_{n-1}^{k-1}x^{k-1}(1-x)^{n-1-(k-1)}.$$

Therefore the left member of the equality can be written in the form

$$nx[C_{n-1}^0(1-x)^{n-1} + C_{n-1}^1 x(1-x)^{n-2} + \ldots + C_{n-1}^{n-1}\cdot x^{n-1}] =$$

$$= nx[x+1-x]^{n-1} = nx.$$

181. Any splitting of the pack indicated in the statement of the problem is equivalent to selecting 16 cards out of the 32 cards that are not aces and two aces out of the four aces. The first selection can be accomplished in C_{32}^{16} ways, and the second in C_4^2 ways. Since every selection of the above 16 cards can be combined with any selection of two aces, the total number of ways in which the pack can be split is equal to $C_{32}^{16}C_4^2$.

182. The sought-for number is equal to the number of permutations of 10 digits taken 5 at a time, i.e. to $10\times9\times8\times7\times6 = 30{,}240$.

183. Imagine that we have an ordered set of n "boxes" which can be filled by pairs of elements. Let us form the partitions and fill, in succession, the boxes by the pairs of elements.

A pair put into the first box can be selected in C_{2n}^2 ways. After the first pair has been selected, we can select the second pair in C_{2n-2}^2 ways, then the third in C_{2n-4}^2 ways and so on. Finally, we obtain a set of $C_{2n}^2 C_{2n-2}^2 C_{2n-4}^2 \ldots C_2^2$ partitions which, however, includes all the partitions differing in the order of the pairs. Consequently, the number of the partitions we are interested in is equal to

$$\frac{C_{2n}^2 C_{2n-2\ldots}^2 C_2^2}{n!} = \frac{2n(2n-1)(2n-2)(2n-3)\ldots2\cdot1}{2^n n!}$$

$$= (2n-1)(2n-3)\ldots3\cdot1.$$

The same result can be obtained by another way of reasoning. Let k_m $(m = 1,2,\ldots)$ be the number of partitions of the desired type when the number of elements equals $2m$. Consider $2n$ elements. Since the order of the pairs is inessential a pair containing the first element can be regarded as the first pair.

The pairs containing the first element can be formed in $2n-1$ ways. After a first pair has been selected, the rest of $2(n-1)$ elements can be partitioned into pairs in k_{n-1} ways. Therefore, $k_n = (2n-1)k_{n-1}$. With the aid of this relation we easily find

$$k_n = (2n-1)(2n-3)\dots 5\cdot 3\cdot 1.$$

184. Out of the total number $n!$ of permutations we have to subtract the number of those in which the elements a and b are adjacent. To form a permutation in which the elements a and b are adjacent we can take one of the permutations [whose number is $(n-2)!$] containing the remaining $n-2$ elements and add the two elements a and b to it so that they are adjacent. This can be obviously done in $2(n-1)$ ways (the factor 2 appears here because a and b can be interchanged). Thus, the number of permutations in which a and b are adjacent is equal to $2(n-2)!(n-1)$, and the number we are interested in is equal to

$$n!-2(n-1)! = (n-1)!(n-2).$$

185. If among these 5 tickets there are exactly two winning tickets, then the remaining three are non-winning. Out of eight winning tickets, one can select two in C_8^2 ways, and out of 50–8=42 non-winning tickets, three tickets can be chosen in C_{42}^3 ways. Each way of selecting two winning tickets can be combined with any choice of three non-winning tickets. Therefore, the total number of ways is equal to

$$C_8^2\cdot C_{42}^3 = \frac{8\times 7}{1\times 2}\,\frac{42\times 41\times 40}{1\times 2\times 3} = 326,240.$$

The number of ways of selecting five tickets so that at least two of them are winning is equal to the sum of the number of ways in which exactly two, exactly three, exactly four and exactly five winning tickets are extracted. Hence, the desired number is equal to

$$C_8^2C_{42}^3 + C_8^3C_{42}^2 + C_8^4C_{42}^1 + C_8^5\cdot$$

$$1 = \frac{8\times 7}{1\times 2}\,\frac{42\times 41\times 40}{1\times 2\times 3} + \frac{8\times 7\times 6}{1\times 2\times 3}\times\frac{42\times 41}{1\times 2} + \frac{8\times 7\times 6\times 5}{1\times 2\times 3\times 4}\times\frac{42}{1} + \frac{8\times 7\times 6\times 5\times 4}{1\times 2\times 3\times 4\times 5}$$

$$= 326,240 + 48,216 + 2,940 + 56 = 377,452.$$

186. *First solution.* For convenience, let us think of the parallel lines as lying one above the other. Suppose that there are n points on the upper line, and m points on the lower one (Fig. 1). Let us break up the set of all joining line segments into the pencils of lines with fixed points on the lower line as vertices. (In Fig. 1 we see such a pencil of segments joining a point A with all the points on the upper line.) Evidently, the number of these pencils is equal to m, and that the number of points of intersection of the segments belonging to two arbitrary pencils is the same for any pair of pencils. If we denote this number by k_n, then the total number of points of intersection of all the segments is equal to the product of k_n by the number of combinations of the m pencils two at a time, i. e. to

$$k_n C_m^2 = k_n \frac{m(m-1)}{2}.$$

To compute the number k_n let us group all the segments joining the n points on the upper line to two points A and B on the lower line into the pairs of segments joining a fixed point on the upper line (for instance, C) to the points A and B. The number of these pairs is equal to n, and there exists exactly one point of intersection of the segments belonging to two pairs (for instance, such is the point of intersection of the diagonals of the trapezoid $ABCD$). Therefore,

$$k_n = C_n^2 = \frac{n(n-1)}{2}.$$

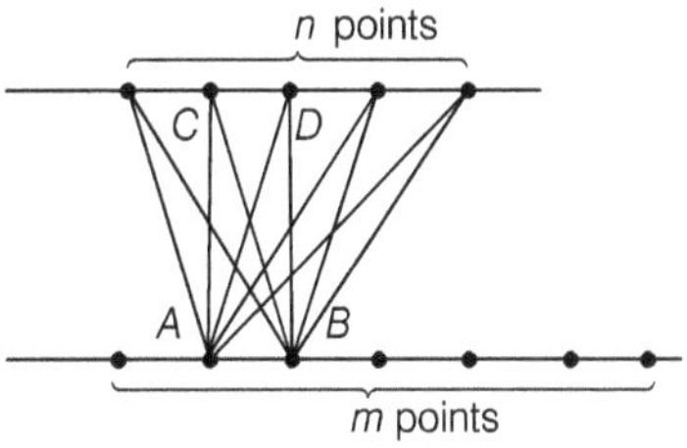

FIG.1

Consequently, the total number of points of intersection of all the segments joining n points on the upper line to m points on the lower line is equal to

$$\frac{n(n-1)}{2}\,\frac{m\,(n-1)}{2}.$$

Second solution. Each point of intersection of the segments can be obtained by selecting two points on the first line (which can be performed in C_m^2 ways) and two points on the second line (which can be performed in C_n^2 ways). Combining all the possible pairs of points we get the total of

$$C_m^2 \cdot C_n^2 = \frac{m(m-1)\,n(n-1)}{4}$$

points of intersection.

187. Each parallelogram is specified by choosing two straight lines of the first family (which can be performed in C_n^2 ways) and two lines of the second family (which can be performed in C_m^2 ways). Thus, the total number of the parallelograms is equal to

$$C_n^2 \cdot C_m^2 = \frac{n\,(n-1)\,m\,(m-1)}{4}.$$

188. Since in the given alphabet every separate character (a dot or a dash) and every pair of the characters denote a letter, the number of ways in which a continuous line consisting of x characters can be read is independent of the particular form of the line and is equal to the total number of all possible partitions of the characters forming the line into the groups of one or two adjacent characters. Let us denote this number by p_n.

Let us now divide all the possible ways of reading the given line consisting of n characters into two sets.

Let the first set comprise the ways in which only the first character of the line is read as a separate letter. The number of ways belonging to the first set is equal to the number of ways in which the rest of the line consisting of $n-1$ characters (remaining after the first character is discarded) can be read, that is to p_{n-1}.

Let the second set comprise the ways in which the first two characters of the line are read as one letter. The number of ways belonging to the second set is equal to the number of ways in which the chain consisting of $n-2$ characters (remaining after the first two characters are discarded) can be read, that is to p_{n-2}.

Since every way of reading the given line belongs either to the first or to the second set, the total number of ways is equal to the sum of ways belonging to the first and second sets, i.e.

$$p_n = p_{n-1} + p_{n-2}. \qquad (1)$$

This equality is a recurrent formula by which one can compute, in succession, $p_1, p_2, \dots p_n$, for any n provided p_1 and p_2 are known. But in the given problem $p_1 = 1$ (for a line consisting of one character there is only one way belonging to the first set) and $p_2 = 2$ (for a line consisting of two characters there are two ways of reading one of which belongs to the first set and the other to the second set).

Using formula (1), we find, in succession,

$$p_3 = p_2 + p_1 = 2 + 1 = 3,$$
$$p_4 = p_3 + p_2 = 3 + 2 = 5,$$
$$p_5 = p_{4+} p_3 = 5 + 3 = 8$$

and so on. Finally, we get

$$p_{12} = 233.$$

6. Problems in Forming Equations

189. Let x be the smaller of the factors. Then the statement of the problem directly implies that

$$x(x+10) - 40 = 39x + 22,$$

that is
$$x^2 - 29x - 62 = 0,$$

whence $x_1 = 31$, $x_2 = -2$. Discarding the negative root we find the sought-for factors which are 31 and 41.

190. Before the first meeting the first cyclist covered $s+a$ km and the second one $s-a$ km where s is the distance between A and B. Consequently, before the second meeting they covered $2s+\frac{1}{k}s$ and $2s-\frac{1}{k}s$ km, respectively.

But if two bodies move with constant speeds, the ratio of the speeds is equal to the ratio of the distances covered by the bodies, provided the times taken are equal. Therefore, for finding s we have the equation

$$\frac{s+a}{s-a}=\frac{2+\frac{1}{k}}{2-\frac{1}{k}}.$$

Hence $t=2ak$ km.

191. If two bodies move with constant speeds, then, for the same path, the ratio of their speeds is the reciprocal of the ratio of the times taken. Let v be the speed of the third car, and t the time of motion of the second car by the moment it was overtaken by the third car. Therefore we have

$$\frac{40}{v}=\frac{t-0.5}{t} \text{ and } \frac{50}{v}=\frac{t+1}{t+1.5}.$$

Dividing termwise the first equation by the second, we find $t=\frac{3}{2}$ hours and then determine $v=60$ km/hr.

192. Let the time period between the start and the meeting be x hours. The distance between the point of meeting and the point B took the cyclist x hours and the pedestrian $x+t$ hours. Since, for equal distances, the times of motion are inversely proportional to the speeds, we can write

$$\frac{x+t}{x}=k,$$

whence we find $x=\frac{t}{k-1}.$

193. Let x be the distance between A and B, and y be the distance between B and C, Then, taking into account that the time of motion is the same in all the cases mentioned in the statement of the problem, we obtain the system of equations

$$\left.\begin{aligned}\frac{x}{3.5}+\frac{y}{4}&=\frac{x+y}{3.75},\\ \frac{x+y}{3.75}&=\frac{14}{60}+\frac{y}{3.75}+\frac{x}{4}.\end{aligned}\right\}$$

Solving this system we find $x=14$ km and $y=16$ km.

194. Let x denote the length of the horizontal path, and y be the length of the uphill portion. Then we can form the following system of equations:

$$\left.\begin{aligned} \frac{y}{3}+\frac{x}{4}+\frac{11.5-(x+y)}{5}&=2\frac{9}{10}, \\ \frac{11.5-(x+y)}{3}+\frac{x}{4}+\frac{y}{5}&=3\frac{1}{10}. \end{aligned}\right\}$$

Adding together the equations, we find $x=4$.

195. Let us denote the distance between the points A and B by l, and the speeds of the motorcyclists by v_1 and v_2. During the time period t the first motorcyclist covered the distance $p+l-q$, and the second the distance $q+l-p$.

Therefore,

$$\left.\begin{aligned} v_1&=\frac{l+p-q}{t}, \\ v_2&=\frac{l+q-p}{t}. \end{aligned}\right\} \qquad (1)$$

On the other hand, the ratio of the speeds is equal to the ratio of the paths covered before the first meeting, i.e.

$$\frac{v_1}{v_2}=\frac{l-p}{p}.$$

Substituting v_1 and v_2 expressed by (1) into the latter relation we get an equation for determining l. Solving it, we find $l=3p-q$. Substituting this value of l into formulas (1) we obtain

$$v_1=\frac{4p-2q}{t}, \quad v_2=\frac{2p}{t}.$$

196. The difference between the delay times of the airplane in the first and second flights which is equal to $\frac{t_1-t_2}{60}$ hours is due to the fact that the distance of d km was covered by the aircraft at different speeds, namely, during the first flight the speed was v km/hr and during the second flight w km/hr (the speeds on the other parts of the flight were equal). Thus, we get the equation

$$\frac{t_1-t_2}{60}=\frac{d}{v}-\frac{d}{w},$$

wherefrom we find that the initial speed of the airplane is equal to

$$w=\frac{60vd}{60\,d+v\,(t_2-t_1)}\,\frac{\text{km}}{\text{hr}}.$$

197. Let us denote the weight of each cut-off piece by x. Suppose that the first piece contained $100a$ % of copper, and the second $100b$ % of copper. Then the weight of copper contained in the first piece after its remainder has been alloyed with the cut-off piece of the other alloy is equal to $a(m-x)+bx$, and the amount of copper in the second piece after its remainder has been alloyed with the cut-off piece of the first alloy is equal to $b(n-x)+ax$. By the hypothesis, we have

$$\frac{a(m-x)+bx}{m}=\frac{b(n-x)+ax}{n}.$$

Solving this equation and taking into account that $a \neq b$ we obtain

$$x=\frac{mn}{m+n}.$$

198. Let the ratio of the weights of the alloyed pieces be $\alpha:\beta$. Then

$$\frac{\frac{\alpha p}{100}+\frac{\beta q}{100}}{\alpha+\beta}=\frac{r}{100}.$$

It follows that

$$\alpha:\beta=(r-q):(p-r).$$

The problem is solvable if either $p>r>q$ or $p<r<q$.

To find the maximum weight of the new alloy let us consider the ratios

$$\frac{P}{|r-q|} \text{ and } \frac{Q}{|p-r|}.$$

If $\frac{P}{|r-q|}=\frac{Q}{|p-r|}$, then the maximum weight is equal to

$$P+Q=\frac{p-q}{r-q}p=\frac{p-q}{p-r}Q.$$

If $\frac{P}{|r-q|}<\frac{Q}{|p-r|}$, the maximum weight is equal to

$$p+\frac{p-r}{r-q}P=\frac{p-q}{r-q}P.$$

If, finally, $\frac{P}{|r-q|}>\frac{Q}{|p-r|}$, then the maximum weight is

$$Q+\frac{r-q}{p-r}Q=\frac{p-q}{p-r}Q.$$

199. Suppose that each worker worked for t days and A earned x roubles while B earned y roubles. From the conditions of the problem deduce the following system of equations :

$$\left.\begin{aligned}(t-1)\frac{x}{t}&=72,\\(t-7)\frac{y}{t}&=64.8,\\(t-1)\frac{y}{t}-(t-7)\frac{x}{t}&=32.4.\end{aligned}\right\}\qquad(1)$$

From the first two equations we find

$$\frac{t-1}{t}=\frac{72}{x},\ \frac{t-7}{t}=\frac{64.8}{y}.$$

Finally, the last equation yields

$$72\frac{y}{x}-64.8\frac{x}{y}=32.4,$$

that is

$$20\left(\frac{y}{x}\right)^2-9\left(\frac{y}{x}\right)-18=0.$$

From the latter equation we find $y=\frac{6}{5}x$ (the negative root is discarded). Now, dividing the second equation of system (1) by the first one and replacing $\frac{y}{x}$ by its value $\frac{6}{5}$ we find

$$\frac{6}{5}\cdot\frac{t-7}{t-1}=\frac{64.8}{7.2},\ \frac{t-7}{t-1}=\frac{3}{4},$$

whence we obtain $t=25$. Consequently,

$$x=75 \text{ roubles},\ y=90 \text{ roubles}.$$

200. Let t_1 be the time elapsed before the first meeting, t_2 be the time elapsed before the second meeting and R be the radius of the circle. During the time t_1 the first body covered the distance vt_1 and the second the distance $\frac{at_1^2}{2}$. The sum of these distances is equal to the circumference of the circle, that is

$$vt_1+\frac{at_12}{2}=2\pi R. \qquad (1)$$

During the time t_2 each body covered the same distance equal to the circumference of the circle, and hence we have

$$vt_2=2\pi R \text{ and } \frac{at_2^2}{2}=2\pi R.$$

Eliminating t_2 from these relations we find $R=\frac{v^2}{\pi a}$. Substituting this value of R into (1) we arrive at a quadratic equation in t_1 of the form

$$\frac{at_1^2}{2}+vt_1-\frac{2v^2}{a}=0.$$

Solving this equation and discarding the negative root (according to the meaning of the problem, we must have $t_1>0$) we finally receive

$$t_1=(\sqrt{5}-1)\frac{v}{a}.$$

201. Let us denote by q_1 and q_2 the capacities of the taps measured in l/min and by v the volume of the tank. The times of filling the tank by each tap alone are, respectively,

$$t_1 = \frac{v}{q_1} \text{ and } t_2 = \frac{v}{q_2}. \tag{1}$$

The first condition of the problem leads to the equation

$$q_1 \cdot \frac{1}{3} t_2 + q_2 \cdot \frac{1}{3} t_1 = \frac{11}{18} v.$$

Using equalities (1) we get the quadratic equation

$$\left(\frac{q_1}{q_2}\right)^2 - \frac{13}{6} \frac{q_1}{q_2} + 1 = 0,$$

whose solutions are $\frac{q_1}{q_2} = \frac{2}{3}$ and $\frac{q_1}{q_2} = \frac{3}{2}$. The second condition of the problem implies that

$$v = (3 \cdot 60 + 36)(q_1 + q_2) = 216 (q_1 + q_2).$$

From (1) we find the sought-for quantities:

$$t_1 = \frac{216 (q_1 + q_2)}{q_1} = 540 \text{ min (9 hours)},$$

$$t_2 = \frac{216 (q_1 + q_2)}{q_2} = 360 \text{ min (6 hours)}.$$

There is a second solution, namely.

$$t_1 = 360 \text{ min}, \ t_2 = 540 \text{ min}.$$

202. Let γ be the specific weight of water and s be the cross-section area of the pipe. Atmospheric pressure p_a is determined by the formula

$$p_a = \gamma c.$$

If p_1 is the pressure under the piston when it is elevated, then, by Boyle and Mariott's law, for the column of air between the piston and the water level we

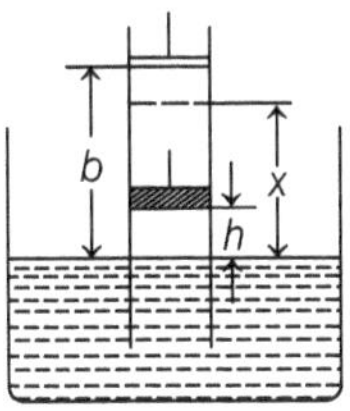

FIG. 2

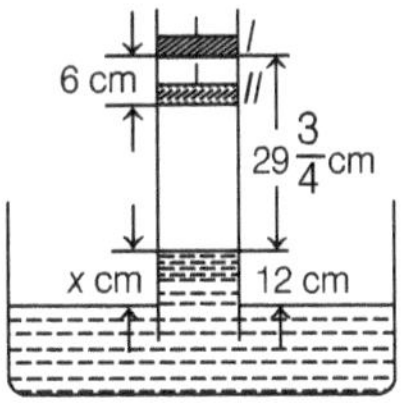

FIG. 3

have $p_1\ (b-x)s = p_a hs$ (see Fig. 2). The equilibrium equation for the column of water is of the form $p_a - p_1 = \gamma x$. This leads to the equation

$$c - \frac{hc}{b-x} = x$$

(after γ has been cancelled out), i. e. to the quadratic equation

$$x^2 - (b+c)x + (b-h)c = 0.$$

Solving the equation we find

$$x = \frac{1}{2}\left[(b+c) - \sqrt{(b-c)^2 + 4hc}\right].$$

203. Let p_1 and p_2 be the air pressures under the piston in positions *I* and *II*, respectively (Fig. 3), and γ be the specific weight of mercury. The equilibrium equation for the columns of mercury 12 cm and x cm high are, respectively,

$$\left.\begin{aligned} 76\gamma - p_1 &= 12\gamma, \\ 76\gamma - p_2 &= x\gamma. \end{aligned}\right\} \qquad (1)$$

Boyle and Mariott's law applied to the column of air below the piston yields the equation

$$p_1 \cdot 29\frac{3}{4} = p_2(36-x).$$

Substituting the expressions of p_1 and p_2 found from (1) into this equation, we obtain the following quadratic equation in x:

$$29\frac{3}{4} \times 64 = (76-x)(36-x),$$

that is

$$x^2 - 112x + 832 = 0.$$

Solving the last equation we find $x = 56 \pm \sqrt{3136-832} = 56 \pm \sqrt{2304}$ $= 56 \pm 48$, and hence $x = 8$ cm.

204. Let the watch gain x minutes a day. Then it will show true time in $\frac{2}{x}$ days. If it were 3 minutes slow at that moment but gained $x + \frac{1}{2}$ minutes more a day, it would show true time in $\frac{3}{x+\frac{1}{2}}$ days.

Hence,

$$\frac{3}{x+\frac{1}{2}} + 1 = \frac{2}{x},$$

whence

$$x^2 + \frac{3}{2}x - 1 = 0.$$

Solving this equation, we find $x = 0.5$.

205. If x is the original sum of money each person deposited and y is the interest paid by the savings bank, then

$$x + x\frac{y}{100}\frac{m}{12} = p,$$

$$x + x\frac{y}{100}\frac{n}{12} = q.$$

Multiplying the first equation by n and the second by m, and subtracting the latter equation from the former, we find

$$x = \frac{pn - qm}{n - m}.$$

Now taking the original system and subtracting the second equation from the first one we get

$$\frac{xy}{1200}(m - n) = p - q,$$

whence we obtain

$$y = \frac{1200\,(p - q)}{qm - pn}\,\%.$$

206. Let v_1 and v_2 be the speeds of the points, and $v_1 > v_2$. The first condition of the problem is expressed by the equation

$$\frac{2\pi R}{v_2} - \frac{2\pi R}{v_1} = t.$$

The second condition means that the distance covered by the point moving in the circle at a higher speed during the time T is by $2\pi R$ longer than that covered by the other point. Thus, we get another equation

$$Tv_1 - Tv_2 = 2\pi R.$$

From the latter equation we find

$$v_2 = v_1 - \frac{2\pi R}{T}.$$

Substituting this expression for v_2 into the first equation we get a quadratic equation for v_1:

$$v_1^2 - \frac{2\pi R}{T}v_1 - \frac{2\pi R}{T}\cdot\frac{2\pi R}{t} = 0.$$

Solving it we find

$$v_1 = \frac{\pi R}{T}\left(\sqrt{1 + \frac{4T}{t}} + 1\right)$$

and then determine

$$v_2 = \frac{\pi R}{T}\left(\sqrt{1 + \frac{4T}{t}} - 1\right).$$

207. Let v be the volume of the solution in the flask and x be the percentage of sodium chloride contained in the solution.

The volume $\frac{v}{n}$ of the solution is poured into the test tube and evaporated until the percentage of sodium chloride in the test tube is doubled. Since the amount of sodium chloride remains unchanged, the volume of the solution in the test tube becomes half as much, and hence the weight of the evaporated water is equal to $\frac{v}{2n}$.

After the evaporated solution is poured back into the flask, the amount of sodium chloride in the flask becomes the same as before. i.e. $v\frac{x}{100}$, and the volume of the solution is reduced by $\frac{v}{2n}$. Thus, we obtain the equation

$$\frac{v\frac{x}{100}}{v-\frac{v}{2n}}=\frac{x+p}{100},$$

wherefrom we find

$$x=(2n-1)p.$$

208. Let the first vessel contain x litres of alcohol, then the second vessel contains $30-x$ litres. After water has been added to the first vessel, one litre of the obtained mixture contains $\frac{x}{30}$ litres of alcohol and $1-\frac{x}{30}$ litres of water. After the resulting mixture is added from the first vessel to the second vessel the latter contains $30-x+\frac{x}{30}x$ litres of alcohol and $\left(1-\frac{x}{30}\right)x$ litres of water.

One litre of the new mixture contains

$1-\frac{x}{30}+\left(\frac{x}{30}\right)^2$ litres of alcohol.

Alter 12 litres of the new mixture is poured out from the second vessel into the first, the first vessel contains

$12\left[1-\frac{x}{30}+\left(\frac{x}{30}\right)^2\right]+\frac{x}{30}(30-x)$ litres of alcohol and the second contains

$$18\left[1-\frac{x}{30}+\left(\frac{x}{30}\right)^2\right] \text{litres of alcohol.}$$

By the hypothesis,

$$18\left[1-\frac{x}{30}+\left(\frac{x}{30}\right)^2\right]+2$$

$$=12\left[1-\frac{x}{30}+\left(\frac{x}{30}\right)^2\right]+x-\frac{x^3}{30},$$

whence we get the equation

$$x^2-30x+200=0.$$

This equation has the roots

$$x_1=20 \text{ and } x_2=10.$$

Hence, the first vessel originally contained either 20 litres of alcohol (and then the second contained 10 l) or 10 litres (and then the second vessel contained 20 l).

209. Let x be the distance between the bank the travellers started from and the place where C left the motor boat. Note that A caught the boat at the same distance from the opposite bank. Indeed, the only distinction between the ways in which A and C crossed the water obstacle is that C started out in the motor boat and then swam and A first swam and then took the motor boat. Since they swam at an equal speed $v\,(v \neq v_1)$ and the crossing took them equal times, the above distances should be equal.

Taking this note into consideration, we easily set up the equation

$$\frac{x+s-2(s-x)}{v_1}=\frac{s-x}{v},$$

its left member expressing the time of motion of the boat from the start to the point where it meets A and its right member being equal to the time of motion of A from the start to that point.

The above equation yields

$$x=\frac{s(v+v_1)}{3v+v_1}.$$

Therefore the duration of the crossing is equal to

$$T=\frac{s-x}{v}+\frac{x}{v_1}=\frac{s}{v_1}\,\frac{v+3v_1}{3v+v_1}.$$

Note. The problem can also be solved without using the equality of the above mentioned distances. But then we have to introduce some new unknowns, and the solution becomes more complicated.

210. Let the sought-for distance be s km and the speed of the train be v km/h. During 6 hours preceding the halt caused by the snow drift the first train covered $6v$ km and the remaining distance of $(s-6v)$ km took it $\frac{5(s-6v)}{6v}$ hours because the speed of the train on that part of the trip was equal to $\frac{6}{5}v$.

The entire trip (including the two-hour wait) lasted $8+\frac{5(s-6v)}{6v}$ hours which exceeds by one hour the interval of $\frac{s}{v}$ hours indicated by the time-table. Thus, we obtain the equation

$$8+\frac{5(s-6v)}{6v}=1+\frac{s}{v}.$$

Reasoning analogously, we set up another equation concerning the second train:

$$\frac{s}{v}+\frac{3}{2}=8+\frac{150}{v}+\frac{5(s-6v-150)}{6v}.$$

From this system of equations we find $s=600$ km.

211. Denoting the speed of the motor boat in still water by v and the speed of the current by w we get the following system of two equations:

$$\left.\begin{aligned}\frac{a}{v+w}+\frac{a}{v-w}&=T,\\ \frac{a}{v-w}=T_0+\frac{a-b}{v+w}+\frac{2b}{v+w}&=T_0+\frac{a+b}{v+w}.\end{aligned}\right\}$$

Solving this system with respect to the unknowns $\frac{1}{v+w}$ and $\frac{1}{v-w}$ and taking their reciprocals we find

$$v+w=\frac{2a+b}{T-T_0}\text{ and }v-w=\frac{a(2a+b)}{T(a+b)+T_0a}.$$

It then follows that

$$v=\frac{1}{2}\left[\frac{2a+b}{T-T_0}+\frac{a(2a+b)}{T(a+b)+T_0a}\right]$$

and

$$w=\frac{1}{2}\left[\frac{2a+b}{T-T_0}-\frac{a(2a+b)}{T(a+b)+T_0a}\right].$$

212. Let x be the time period during which the second tap was kept open and $v(w)$ be the capacity of the first (second) tap measured in m^3/hr. We have

$$\left.\begin{aligned}v(x+5)+wx&=425,\\ 2vx&=w(x+5),\\ (v+w)17&=425.\end{aligned}\right\}$$

From the second and third equations we get

$$v = 25\frac{x+5}{3x+5}, \quad w = \frac{50x}{3x+5}.$$

Substituting these expressions into the first equation we find

$$3x^2 - 41x - 60 = 0,$$

whence $x = 15$ hours (the negative root is discarded).

213. Let the sought-for speed of the train be v km/hr and the scheduled speed be v_1 km/hr. The first half of the way took the train $\frac{10}{v_1}$ hours and the second half of the way together with the halt took it $\frac{10}{v_1+10} + \frac{1}{20}$ hours in the first trip and $\frac{10}{v} + \frac{1}{12}$ hours in the second trip. But both times the train arrived at B on schedule and therefore

$$\frac{10}{v_1} = \frac{10}{v_1+10} + \frac{1}{20}, \quad \frac{10}{v_1} = \frac{10}{v} + \frac{1}{12}.$$

From the first equation we can find v_1. We have

$$10\left(\frac{1}{v_1} - \frac{1}{v_1+10}\right) = \frac{1}{20}, \quad \frac{100}{v_1(v_1+10)} = \frac{1}{20},$$

that is $\quad v_1^2 + 10v_1 - 2000 = 0,$

and the latter equation has the only one positive root $v_1 = 40$. From the second equation we find that $v = 60$ km/hr.

214. Let the distance AB be equal to s km, and the speeds of the first and second airplanes be respectively equal to v_1 and v_2. Then, by the conditions of the problem, we have the following system of three equations:

$$\left.\begin{aligned} \frac{s}{2v_1} + \frac{a}{v_1} &= \frac{s}{2v_2} - \frac{a}{v_2}, \\ \frac{s}{2v_2} - \frac{s}{2v_1} &= b, \\ \frac{3s}{4v_1} - b &= \frac{s}{4v_2}. \end{aligned}\right\}$$

Let us put $\quad \frac{s}{2v_1} = x, \ \frac{s}{2v_2} = y.$

From the second and third equations we find $x = \frac{3}{2}b$ and $y = \frac{5}{2}b$, and the first equation yields $a\left(\frac{1}{v_1} + \frac{1}{v_2}\right) = b$. But $\frac{v_2}{v_1} = \frac{x}{y} = \frac{3}{5}$, and now we readily find that $v_1 = \frac{8a}{3b}$, $v_2 = \frac{8a}{5b}$ and $s = 8a$.

215. Let u be the speed of the motor boat in still water and v be the speed of the current. Then we have the following system:

$$\left.\begin{aligned}\frac{96}{u+v}+\frac{96}{u-u}&=14,\\ \frac{24}{v}=\frac{96}{u+v}+\frac{72}{u-v}&.\end{aligned}\right\}$$

To solve it let us put $\frac{u}{v}=z$. Multiplying both members of the second equation by v we find

$$24=\frac{96}{z+1}+\frac{72}{z-1}.$$

Reducing the terms of this equation to a common denominator and discarding it we obtain the quadratic equation

$$24z^2-168z=0,$$

whose roots are $z=0$ and $z=7$. Since $z\neq 0$, we must take $z=7$. Hence,

$$u=7\,v.$$

Substituting $u=7v$ into the first equation of the system we derive

$$\frac{96}{8v}+\frac{96}{6v}=14,$$

whence we find

$$v=2\text{ km/hr},\quad u=14\text{ km/hr}.$$

216. The distance covered by a body moving with constant acceleration a during t sec is determined by the formula

$$s=v_0t+\frac{at^2}{2}.$$

To find v_0 and a for each body we must substitute the given numerical data into this formula.

(1) For the first body we have

$$25=v_0+\frac{a}{2}\quad\text{for } t=1$$

and

$$50\frac{1}{3}=2v_0+2a\quad\text{for } t=2,$$

whence $a=\frac{1}{3}$, $v_n=25-\frac{1}{6}$ and $s_1=24\frac{5}{6}t+\frac{t^2}{6}$.

(2) For the second body we have

$$30=v_0+\frac{a}{2}\text{ for } t=1$$

and

$$59\frac{1}{2} = 2v_0 + 2a \text{ for } t = 2,$$

whence $a = -\frac{1}{2}$, $v_0 = 30 + \frac{1}{4}$

and $$s_2 = 30\frac{1}{4}t - \frac{t^2}{4}.$$

For the moment when the first body catches up with the second we have $s_1 = s_2 + 20$ which results in a quadratic equation for determining t of the form

$$t^2 - 13t - 48 = 0.$$

Solving it we find $t = 16$, the negative root being discarded.

217. Let v denote the relative speed of the boat. Then the time of motion of the boat is equal to

$$t = \frac{10}{v+1} + \frac{6}{v-1}.$$

By the hypothesis, we have

$$3 \le \frac{10}{v+1} + \frac{6}{v-1} \le 4. \qquad (1)$$

It is necessary that $v > 1$, since otherwise the boat cannot move upstream.

Let us pass from the system of inequalities (1) to an equivalent system of inequalities of the form

$$3(v^2 - 1) \le 16v - 4 \le 4(v^2 - 1).$$

Thus, the two inequalities

$$3v^2 - 16v + 1 \le 0$$

and

$$4v^2 - 16v \ge 0.$$

must hold simultaneously. The first inequality is satisfied if

$$\frac{8 - \sqrt{61}}{3} \le v \le \frac{8 + \sqrt{61}}{3}.$$

The second inequality is satisfied if $v < 0$ or $v > 4$. But since $v > 1$, we finally obtain

$$4 \le v \le \frac{8 + \sqrt{61}}{3}.$$

218. Let x be the volume of water in the vessel A before pouring the water from A into B. Then the original volume of water in the vessels B and C is equal to $2x$ und $3x$ respectively, and the total volume is equal to $x + 2x + 3x = 6x$.

After the water has been poured from A into B and from B into C for the first time, the water level in all three vessels becomes the same, and therefore the volumes of water in them are in the ratio equal to that of the areas of the bases which is 1:4:9. Therefore, after the first pouring the volumes of water in the vessels A, B and C are respectively equal to

$$1\cdot\frac{6x}{1+4+9}=\frac{3}{7}x,\quad 4\cdot\frac{6x}{1+4+9}=\frac{12}{7}x.$$

and

$$9\cdot\frac{6x}{1+4+9}=\frac{27}{7}x.$$

After the second pouring from C into B these volumes assume the values

$$\frac{3}{7}x,\ \frac{12}{7}x+128\frac{4}{7}\ \text{ and }\ \frac{27}{7}x-128\frac{4}{7},$$

respectively. After the third pouring from B into A the volume of water in A becomes equal to x–100, and in B equal to

$$\frac{1}{2}(x-100)\cdot 4=2(x-100).$$

Adding together the volumes of water in all the vessels we obtain the following linear equation with respect to x:

$$(x-100)+2(x-100)+\frac{27}{7}x-128\frac{4}{7}=6x.$$

Solving this equation we find $x = 500$.

Thus we find the original amount of water in each vessel:

A contains 500 litres,

B contains 1000 litres,

C contains 1500 litres.

219. Let the desired number have the form $xyzt$ where the letters x, y, z and t denote the digits in the corresponding decimal places. By the conditions of the problem, we obtain the following system of equations:

$$\left.\begin{aligned} x^2+t^2&=13,\\ y^2+z^2&=85,\\ xyzt-1089&=tzyx.\end{aligned}\right\}\qquad(1)$$

The rules of subtraction of decimal numbers imply that in the third equation of the above system t is equal either to 9 or to

$$(10+t)-9=x,$$

i. e.

$$x=t+1.\qquad(2)$$

But from the first equation of system (1) it follows that $t < 4$ and therefore (2) takes place. Then from the first equation of system (1) we get the equation for determining t:

$$(t+1)^2 + t^2 = 13,$$

whence we find

$$t = 2.$$

From (2) it then follows that $x = 3$, and the third equation of system (1) takes the form

$$3yz^2 - 1089 = 2zy3. \qquad (3)$$

Now let us note that $z < 9$ because if $z = 9$, then (3) implies that $y = 0$ and therefore the second equation of system (1) is not fulfilled. From (3) we find

$$(z-1+10) - 8 = y,$$

i. e.

$$z = y - 1. \qquad (4)$$

Finally, from the second equation of system (1) and from (4) we determine $z = 6$, $y = 7$. Thus, the sought-for number is 3762.

220. Let us begin with finding the distance x between the start of motion and the first meeting. The equation for the times of motion of both points has the form

$$\frac{a+x}{v} - t = \frac{x}{w},$$

whence

$$x = \frac{(a-vt)\,w}{v-w}.$$

The time from the start of motion to the first meeting is equal to

$$t_1 = \frac{a+x}{v}.$$

Substituting the above value of x into this expression we get

$$t_1 = \frac{a-wt}{v-w}.$$

Let τ be the time interval between two successive meetings. Then

$$v\tau - w\tau = l,$$

which results in $$\tau = \frac{l}{v-w}.$$

The successive meetings will thus occur at the moments of time

$$t_1, t_1+\tau, t_1+2\tau, \ldots .$$

The moment of the nth meeting is

$$t_n = \frac{a - wt + l(n-1)}{v_1 - w}.$$

221. Let γ_1 be the specific weight of the first component of the alloy, γ_2 be the specific weight of the second component and γ be that of water. Suppose that the weight of the first component is x. According to the Archimedes principle, when immersed in water, the alloy loses in its weight a portion of

$$\left(\frac{x}{\gamma_1}+\frac{P-x}{\gamma_2}\right)\gamma.$$

Analogously, for the components the losses in weight are equal to

$$\frac{P}{\gamma_1}\gamma \text{ and } \frac{P}{\gamma_2}\gamma.$$

These losses are given: they are equal to B and C respectively. Consequently, we have

$$\frac{\gamma}{\gamma_1}=\frac{B}{P}, \quad \frac{\gamma}{\gamma_2}=\frac{C}{P}.$$

Thus, the loss of weight of the alloy is

$$A=\frac{B}{P}x+\frac{C}{P}(P-x).$$

Hence,

$$x=\frac{A-C}{B-C}P.$$

For the problem to be solvable it is necessary that $B \neq C$. Furthermore, the fact that $\frac{x}{p}$ is a number lying between 0 and 1 implies the inequality

$$0<\frac{A-C}{B-C}<1.$$

It follows that either $B > A > C$ or $C > A > B$. Therefore, for the problem to be solvable it is necessary and sufficient that the number A lie between the numbers B and C.

222. Let us denote the distance from the point A to the mouth of the river by s, the distance between the mouth of the river and the point B across the lake by s_1, the speed of the towboat (without towing) by v and the speed of the current by v_1. It is necessary to determine the quantity $\frac{2s_1}{v}=x$.

The conditions of the problem enable us to set up the three equations

$$\left.\begin{aligned} \frac{s}{v+v_1}+\frac{x}{2}&=61, \\ \frac{s}{v-v_1}+\frac{x}{2}&=79, \\ \frac{s}{v_1}+x&=411. \end{aligned}\right\}$$

From the first equation we obtain

$$\frac{v+v_1}{s}=\frac{2}{112-x}, \qquad (1)$$

from the second equation we find

$$\frac{v-v_1}{s}=\frac{2}{158-x} \qquad (2)$$

and from the third equation we get

$$\frac{v_1}{s}=\frac{1}{411-x}. \qquad (3)$$

Subtracting equality (2) from equality (1) and using equality (3) we obtain the following equation in x:

$$\frac{1}{122-x}-\frac{1}{158-x}=\frac{1}{411-x},$$

that is

$$x^2-244x+4480=0.$$

Solving this equation we find $x_1=20,\ x_2=224.$

It is obvious that the value $x_2=224$ should be discarded because the left member of equation (1) cannot be negative.

223. Let the distance AB be denoted by s, the distance BC by s_1, the speed of the boat by v and the speed of the current by v_1 (s and s_1 are supposed to be expressed in the same units of length and v and v_1 in those units per hour).

For the motion of the boat from A to C downstream we have

$$\frac{s}{v}+\frac{s_1}{v+v_1}=6. \qquad (1)$$

For the boat going upstream from C to A we have

$$\frac{s_1}{v-v_1}+\frac{s}{v}=7. \qquad (2)$$

If between A and B the current is the same as between B and C, then the trip from A to C takes

$$\frac{s+s_1}{v+v_1}=5.5 \text{ hours} \qquad (3)$$

Now we have to determine the ratio $\frac{s+s_1}{v-v_1}$.

Reducing equations (1), (2) and (3) to a common denominator and multiplying both members of equation (3) by $v \neq 0$, we get the system

$$\left.\begin{aligned} (s+s_1)v &= 6v(v+v_1) - sv_1, \\ (s+s_1)v &= 7v(v-v_1) + sv_1, \\ (s+s_1)v &= 5.5\,(v+v_1)v. \end{aligned}\right\} \quad (4)$$

Adding together the first two equations and using the third one we obtain $\quad 2\,(s+s_1)v = v(13v - v_1) = 11v(v+v_1)$,

whence we find $v = 6v_1$. But from the third equation of system (4) we have

$$\frac{s+s_1}{v_1} = 7 \times 5.5. \text{ Consequently,}$$

$$\frac{s+s_1}{v-v_1} = \frac{s+s_1}{5v_1} = 7.\ 7 \text{ hours.}$$

224. Let v be the volume of the vessel, α_1 be the percentage of the acid in it after the first mixing, α_2 the percentage of the acid after the second mixing and so on. We have

$$\left.\begin{aligned} &\frac{(v-a)p+aq}{v} + \alpha_1, \\ &\frac{(v-a)\alpha_1 + aq}{v} = \alpha_2, \\ &\cdots\cdots\cdots\cdots\cdots\cdots \\ &\frac{(v-a)\alpha_{k-2} + aq}{v} = \alpha_{k-1}, \\ &\frac{(v-a)\alpha_{k-1} + aq}{v} = r. \end{aligned}\right\}$$

Multiplying the sth equality by $\left(\frac{v-a}{v}\right)^{k-s}$ $(s = 1, 2, \ldots., r)$ and adding together the results we obtain

$$\left(\frac{v-a}{v}\right)^k p + \frac{a}{v} q \left[1 + \frac{v-a}{v} + \left(\frac{v-a}{v}\right)^2 + \ldots + \left(\frac{v-a}{v}\right)^{k-1}\right] = r.$$

whence it follows that

$$\left(\frac{v-a}{v}\right)^k p + \frac{a}{v} q \frac{\left(\frac{v-a}{v}\right)^k - 1}{\frac{v-a}{v} - 1} = r.$$

Consequently, $\left(1 - \frac{a}{v}\right)^k (p-q) = r - q.$

Answer:

$$v = \frac{a}{1 = \sqrt[k]{\frac{r-q}{p-q}}}.$$

225. At the end of the first year the deposit increased by $\frac{Ap}{100}$ roubles and the depositor took out B roubles. Therefore, at the beginning of the second year the deposit was equal (in roubles) to

$$P_1 = A\left(1+\frac{p}{100}\right)-B.$$

At the end of the second year the deposit was equal to

$$P_2 = P_1\left(1+\frac{p}{100}\right)-B = A\left(1+\frac{p}{100}\right)^2 - B\left[1+\left(\frac{p}{100}+1\right)\right]$$

and at the end of the third year it was

$$P_3 = Ak^3 - B(1+k+k^2)$$

where

$$k = 1+\frac{p}{100}.$$

Obviously, at the end of the nth year the deposited sum became equal to

$$P_n = Ak^n - B(1+k+k^2+\ldots+k^{n-1}),$$

i.e.

$$P_n = \frac{Ap-100B}{p}\left(1+\frac{p}{100}\right)^n + \frac{100B}{p}.$$

To solve the problem we must find n such that $P_n \geq 3A$. Then

$$n \geq \frac{\log(3Ap-100B)-\log(Ap-100B)}{\log\left(1+\frac{p}{100}\right)}. \qquad (1)$$

The meaning of the problem indicates that the deposited sum must increase, and therefore

$$Ap > 100B.$$

Furthermore, we have $p > 0$, $A > 0$ and $B > 0$ and hence the expression on the right-hand side of inequality (1) makes sense.

226. The amount of wood in the forestry at the end of the first year is equal to

$$a\left(1+\frac{p}{100}\right)-x = a_1,$$

at the end of the second to

$$a_1\left(1+\frac{p}{100}\right)-x = a_2.$$

at the end of the third year to

$$a_2\left(1+\frac{p}{100}\right)-x = a_3$$

and so on. Lastly, at the end of the nth year the amount of wood is equal to

$$a_{n-1}\left(1+\frac{p}{100}\right)-x=a_n=aq.$$

Now we can find x. Putting, for brevity, $1+\frac{p}{100}=k$, we get from the last equation the expression $x=ka_{n-1}-aq$. Expressing a_{n-1} from the foregoing equation we obtain

$$x=k\,(ka_{n-2}-x)-aq=k^2a_{n-2}-kx-aq.$$

But

$$a_{n-2}=ka_{n-3}-x.$$

Hence,

$$x=k^3a_{n-3}-k^2x-kx-aq.$$

Proceeding in the same way, we finally express a_2 in terms of a_1 and obtain the following equation with respect to x:

$$x=k^na-x\,(k^{n-1}+k^{n-2}+\ldots+k)-aq.$$

It follows that

$$x=a\,\frac{k^n-q}{k^n-1}\,(k-1)=a\,\frac{\left(1+\frac{p}{100}\right)^n-q}{\left(1+\frac{p}{100}\right)^n-1}\,\frac{p}{100}.$$

227. Before pouring the concentration qi $(i=1,2,\ldots,n)$ of alcohol was

$q_1=1$ in the first vessel,

$q_2=\frac{1}{k}$ the second vessel,

. . . .

$q_n=\frac{1}{k^{n-1}}$. in the nth vessel.

After all the manipulations the concentrations became respectively equal to p_1, p_2, ..., p_n. Then $p_1=1$, and P_i for $i>1$ is determined from the equation

$$pi=\frac{qi\frac{v}{2}+pi-1\frac{v}{2}}{v}=\frac{qi+pi-1}{2}\ (i=2,\ldots,n).$$

We obtain this equation by dividing the amount

$$qi-\frac{v}{2}+pi-1\frac{v}{2}$$

of alcohol contained in the ith vessel after it has been filled from the $(i-1)$th vessel, by the volume v of the vessel.

Thus,

$$p_2 = \frac{q_2 + p_1}{2}, \quad p_3 = \frac{q_3 + p_2}{2}, \ \dots \ p_n = \frac{q_n + p_{n-1}}{2}.$$

Hence,

$$p_n = \frac{q_n + p_{n-1}}{2} = \frac{q_n + \dfrac{q_{n-1} + p_{n-1}}{2}}{2}$$

$$= \frac{q_n}{2} + \frac{q_{n-1}}{2^2} + \frac{1}{2^2} p_{n-2}$$

$$= \frac{q_n}{2} + \frac{q_{n-1}}{2^2} + \frac{1}{2^2} \frac{q_{n-2} + p_{n-3}}{2}$$

$$= \frac{q_n}{2} + \frac{q_{n-1}}{2^2} + \frac{q_{n-2}}{2^3} + \frac{p_{n-3}}{2^3} = \dots$$

$$\dots = \frac{q_n}{2} + \frac{q_{n-1}}{2^2} + \dots + \frac{q_2}{2^{n-1}} + \frac{p_1}{2^{n-1}}$$

$$= \frac{1}{2k^{n-1}} + \frac{1}{2^2 k^{n-2}} + \dots + \frac{1}{2^{n-1} k} + \frac{1}{2^{n-1}}.$$

For $k \neq 2$ the last sum is equal to

$$p_n = \frac{1}{2k} \frac{\dfrac{1}{k^{n-1}} - \dfrac{1}{2^{n-1}}}{\dfrac{1}{k} - \dfrac{1}{2}} + \frac{1}{2^{n-1}} = \frac{2^{n-1} - k^{n-1}}{(2k)^{n-1}(2-k)} + \frac{1}{2^{n-1}}.$$

For $k = 2$ it equals

$$p_n = \frac{n-1}{2^n} + \frac{1}{2^{n-1}} = \frac{n-1}{2^n} + \frac{2}{2^n} = \frac{n+1}{2^n}.$$

228. The quotient is expressed by the fraction of the form $\dfrac{p}{p^2-1}$ where p is a positive integer. The conditions of the problem are written in the form of the inequalities

$$\frac{p+2}{p^2+1} > \frac{1}{3} \text{ and } 0 < \frac{p-3}{p^2-4} < \frac{1}{10}.$$

We now transform the first inequality to the form

$$3(p+2) > p^2 + 1, \text{that is } 0 > p^2 - 3p - 5.$$

Solving the quadratic equation $p^2 - 3p - 5 = 0$ we obtain

$$p_{1,2} = \frac{3 \pm \sqrt{29}}{2}.$$

From the inequality $0 > p^2 - 3p - 5$ we get $p_2 < p < p_1$, But $p_2 < 0$ and $p > 0$, therefore

$$0 < p < p_1 = \frac{3 + \sqrt{29}}{2}.$$

It is readily seen that p_1 lies between 4 and 4.5. Consequently, it follows from the latter inequality that p as integer can assume only one of the four values $p = 1, 2, 3, 4$. Substituting these values into the second inequality

$$0 < \frac{p-3}{p^2-4} < \frac{1}{10},$$

we find that $p \neq 1, p \neq 2$ and $p \neq 3$. Thus, $p = 4, \frac{p}{p^2-1} = \frac{4}{15}$.

7. Miscellaneous Problems

229. We have

$$\frac{1}{n(n+1)} + \frac{1}{(n+1)(n+2)} + \ldots + \frac{1}{(n+k-1)(n+k)} =$$

$$= \left(\frac{1}{n} - \frac{1}{n+1}\right) + \left(\frac{1}{n+1} - \frac{1}{n+2}\right) + \ldots +$$

$$+ \left(\frac{1}{n+k-1} - \frac{1}{n+k}\right) = \frac{1}{n} - \frac{1}{n+k} = \frac{k}{n(n+k)}.$$

230. Let first $x \neq a$. Multiplying and dividing the product in question by $x - a$ and applying, in succession, the formula for the difference of the squares of two numbers we obtain

$$\frac{(x-a)(x+a)(x^2+a^2)(x^4+a^4)\ldots(x^{2^{n-1}}+a^{2^{n-1}})}{x-a}$$

$$= \frac{(x^2-a^2)(x^2+a^2)(x^4+a^4)\ldots(x^{2^{n-1}}+a^{2^{n-1}})}{x-a}$$

$$= \frac{(x^4-a^{4)}(x^4+a^4)\ldots x^{2^{n-1}}+a^{2^{n-1}}}{x-a}$$

$$= \frac{(x^8-a^8)\ldots(x^{2^{n-1}}+a^{2^{n-1}})}{x-a} = \frac{x^{2^n}-a^{2^n}}{x-a}.$$

Let now $x = a$. Then the product is equal to

$$2a \cdot 2a^2 \cdot 2a^4 \ldots 2a^{2^{n-1}} = 2^n a^{1+2+2^2+\ldots 2^{n-1}}$$

$$= 2^n a^{\frac{2^n-1}{2-1} = 2^n a\, 2^n - 1} = 2^n a^{(2n-1)}$$

231. Multiply and divide the given expression by the product

$$(x+a)(x^2+a^2)(x^4+a^4)\ldots(x^{2^{n-1}}+a^{2^{n-1}}),$$

which is different from zero for all real $x \neq -a$. It is readily seen that the result can be written as follows:

$$\frac{(x^3+a^3)(x^6+a^6)(x^{12}+a^{12})\dots(x^{3\cdot 2^{n-1}}+a^{3\cdot 2^{n-1}})}{(x+a)(x^2+a^2)(x^4+a^4)\dots(x^{2^{n-1}}+a^{2^{n-1}})}.$$ The numerator and denominator of this fraction are products similar to that in the foregoing problem. Therefore, multiplying the numerator and denominator by the product $(x-a)(x^3-a^3)$ we transform the expression to the form

$$\frac{x^{3\cdot 2^n}-a^{3\cdot 2^n}}{x^3-a^3}\cdot\frac{x-a}{x^{2^n}-a^{2^n}}=\frac{x^{2^{n+1}}+a^{2^n}x^{2^n}+a^{2^{n+1}}}{x^2+ax+a^2}.$$

This method is inapplicable for $x=\pm a$. But in these cases a simple computation shows that for $x=-a$ the product is equal to $3^n a^{2(2^n-1)}$ and for $x=a$ it is equal to $a^{2(2^n-1)}$.

232. It is obvious that

$$S_k-S_{k-1}=b_k \quad (k=2,\ 3,\ 4,\ \dots,\ n) \tag{1}$$

and

$$S_1=b_1 \tag{2}$$

Substituting the values of $b_1,\ b_2,\dots,\ b_n$ obtained from (1) and (2) into the sum

$$a_1b_1+a_2b_2+\dots+a_nb_n,$$

we get

$$a_1b_1+a_2b_2+\dots+a_nb_n=a_1S_1+a_2(S_2-S_1)+a_3(S_3-S_2)+$$
$$+\dots+a_n(S_n-S_{n-1})=S_1(a_1-a_2)$$
$$+S_2(a_2-a_3)+\dots+S_{n-1}(a_{n-1}-a_n)+a_nS_n.$$

233. Multiply both members of the equality by 2 and transpose its right member to the left. After simple transformations we get

$$2(a^2+b^2+c^2-ab-ac-bc)=a^2-2ab+b^2+a^2-2ac+c^2+b^2-2bc+c^2$$
$$=(a-b)^2+(a-c)^2+(b-c)^2=0.$$

Since a, b and c are real, the latter relation is only possible if $a=b=c$.

234. Let us multiply $a^2+b^2+c^2-bc-ca-ab$ by $a+b+c$.

Carrying out simple computations we find that the product is equal to $a^3+b^3+c^3-3abc$, that is, according to the condition of the problem, it is equal to zero. Hence, the assertion stated in the problem is true.

235. Since $p\neq 0$ and $q\neq 0$ we can write

$$\left(\frac{a_1}{p}\right)^2+\left(\frac{a_2}{p}\right)^2+\dots+\left(\frac{a_n}{p}\right)^2=1,$$

$$\left(\frac{b_1}{q}\right)^2+\left(\frac{b_2}{q}\right)^2+\dots+\left(\frac{b_n}{q}\right)^2=1,$$

$$\frac{a_1}{p}\frac{b_1}{q}+\frac{a_2}{p}\frac{b_2}{q}+\ldots+\frac{a_n}{p}\frac{b_n}{q}=1.$$

Adding the first two of these equalities termwise and subtracting the doubled third equality we find

$$\left(\frac{a_1}{p}-\frac{b_1}{q}\right)^2+\left(\frac{a_2}{p}-\frac{b_2}{q}\right)^2+\ldots+\left(\frac{a_n}{p}-\frac{b_n}{q}\right)^2=0.$$

Taking into account that all the quantities involved are real we conclude that

$$\frac{a_1}{p}-\frac{b_1}{q}=0,\ \frac{a_2}{p}-\frac{b_2}{q}=0,\ \ldots,\ \frac{a_n}{p}-\frac{b_n}{q}=0,$$

which immediately implies the assertion of the problem.

235. Put $p_n=a_n-a_{n-1}$. Then the statement of the problem implies the formula $p_n=p_{n-1}+1$, showing that the numbers p_n form an arithmetic progression with unity is common difference. Therefore, $p_n=p_2+n-2$. Now we find

$$a_n=(a_n-a_{n-1})+(a_{n-1}-a_{n-2})+\ldots+(a_2-a_1)+a_1=$$
$$=p_n+p_{n-1}+\ldots+p_2+a_1=(n-1)p_2+(n-2)+(n-3)+$$
$$+\ldots+1+a_1=(n-1)(a_2-a_1)+a_1+\frac{(n-2)(n-1)}{2},$$

and, finally,

$$a_n=(n-1)a_2-(n-2)a_1+\frac{(n-2)(n-1)}{2}.$$

237. *First solution.* The given relation can be written in the two forms

$$a_n-\alpha a_{n-1}=\beta(a_{n-1}-\alpha a_{n-2})$$

and

$$a_n-\beta a_{n-1}=\alpha(a_{n-1}-\beta a_{n-2}).$$

Putting $a_n-\alpha a_{n-1}=u_n$

and $a_n-\beta a_{n-1}=v_n$ we find that

$$u_n=\beta u_{n-1},\ v_n=\alpha v_{n-1},$$

whence it follows that

$$u_n=\beta^{n-2}u_2,\ v_n=\alpha^{n-2}v_2$$

or

$$a_n-\alpha a_{n-1}=\beta^{n-2}(a_2-\alpha a_1),$$
$$a_n-\beta a_{n-1}=\alpha^{n-2}(a_2-\beta a_1).$$

Eliminating a_{n-1} from these relations we finally obtain

$$a_n=\frac{\beta^{n-1}-\alpha^{n-1}}{\beta-\alpha}.\ a_2-\alpha\beta\frac{\beta^{n-2}-\alpha^{n-2}}{\beta-\alpha}.a_1.$$

Second solution. Making n in the original relation take on consecutive values
3, 4, ... we find

$$a_3 = (\alpha+\beta)a_2 - \alpha\beta a_1 = \frac{\alpha^2-\beta^2}{\alpha-\beta}a_2 - \alpha\beta\frac{\alpha-\beta}{\alpha-\beta}a_1$$

and

$$a_4 = (\alpha+\beta)a_3 - \alpha\beta a_2 = (\alpha+\beta)\frac{\alpha^2-\beta^2}{\alpha-\beta}a_2 - \alpha\beta\frac{\alpha^2-\beta^2}{\alpha-\beta}a_1 -$$

$$-\alpha\beta\frac{\alpha-\beta}{\alpha-\beta}a_2 = \frac{\alpha^3-\beta^3}{\alpha-\beta}a_2 - \alpha\beta\frac{\alpha^2-\beta^2}{\alpha-\beta}a_1.$$

The general formula

$$a_n = \frac{\alpha^{n-1}-\beta^{n-1}}{\alpha-\beta}a_2 - \alpha\beta\frac{\alpha^{n-2}-\beta^{n-2}}{\alpha-\beta}a_1$$

can now be easily proved by induction.

238. We have $x_1 + x_2 = 3a$, $x_1x_2 = a^2$. Therefore

$$x_1^2 + x_2^2 = (x_1+x_2)^2 - 2x_1x_2 = 7a^2 = \frac{7}{4}$$

whence $a^2 = \frac{1}{4}$. Hence, there are two possible values of a, namely $a_1 = \frac{1}{2}$ and $a_2 = -\frac{1}{2}$.

239. We find

$$y_1 = (x_1+x_2)^2 - 2x_2x_2 = p^2 - 2q,$$
$$y_2 = (x_1+x_2)^3 - 3(x_1+x_2)x_1x_2 = -p^3 + 3pq.$$

The coefficients of the quadratic equation $y^2 + ry + s = 0$ with y_1 and y_2 are respectively equal to

$$r = -(y_1+y_2) = p^3 - p^2 - 3pq + 2q$$

and $$s = y_1y_2 = (p^2-2q)(-p^3+3pq).$$

240. We have $x_1 + x_2 = -\frac{b}{a}$ and $x_1x_2 = \frac{c}{a}$. With the aid of these formulas we find

$$\frac{1}{x_1^2} + \frac{1}{x_2^2} = \frac{(x_1+x_2)^2 - 2x_1x_2}{x_1^2x_2^2} = \frac{b^2-2ac}{c^2}$$

and

$$x_1^4 + x_1^2x_2^2 + x_2^1 = (x_1^2+x_2^2)^2 - x_1^2x_2^2$$
$$= \left(\frac{b^2}{a^2} - 2\frac{c}{a}\right)^2 - \left(\frac{c}{a}\right)^2.$$

241. Let

$$(a_1 + b_1x)^2 + (a_2 + b_2x)^2 + (a_3 + b_3x)^2 = (A + Bx)^2, \quad (1)$$

for all x where $B \neq 0$. Putting $x = -\frac{A}{B}$ we get

$\left(a_1 - b_1\frac{A}{B}\right)^2 + \left(a_2 - b_2\frac{A}{B}\right)^2 + \left(a_3 - b_3\frac{A}{B}\right)^2 = 0.$ All the quantities involved being real, we thus have the three equalities

$$a_1 = \lambda b_1,\ a_2 = \lambda b_2,\ a_3 = \lambda b_3, \quad (2)$$

where $\lambda = \frac{A}{B}$. Besides, the condition

$$b_1^2 + b_2^2 + b_3^2 \neq 0 \quad (3)$$

should hold because, if otherwise, all the three numbers b_1, b_2 and b_3 were equal to zero, and then the left member of (1) were independent of x.

Let now, conversely, conditions (2) and (3) be fulfilled. Then

$$\begin{aligned}(a_1 + b_1x)^2 + (a_2 + b_2x)^2 + (a_3 + b_3x)^2 &= \\ &= b_1^2(\lambda + x)^2 + b_2^2(\lambda + x)^2 + b_3^2(\lambda + x)^2 = \\ &= (\lambda\sqrt{b_1^2 + b_2^2 + b_3^2} + \sqrt{b_1^2 + b_2^2 + b_3^2}x)^2,\end{aligned}$$

and, consequently, the sum indicated in the problem is a square of a polynomial of the first degree. Thus, conditions (2) and (3) are necessary and sufficient.

242. Let us denote the roots of the equation by x_1 and x_2. Then $x_1 + x_2 = -p$ and $x_1x_2 = q$.

If x_1 and x_2 are negative then, obviously, $p > 0$ and $q > 0$. But if $x_1 = \alpha + i\beta$ where $\alpha < 0$ and $\beta \neq 0$, then $x_2 = \alpha - i\beta$, and we see that

$$p = -x_1 - x_2 = -2\alpha > 0$$

and

$$q = x_1x_2 = \alpha^2 + \beta^2 > 0.$$

Conversely, let it be known that $p > 0$ and $q > 0$. Then, if x_1 and x_2 are real, from the equality $x_1x_2 = q$ it follows that x_1 and x_2 are of the same sign, and the equality $x_1 + x_2 = -p$ implies that the roots are negative. But if $x_1 = \alpha + i\beta$,

$x_2 = \alpha - i\beta$ and $\beta \neq 0$, then $x_1 + x_2 = -p = 2\alpha$, and, consequently, is negative.

243. The roots of the equation $x^2 + px + q = 0$ being positive, the discriminant D of the equation satisfies the condition

$$D = p^2 - 4p \geq 0, \quad (1)$$

and the coefficients p and q satisfy the inequalities

$$p = -x_1 - x_2 < 0 \quad (2)$$

and

$$q = x_1 x_2 > 0, \qquad (3)$$

Let y_1 and y_2 be the roots of the equation

$$qy^2 + (p - 2rq)y + 1 - pr = 0. \qquad (4)$$

The discriminant of this equation is equal to

$$D_1 = 4r^2q^2 + p^2 - 4q$$

and, by virtue of (1), it is non-negative for all r. Consequently, y_1 and y_2 are real for all r. Taking into account (2) and (3), and applying Vieta's theorem we get, for $r \geq 0$, the inequality

$$y_1 y_2 = \frac{1 - pr}{q} > 0, \qquad (5)$$

and, hence, y_1 and y_2 are of the same sign. Furthermore, we have

$$y_1 + y_2 = -\frac{p - 2rq}{q} > 0 \qquad (6)$$

and, hence, y_1 and y_2 are positive for $r \geq 0$ which is what we set out to prove.

It is obvious that the assertion remains true if we require that the inequalities

$$1 - pr > 0 \text{ and } p - 2rq < 0,$$

hold simultaneously, that is

$$r > \frac{1}{p} \qquad (7)$$

and

$$r > \frac{p}{2q}. \qquad (8)$$

Thus, for negative r satisfying conditions (7) and (8) the roots y_1 and y_2 are positive. If these conditions are not observed, one (or both) roots of equation (4) is nonpositive.

244. Let us first suppose that $p \neq 3$. For the roots of a quadratic equation with real coefficients to be real it is necessary and sufficient that the discriminant D of this equation be non-negative. We have

$$D = 4p^2 - 24p(p - 3) = 4p(18 - 5p)$$

and therefore the condition $D \geq 0$ holds for

$$0 \leq p \leq 3.6. \qquad (1)$$

The real roots x_1 and x_2 are positive if and only if their sum and product are positive, i.e.

$$x_1 + x_2 = \frac{2p}{p - 3} > 0, \quad x_1 x_2 = \frac{6p}{p - 3} > 0. \qquad (2)$$

The system of inequalities (1), (2) is satisfied for

$$3 < p \le 3.6.$$

It should also be noted that for $p = 3$ the equation under consideration has the unique root $x = 3 > 0$. Therefore, all the sought-for values are determined by the condition

$$3 \le p \le 3.6.$$

245. We shall prove the assertion by contradiction. Let us suppose that $a \ne 0$. Then for the roots x_1 and x_2, we have

$$x_{1,2} = \frac{-b \pm \sqrt{b^2 - 4a(c + \lambda)}}{2a}.$$

Now there are two possible cases here:

(1) Let $a > 0$. Then λ is chosen so that the inequality

$$\lambda > \frac{b^2}{4a} - c$$

is fulfilled. In this case we obviously have $b^2 - 4a(c + \lambda) < 0$ and, hence, the given equation has nonreal roots.

(2) Let $a < 0$. Then if $\lambda > -c$, we have

$$-b + \sqrt{b^2 - 4a(c + \lambda)} > 0,$$

and, hence, the root $\dfrac{-b + \sqrt{b^2 - 4a(c + \lambda)}}{2a}$ is negative. Thus, both assumptions lead to a contradiction. The assertion has thus been proved.

246. The roots $x_{1,\,2}$ of the equation $x^2 + x + 1 = 0$ satisfy the equation $x^3 - 1 = 0$ as well. Therefore, $x_{1,2}^{3m} = x_{1,2}^{3n} = x_{1,2}^{3p} = 1$ which implies the assertion .

247. Substituting y expressed from the second equation into the first we get the equation

$$2ax^2 + 2(a\lambda + 1)x + a\lambda^2 = 0, \qquad (1)$$

which, by the hypothesis, has real roots for all values of λ. Let us show that then $a = 0$. Suppose the contrary. Then for the discriminant D of the quadratic equation (1) the following inequality

$$D = 4(a\lambda + 1)^2 - 8a^2\lambda^2 \ge 0 \qquad (2)$$

holds for all λ. However, the left member of inequality (2) has the form

$$-4a^2\lambda^2 + 8\lambda + 1$$

and is negative for all sufficiently large absolute values of λ. For instance, if $\lambda = \dfrac{10}{a}$, the left member of equation (1) is equal to -321. Thus we arrive at a contradiction.

248. The equation in question takes the form

$$x^2-(p+q+2a^2)x+pq+(p+q)a^2=0$$

after reducing the fractions to a common denominator and discarding it. Computing the discriminant D of this quadratic equation we get

$$D=(p+q+2a^2)^2-4\,[pq+(p+q)a^2]=(p-q)^2+4a^4.$$

Since $D \geq 0$ for all real a, p and q, the quadratic equation has real roots, and hence the same is true for the original equation.

249. Consider the discriminant of the given quadratic equations:

$$\begin{aligned} D &= (b^2+a^2-c^2)^2-4a^2b^2 \\ &= (b^2+a^2-c^2-2ab)\,(b^2+a^2-c^2+2ab)= \\ &= [(a-b)^2-c^2][(a+b)^2-c^2]. \end{aligned}$$

Since $a+b>c$ and $|a-b|<c$, we have $(a+b)^2>c^2$ and $(a-b)^2<c^2$. Consequently, $D<0$.

250. By Vieta's formulas (see page 10) we have

$$x_1+x_2+x_3=2, x_1x_2+x_2x_3+x_3x_1=1, x_1x_2x_3=-1.$$

Using these equalities we obtain

$$\begin{aligned} &y_1+y_2+y_3=x_1x_2+x_2x_3+x_3x_1=1, \\ &y_1y_2+y_2y_3+y_3y_1=x_1x_2x_3\,(x_1+x_2+x_3)=-2, \\ &y_1y_2y_3=(x_1x_2x_3)^2=1. \end{aligned}$$

Consequently, the new equation is

$$y^3-y^2-2y-1=0.$$

251. On the basis of Vieta's formulas, we have

$$x_1+x_2+x_3=1,\ x_1x_2+x_2x_3+x_3x_1=0,\ x_1x_2x_3=1.$$

By virtue of these equalities, we write

$$y_1+y_2+y_3=2(x_1+x_2+x_3)=2.$$

Since $\quad y_1=1-x_1$

and $\quad y_2=1-x_2,\ y_3=1-x_3,$

we have

$$\begin{aligned} y_1y_2+y_2y_3+y_3y_1 &= (1-x_1)\,(1-x_2)+(1-x_2)\,(1-x_3)+ \\ &\quad +(1-x_3)(1-x_1)=3-2\,(x_1+x_2+x_3) \\ &\quad +x_1x_2+x_2x_3+x_1x_3=1, \end{aligned}$$

and, finally, $\quad y_1y_2y_3=(1-x_1)\,(1-x_2)\,(1-x_3)=-1.$

The new equation, therefore, has the form

$$y^3-2y^2+y+1=0.$$

252. Let

$$x_1 = p-d, x_2 = p, x_3 = p+d.$$

Then $x_1+x_2+x_3 = 3p$. On the other hand, by Vieta's formulas, we have $x_1+x_2+x_3 = -a$ whence we find $3p = -a$ and, hence,

$$x_2 = p = -\frac{a}{3}.$$

Substituting this root into the equation we obtain

$$\left(-\frac{a}{3}\right)^3 + \left(-\frac{a}{3}\right)^2 + b\left(-\frac{a}{3}\right) + c = 0,$$

which yields

$$c = -\frac{2}{27}a^3 + \frac{1}{3}ab.$$

253. Let $x_1 > 0, x_2 > 0$ and $x_3 > 0$ be the roots of the given equation. Following the hint, we consider the expression

$$(x_1+x_2-x_3)(x_2+x_3-x_1)(x_3+x_1-x_2). \qquad (1)$$

For the triangle with line segments of the lengths x_1, x_2, x_3 as sides to exist, it is necessary and sufficient as was proved in the solution of Problem 106 that the condition

$$(x_1+x_2-x_3)(x_2+x_3-x_1)(x_3+x_1-x_2) > 0 \qquad (2)$$

be fulfilled.

To obtain the condition required in the problem let us express the left member of (2) in terms of p, q and r. For this purpose we make use of the relations

$$x_1+x_2+x_3 = -p,\ x_1x_2+x_1x_3+x_2x_3 = q,$$
$$x_1x_2x_3 = -r$$

connecting the roots and coefficients of the equation. Condition (2) is now written in the form

$$(-p-2x_3)(-p-2x_1)(-p-2x_2) > 0;$$

whence it follows that

$$-p^3 - 2p^2(x_1+x_2+x_3) - 4p$$
$$(x_1x_2+x_1x_3+x_2x_3) - 8x_1x_2x_3 > 0$$

and, hence,

$$p^3 - 4pq + 8r > 0.$$

254. Let x_0 be a common root of the equations. Substituting x_0 into both equations and subtracting one equation from the other we find

$$x_0 = \frac{q_2-q_1}{p_1-p_2} \neq 0.$$

Let x^2+ax+b be the quotient obtained by dividing the trinomial $x^3+p_1x+q_1$ by $x-x_0$. Then

$$x^3+p_1x+q_1 = (x-x_0)(x^2+ax+b).$$

Equating the coefficients in x^2 and the constant terms in this identity we find $a = x_0$ and $b = -\frac{q_1}{x_0}$, whence it follows that the other two roots of the first equation are determined by the formula

$$x_{2,3}^{(1)} = \frac{-x_0 \pm \sqrt{x_0^2 + \frac{4q_1}{x_0}}}{2}$$

and of the second equation by the formula

$$x_{2,3}^{(2)} = \frac{-x_0 \pm \sqrt{x_0^2 + \frac{4q_2}{x_0}}}{2}.$$

255. It can easily be verified that for $\lambda = 0$ the equations have no roots in common. Let x_0 be a common root of the equations for some $\lambda \neq 0$. Then we can write

$$\left.\begin{aligned} \lambda x_0^3 - x_0^2 - x_0 - (\lambda + 1) = 0, \\ \lambda x_0^2 - x_0 - (\lambda + 1) = 0. \end{aligned}\right\} \quad (1)$$

Multiplying the second equality by x_0 and subtracting it from the first we find

$$x_0 = \frac{\lambda + 1}{\lambda}. \quad (2)$$

Thus, if there is a common root, then it is connected with λ by formula (2).

It can now be readily verified that the fraction $\frac{\lambda + 1}{\lambda}$ in fact satisfies both equations (it is obviously sufficient to establish this fact only for the second equation). Thus, both equations (1) have a common root for all $\lambda \neq 0$, the root being determined by formula (2).

256. *First solution.* Let x_1, x_2 and x_3 be the roots of the polynomial $P(x)$.

According to Vieta's theorem, we have

$$x_1 + x_2 + x_3 = 0,\ x_1x_2 + x_1x_3 + x_2x_3 = p,$$

whence it readily follows that

$$x_1^2 + x_2^2 + x_3^2 + 2p = 0.$$

Since x_1, x_2 and x_3 are real and different from zero (because $q \neq 0$), we have $x_1^2 + x_2^2 + x_3^2 > 0$ and, hence, $p < 0$.

Second solution. It is apparent that among the three roots of the polynomial $P(x)$ there are two unequal ones. Indeed, if otherwise, we must have $P(x) \equiv (x - x_0)^3$ which is obviously not the case.

Now let x_1 and x_2 be two unequal roots of the polynomial, and let $x_1 < x_2$, Suppose the contrary, that is $p \geq 0$. Then $x_1^3 < x_2^3$ and $px_1 \leq px_2$. Then it follows that

$$P(x_1) = x_1^3 + px_1 + q < x_2^3 + px_2 + q = 0,$$

because $P(x_2) = 0$. We arrive at a conclusion that $P(x_1) < 0$ which contradicts the fact that x_1 is a root of $P(x)$. Consequently, $p < 0$.

257. Let x_1, x_2 and x_3 be the roots of the given equation. By virtue of Vieta's formulas, we have

$$x_1x_3 + x_1x_3 + x_2x_3 = 0, \qquad (1)$$

$$x_1x_2x_3 = b > 0. \qquad (2)$$

Let us first suppose that all the three roots are real. Then from condition (2) it follows that atleast one of them is positive. If in this case we suppose that two roots are positive, then formula (2) implies that the third roots is also positive, which contradicts condition (1). Thus, if all the root are real, the problem has been solved.

Let now x_1 be a nonreal root of the equation, then, as is known, the equation also has the conjugate complex root $x_2 = \overline{x_1}$. Since in this case $x_1x_2 = x_1\overline{x_2} > 0$,

we conclude from equality (2) that $x_3 = \dfrac{b}{x_1\overline{x_1}} > 0.$

The assertion has thus been completely proved.

258. Let α, β and γ_1, be the roots of the first equation and α, β and γ_2 the roots of the second. By virtue of Vieta's formulas, we have

$$\alpha + \beta + \gamma_1 = -a, \qquad (1)$$

$$\alpha\beta\gamma_1 = -18, \qquad (2)$$

$$\alpha + \beta + \gamma_2 = 0, \qquad (3)$$

$$\alpha\beta\gamma_2 = -12 \qquad (4)$$

Now we obtain $\gamma_1 - \gamma_2 = -a$ (5)

from equations (1) and (3) and

$$\frac{\gamma_1}{\gamma_2} = \frac{3}{2} \qquad (6)$$

from equations (2) and (4). Solving (5) and (6) as simultaneous equations we find

$$\gamma_1 = -3a, \gamma_2 = -2a. \qquad (7)$$

Thus, if for some a and b the equations have two common roots, their third roots are determined by formula (7). Substituting $\gamma_1 = -3a$ into the first equation and $\gamma_2 = -2a$ into the second we obtain

$$-18a^3 + 18 = 0$$

and

$$-8a^3 - 2ab + 12 = 0.$$

Solving these equations we see that there can be only one pair of real values satisfying the condition of the problem, namely

$$a = 1, b = 2. \qquad (8)$$

Substituting these values into the equations we readily find

$$x^3 + x^2 + 18 = (x+3)(x^2 - 2x + 6)$$

and

$$x^3 + 2x + 12 = (x+2)\,(x^2 - 2x + 6).$$

Consequently, for the above values of a and b the equations have in fact two common roots. These roots are determined by the formula

$$x_{1,2} = -1 \pm \sqrt{-5}.$$

259. Let us denote the left-hand side of the equality by A. We have

$$A^3 = 20 + 14\sqrt{2} + 3\sqrt[3]{(20+14\sqrt{2})^2}\,\sqrt[3]{20-14\sqrt{2}} + $$
$$+ 3\sqrt[3]{20+14\sqrt{2}}\,\sqrt[3]{(20-14\sqrt{2})^2} + 20 - 14\sqrt{2} =$$
$$= 40 + 3\sqrt[3]{400 - 2\times 14^2}\;A = 40 + 6A.$$

Thus, the left-hand side of the equality to be proved satisfies the cubic equation

$$x^3 - 6x - 40 = 0 \qquad (1)$$

It can be easily checked that equation (1) is satisfied by $x = 4$. Dividing the left-hand side of equation (1) by $x - 4$ we get the equation for finding the other two roots

$$x^2 + 4x + 10 = 0.$$

This equation has nonreal roots because its discriminant is negative:

$$D = -24 < 0.$$

Thus, equation (1) has only one real root $x = 4$, and since A is a priori a real number, we have $A = 4$ which is what we set out to prove.

260. As is easily seen, the expression in question vanishes if any two of the numbers a, b and c are equal. Then, by Bézout's theorem, it is divisible by each of the differences

$$(b-c), (c-a) \text{ and } (a-b).$$

Therefore it seems natural to suppose that the given expression is the product of these factors. Indeed, we have

$$a^2(c-b) + b^2(a-c) + c^2(b-a)$$
$$= a^2c - a^2b + b^2a - b^2c + c^2b - c^2a$$
$$= a^2(c-b) - a(c^2 - b^2) + bc\,(c-b)$$
$$= (c-b)\,[a^2 - ac - ab + bc]$$
$$= (c-b)\,[a\,(a-c) - b(a-c)]$$
$$= (c-b)\,(b-a)\,(c-a) \qquad (1)$$

and thus the assumption turns out to be true. Since a, b and c are pairwise different, the assertion has been proved.

261. Note that for $x=-y$ the given expression turns into zero. Consequently, by Bézout's theorem, it is divisible by $x+y$. To perform the division let us represent $x+y+z$ in the form of a sum of two summands: $(x+y)$ and z. Cubing the sum, we get

$$[(x+y)+z]^3 - x^3 - y^3 - z^3$$
$$= (x+y)^3 + 3(x+y)^2 z + 3(x+y)z^2 - x^3 - y^3$$
$$= 3(x+y)[z^2 + z(x+y) + xy].$$

The quadratic trinomial with respect to z in the square brackets on the right-hand side is readily factorized because its roots are obviously $-x$ and $-y$.

Hence, we obtain $(x+y+z)^3 - x^3 - y^3 - z^3 = 3(x+y)(z+x)(z+y)$.

262. Multiplying both members of the given equality by $abc(a+b+c)$ we transform it to the form $(ab+bc+ac)(a+b+c) - abc = 0$.

Removing the brackets we get

$$a^2b + 2abc + a^2c + ab^2 + b^2c + bc^2 + ac^2 = 0.$$

The left member of this equality is readily factorized:

$$a^2(b+c) + ab(c+b) + ac(b+c) + bc(b+c)$$
$$= (b+c)(a^2 + ab + ac + bc) = (b+c)(a+b)(a+c).$$

Since the last product is equal to zero, we conclude that atleast one of the factors is equal to zero which implies the desired assertion.

263. Let α and β be the roots of the quadratic trinomial $x^2 + px + q$. If the binomial $x^4 - 1$ is divisible by this trinomial, then α and β are the roots of the binomial as well. It is easily seen, that the converse is also true: if α and β are the roots of the binomial $x^4 - 1$, then it is divisible by $x^2 + px + q$ *. The binomial $x^4 - 1$ has the roots 1, −1, i and $-i$ and therefore we can write the factorization

$$(x^4 - 1) = (x-1)(x+1)(x-i)(x+i). \qquad (1)$$

What was said above implies that the trinomials we are interested in may only be products of two of the factors on the right-hand side of (1). Forming all possible permutations we find $C_4^2 = 6$ trinomials:

$$(x-1)(x+1) = x^2 - 1,$$
$$(x-1)(x-i) = x^2 - (1+i)x + i,$$
$$(x-1)(x+i) = x^2 - (1-i)x - i,$$
$$(x+1)(x-i) = x^2 + (1-i)x - i,$$
$$(x+1)(x+i) = x^2 + (1+i)x + i,$$
$$(x-i)(x+i) = x^2 + 1.$$

These obviously are all the sought-for trinomials.

---

* If, in this argument, $\alpha = \beta$, the number α must be a multiple root of the dividend as well.

264. Representing the given polynomial in the form $x^n - 1 - ax(x^{n-2} - 1)$ we divide it by the difference $(x-1)$ using the formula

$$\frac{x^{k+1} - 1}{x-1} = 1 + x + \ldots x^k. \quad (1)^*$$

Performing the division we see that the quotient is the polynomial

$$x^{n-1} + x^{n-2} + \ldots + x + 1 + 1 - ax(x^{n-3} + x^{n-4} + \ldots + x + 1).$$

For the latter polynomial to be divisible by $x-1$, it is necessary and sufficient that (according to Bézout's theorem) the following equality be fulfilled: $n - a(n-2) = 0$.

Therefore, the polynomial given in the problem is divisible by $(x-1)^2$ for any natural $n > 2$ and $a = \frac{n}{n-2}$.

265. The conditions of the problem imply that

$$\left.\begin{aligned} p(a) &= A, \\ p(b) &= B, \\ p(c) &= C. \end{aligned}\right\} \quad (1)$$

Dividing the polynomial $p(x)$ by $(x-a)(x-b)(x-c)$ we represent it in the form $p(x) = (x-a)(x-b)(x-c)\,q(x) + r(x)$. (2)

It is obvious that $r(x)$ is a polynomial of degree not higher than the second. Writing it in the form

$$r(x) = lx^2 + mx + n, \quad (3)$$

we substitute, in succession, the values $x = a$, $x = b$ and $x = c$ into identity (2).

By virtue of equality (1), we arrive at the following system of equations for defining the coefficients l, m and n of polynomial (3):

$$\left.\begin{aligned} la^2 + ma + n &= A, \\ lb^2 + mb + n &= B, \\ lc^2 + mc + n &= C. \end{aligned}\right\} \quad (4)$$

Solving this system we find

$$l = \frac{(A-B)(b-c) - (B-C)(a-b)}{(a-b)(b-c)(a-c)},$$

$$m = \frac{(A-B)(b^2-c^2) - (B-C)(a^2-b^2)}{(a-b)(b-c)(c-a)},$$

$$n = \frac{a^2(Bc - Cb) + a(Cb^2 - Bc^2) + A(Bc^2 - Cb^2)}{(a-b)(b-c)(c-a)}.$$

* Formula (1) can be easily verified by division but it should be noted that it simply coincides with the formula for the sum of k terms of a geometric progression with common ratio x.

Note. For $x = a$, $x = b$ and $x = c$ the sought-for polynomial $r(x)$ takes on the values A, B and C, respectively. It can easily be verified that the polynomial (of degree not higher than the second) given below is one possessing this property:

$$A\frac{(x-b)(x-c)}{(a-b)(a-c)} + B\frac{(x-a)(x-c)}{(b-a)(b-c)} + C\frac{(x-a)(x-b)}{(c-a)(c-b)}. \tag{5}$$

System (4) having only one solution, there exists only one polynomial possessing the above property. Consequently, $r(x)$ coincides with polynomial (5).

266. The formula is obviously true for $n = 1$. Let us suppose that it is true for a certain n and prove that then it is true for $n + 1$ as well. Denoting the sum standing on the left-hand side of the formula to be proved by S_n, we can write

$$S_{n+1} = S_n + \frac{(n+1)(n+2)}{2} = \frac{(n+1)(n+2)(n+3)}{6}$$
$$= \frac{(n+1)[(n+1)+1][(n+1)+2]}{6}.$$

Thus, it follows by induction that the formula is valid for any natural n.

267. Let S_n be the sum on the left-hand side of the formula. For $n = 1$ both sides of the formula coincide. Let us show that if the formula holds for some n, then it is also true for $n + 1$. We have

$$S_{n+1} = S_n + (n+1)^2 = \frac{n(n+1)(2n+1)}{6} + (n+1)^2$$
$$= \frac{(n+1)(2n^2+7n+6)}{6} = \frac{(n+1)[2n(n+2)+3(n+2)]}{6}$$
$$= \frac{(n+1)[(n+1)+1][2(n+1)+1]}{6}.$$

Consequently, the formula holds for any natural n.

268. The validity of the assertion is readily established for $n = 1$. Suppose that the formula be true for some $n \geq 1$. Let S_n be the sum on the left-hand side of the formula, We have

$$S_{n+1} = S_n + \frac{1}{(n+1)(n+2)(n+3)}$$
$$= \frac{n(n+3)}{4(n+1)(n+2)} + \frac{1}{(n+1)(n+2)(n+3)}.$$

It follows that $S_{n+1} = \dfrac{n^3+6n^2+9n+4}{4(n+1)(n+2)(n+3)} = \dfrac{(n+1)(n^2+5n+4)}{4(n+1)(n+2)(n+3)}$

$$= \frac{(n+1)[(n+1)+3]}{4[(n+1)+1][(n+1)+2]}$$

Hence, the formula is true for any natural n.

269. The formula is obviously true for $n = 1$. Suppose that it is true for some $n \geq 1$, i.e.,

$$(\cos\varphi + i\sin\varphi)^n = \cos n\varphi + i\sin n\varphi. \quad (1)$$

To prove that the formula holds for $n+1$ let us multiply both members of (1) by $\cos\varphi + i\sin\varphi$. According to the rule for multiplying complex numbers we obtain

$$(\cos\varphi i + \sin\varphi)^{n+1} = (\cos n\varphi + i\sin n\varphi)(\cos\varphi + i\sin\varphi)$$
$$= (\cos n\varphi\cos\varphi - \sin n\varphi\sin\varphi)$$
$$+ i(\cos n\varphi\sin\varphi + \sin n\varphi\cos\varphi)$$
$$= \cos(n+1)\varphi + i\sin(n+1)\varphi.$$

Consequently, the formula is true for any natural n.

270. Apparently, $a + b = 1$ and $ab = -1$. Using this, we can write

$$a_n = a_n(a+b) = \frac{a^{n+1} - ab^n + a^n b - b^{n+1}}{\sqrt{5}}$$
$$= \frac{a^{n+1} - b^{n+1}}{\sqrt{5}} - \frac{a^{n-1} - b^{n-1}}{\sqrt{5}},$$

that is $a_n = a_{n+1} - a_{n-1}$ which implies

$$a_{n+1} = a_n + a_{n-1}.$$

It follows that if for some n the numbers a_{n-1} and a_n are positive integers, then a_{n+1} is also a positive integers. Consequently by induction, a_{n+2}, a_{n+3} etc. are positive integers. But we have $a_1 = 1$ and $a_2 = 1$, and hence all a_n are positive integers for $n > 2$.

271. For $r = 1$ the inequality is true. Let us suppose that it is true for some a. Multiplying both members by $1 + a_{n+1} > 0$ we find

$$(1+a_1)(1+a_2)\ldots(1+a_n) \geq (1 + a_1 + a_2 + \ldots + a_n)(1 + a_{n+1})$$
$$= 1 + a_1 + a_2 + \ldots + a_n + a_{n+1} + a_1 a_{n+1} + a_2 a_{n+1} + \ldots + a_n a_{n+1}.$$

We have $a_1 a_{n+1} + a_2 a_{n+1} + \ldots + a_n a_{n+1} > 0$ and therefore the inequality is true for $n + 1$ as well.

272. Let us first of all verify that the formula holds for $n = 1$. Indeed, for $n = 1$. It takes the form

$$(a+b)_1 = C_1^0(a)_0(b)_1 + C_1^1(a)_1(b). \quad (1)$$

If now we use the definition for the generalized nth power of a number, it becomes evident that both members of formula (1) are equal to $a + b$ and, consequently, the equality is in fact true.

Now suppose that the formula is true for some n and prove that it is then true for $n+1$ as well. The definition of the generalized nth power implies that

$$(a+b)_{n+1} = (a+b)_n(a+b-n)$$
$$= [C_n^0(a)_0(b)_n + C_n^1(a)_1(b)_{n-1} + \ldots$$
$$\ldots + C_n^k(a)_k(b)_{n-k} + \ldots + C_n^n(a)_n(b)_0](a+b-n).$$

Removing the square brackets, we transform each of the $n+1$ summands according to the formula

$$C_n^k(a)_k(b)_{n-k}(a+b-n)=C_n^k(a)_k$$
$$(b)_{n-k}[(a-k)+(b-n+k)]=$$
$$=C_n^k(a)_k(a-k)(b)_{n-k}+C_n^k(a)_k(b)_{n-k}(b-n+k)=$$
$$=C_n^k(a)_{k+1}(b)_{n-k}+C_n^k(a)_k(b)_{n-k+1}\ (k=0,1,\ldots,n).$$

This results in

$$(a+b)_{n+1}=C_n^0(a)_1(b)_n+C_n^0(a)_0(b)_{n+1}+C_n^1(a)_2(b)_{n-1}$$
$$+C_n^1(a)_1(b)_n+\ldots C_n^k(a)_{k+1}(b)_{n-k}$$
$$+C_n^k(a)_k(b)_{n-k+1}+\ldots+C_n^n(a)_{n+1}(b)_0+C_n^n(a)_n(b)_1.$$

Collecting like terms, we obtain

$$(a+b)_{n+1}=C_n^0(a)_0(b)_{n+1}+(C_n^0+C_n^1)(a)_1(b)_n+$$
$$+(C_n^1+C_n^2)(a)_2(b)_{n-1}+\ldots+(C_n^k+C_n^{k+1})(a)_{k+1}(b)_{n-k}+\ldots$$
$$+(C_n^{n-1}+C_n^n)(a)_n(b)_1+C_n^n(a)_{n+1}(b)_0.$$

Furthermore, using then the fact that $C_n^0=C_{n+1}^0=1$, $C_n^n=C_{n+1}^{n+1}=1$, and the identity $C_n^k+C_n^{k+1}=C_{n+1}^{k+1}$, which is easily verified, we obtain

$$(a+b)_{n+1}=C_{n+1}^0(a)_0(b)_{n+1}+C_{n+1}^1(a)_1(b)_n++C_{n+1}^2(a)_2(b)_{n-1}+\ldots$$
$$+C_{n+1}^{k+1}(a)_{k+1}(b)_{n-k}C_{n+1}^n(a)_n(b)_n+C_{n+1}^{n+1}(a)_{n+1}(b)_0.$$

Hence, we have proved that if the given formula is true for some n, then it is true for $n+1$ as well. But it holds for $n=1$, and consequently, we conclude, by induction, that it holds for all natural n.

273. Let $r(t)$ be the distance between the trains at the moment t. Then

$$r^2(t)=(a-v_1t)^2+(b-v_2t)^2=(v_1^2+v_2^2)t^2-2$$
$$(av_1+bv_2)t+a^2+b^2.$$

Note that if $r^2(t)$ attains its least value for $t=t_0$, then $r(t)$ also attains the least value for $t=t_0$, the converse also being true. The problem is thus reduced to finding the least value of the quadratic trinomial $r^2(t)$.

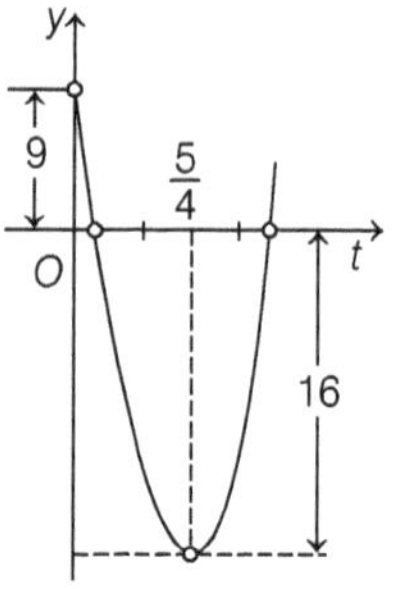

FIG. 4

According to formula (4), page 43, the least value of $r^2(t)$ (and, hence, of $r(t)$) is attained at the moment

$$t_0=\frac{av_1+bv_2}{v_1^2+v_2^2}.$$

Now using formula (3) we find the least distance between the trains:

$$r(t_0)=\sqrt{\frac{4(a^2+b^2)(v_1^2+v_2^2)-4(av_1+bv_2)^2}{4(v_1^2+v_2^2)}}=\frac{|av_2-bv_1|}{\sqrt{v_1^2+v_2^2}}.$$

274. At the moment t the car is at a distance of $40t$ km from the point A, and the motorcycle at a distance of $\frac{32}{2}t^2+9$ km from that point. Consequently, the distance between them is equal to the absolute value of the expression $16t^2+9-40t$. Denoting this distance by $y(t)$, we can plot the graph of the quadratic trinomial $y(t)$(See Fig. 4). The graph is a parabola intersecting the t-axis at the points $t_1=\frac{1}{4}$ and $t_2=2\frac{1}{4}$. The graph clearly shows that if $0\le t\le 2$, the greatest in its absolute value ordinate y corresponds to the vertex of the parabola. The latter lies on the axis of symmetry which intersects the t-axis at the point
$$t_0=\frac{t_1+t_2}{2}=\frac{5}{4}.$$
Thus, the distance attains its greatest value in an hour and a quarter after the start of the motion and is equal to 16 km.

275. Denote the expression in question by y and transform it in the following way: $y=\log_2^4 x+12\log_2^2 x(\log_2 8-\log_2 x)$
$$=\log_2^2 x(\log_2^2 x-12\log_2 x+36)=\log_2^2 x(6-\log_2 x)^2.$$
Let us put $\log_2 x=z$, then $0\le z\le 6$. The problem is thus reduced to finding the greatest value of the variable
$$y=z^2(6-z)^2.$$
It is sufficient to find the greatest value of $z(6-z)$ for $0\le z\le 6$ because the greater a positive number, the greater its square. The quadratic trinomial $z(6-z)=-(z-3)^2+9$ attains its greatest value for $z=3$. Thus, the sought-for greatest value is attained for $z=3$ and is equal to 81.

276. *First solution.* It is obviously sufficient to consider only positive values of x. According to the well-known inequality (3), page 20, we have
$$\frac{ax^2+b}{2}\le\sqrt{ax^2b}=x\sqrt{ab}. \qquad (1)$$
Consequently, for all $x>0$,
$$y=\frac{x}{ax^2+b}\le\frac{x}{2x\sqrt{ab}}=\frac{1}{2\sqrt{ab}}. \qquad (2)$$
Relation (1) turns into an equality when $ax^2=b$, and consequently for
$$x_0=\sqrt{\frac{b}{a}} \text{ we have}$$
$$y_0=\frac{1}{2\sqrt{ab}}. \qquad (3)$$
By virtue of (2), this is just the greatest value of the function.

Second solution. Solving the equation

$$y = \frac{x}{ax^2 + b} \qquad (4)$$

for x we obtain

$$x = \frac{1 \pm \sqrt{1 - 4aby^2}}{2ay}. \qquad (5)$$

Formula (5) implies that the inequality $1 - 4aby^2 \geq 0$ must be fulfilled for all real x. Hence

$$y \leq \frac{1}{2\sqrt{ab}}. \qquad (6)$$

Function (4) attains the value $y_0 = \dfrac{1}{2\sqrt{ab}}$ for a real value of $x = x_0$ (from (5) we find that $x_0 = \sqrt{\dfrac{b}{a}}$), and therefore, by virtue of (6), this value is the greatest. (7)

277. Performing some simple transformations we get

$$\frac{x^2+1}{x+1} = x - 1 + \frac{2}{x+1} = -2 + \left[x + 1 + \frac{2}{x+1}\right].$$

By virtue of inequality (3), page 20, we have

$$x + 1 + \frac{2}{x+1} \geq 2\sqrt{(x+1)\frac{2}{(x+1)}} = 2\sqrt{2}, \qquad (1)$$

and the sign of equality in (1) only appears if

$$1 + x = \frac{2}{x+1}, \text{ i.e. for } x_0 = \sqrt{2} - 1.$$

Thus, for all $x_0 \geq 0$ we have

$$\frac{x^2+1}{x+1} \geq -2 + 2\sqrt{2}. \qquad (2)$$

and the sign of equality in the latter formula takes place for

$$x = \sqrt{2} - 1.$$

278. Let us take a number scale and mark on it the points A, B, C and D corresponding to the numbers a, b, c and d. Let M denote a point with variable abscissa x (Fig. 5). There can be the following five cases here:

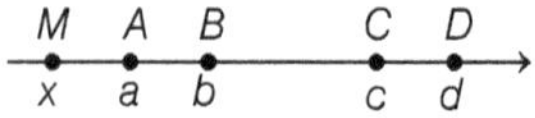

FIG. 5

(1) If $x \leq a$, then we have

$$\varphi(x) = MA + MB + MC + MD = AB + 2MB + 2BC + CD,$$

which clearly shows that $\varphi(x)$ attains the least value when the point M coincides with the point A and that this value is equal to

$$3AB + 2BC + CD.$$

(2) If $a \le x \le b$, then

$$\varphi(x) = AM + MB + MC + MD$$
$$= AB + 2MB + 2BC + CD.$$

In this case the least value is attained by the function $\varphi(x)$ when the point M coincides with the point B, this value being equal to

$$AB + 2BC + CD.$$

(3) If $b \le x \le c$, then for these values of x the function $\varphi(x)$ is constant and is equal to

$$AB + 2BC + CD.$$

(4) If $c \le x < d$, then the least value of the function $\varphi(x)$ is attained at the point $x = c$, and it is also equal to

$$AB + 2BC + CD.$$

(5) If $x \ge d$, then the least value of the function $\varphi(x)$ is equal to

$$AB + 2BC + 3CD.$$

Comparing the results thus obtained we see that the least value of the function $\varphi(x)$ is equal to $AB + 2BC + CD$, that is to

$$b - a + 2(c - b) + d - c = d + c - b - a.$$

This value the function $\varphi(x)$ takes on provided $b \le x \le c$.

279. Let r be the modulus and φ the argument of the complex number $z\,(r \ge 0, 0 \le \varphi \le 2\pi)$. Then $z = r(\cos\varphi + i\sin\varphi)$ and the given equation takes the form

$$r^2(\cos 2\varphi + i\sin 2\varphi) + r = 0.$$

It follows that either $r = 0$ and $z = z_1 = 0$ or $r\cos 2\varphi + 1 + ir\sin 2\varphi = 0$, and, consequently,

$$\left.\begin{aligned} \sin 2\varphi &= 0, \\ r\cos 2\varphi + 1 &= 0. \end{aligned}\right\}$$

The first equation is satisfied by the values $\varphi = 0, \frac{\pi}{2}, \pi, \frac{3\pi}{2}$, and since by virtue of the second equation we have $\cos 2\varphi < 0$, only the values $\varphi = \frac{\pi}{2}$ and $\varphi = \frac{3\pi}{2}$ must be taken. In both cases we find from the second equation the value $r = 1$, which yields two more solutions:

$$z_2 = 1\left(\cos\frac{\pi}{2} + i\sin\frac{\pi}{2}\right) = i,$$

$$z_3 = 1\left(\cos\frac{3\pi}{2} + i\sin\frac{3\pi}{2}\right) = -i.$$

280. Let us represent z in the form $z = x + iy$. Then the equation $\left|\frac{z-4}{z-8}\right| = 1$ takes the form

$$(x-4)^2 + y^2 = (x-8)^2 + y^2.$$

It follows that $x = 6$ and, hence, $z = 6 + iy$. Substitute this value into the equation $\left|\frac{z-12}{z-8i}\right| = \frac{5}{3}$. Then after simplification the equation takes the form $$y^2 - 25y + 136 = 0.$$

This yields

$$y_{1,2} = \frac{25 \pm \sqrt{625 - 4 \times 135}}{2} = \frac{25 \pm \sqrt{625 - 544}}{2} = \frac{25 \pm 9}{2},$$

i.e. $y_1 = 17$ and $y_2 = 8$.

Answer: $z_1 = 6 + 17i$; $z_2 = 6 + 8i$.

281. For brevity, put $\frac{1+i}{2} = z$. The product $(1+z)(1+z^2)(1+z^{2^2})\dots(1+z^{2^n})$ has the same form as the product in Problem 230. Let us denote this product by P.

Proceeding as in Problem 230, we find

$$P = \frac{1 - z^{2^{n+1}}}{1 - z}.$$

Now we must substitute $\frac{1+i}{2}$ for z into the above formula. We have

$$\frac{1}{1-z} = \frac{1}{1 - \frac{1+i}{2}} = \frac{2}{1-i} = \frac{2(1+i)}{(1-i)(1+i)} = 1 + i.$$

Furthermore, we find

$$1 - z^{2^{n+1}} = 1 - \left(\frac{1+i}{2}\right)^{2^{n+1}} = 1 - \left[\left(\frac{1+i}{2}\right)^2\right]^{2^n} = 1 - \left(\frac{i}{2}\right)^{2^n}. \quad (1)$$

Note that for $n \geq 2$ we have $i^{2^n} = (i^4)^{2^{n-2}} = 1$.

Hence, by virtue of (1), for $n \geq 2$ we have $1 - z^{2^{n+1}} = 1 - \frac{1}{2^{2^n}}$ and

$$P = (1+i)\left(1 - \frac{1}{2^{2^n}}\right).$$

For $n = 1$ we obtain

$$1 - z^{2^{n+1}} = 1 - \left(\frac{i}{2}\right)^2 = \frac{5}{4}.$$

Answer: $$P = (1+i)\frac{5}{4}.$$

282. As is known, the addition and subtraction of complex numbers can be performed geometrically according to the well-known parallelogram law. Therefore the modulus of a difference of two complex numbers $|z'-z''|$ is equal to the distance between the corresponding points of the complex z-plane. Consequently, the condition $|z-25i| \leq 15$ is satisfied by the points of the complex plane lying inside and on the circumference of the circle of radius 15 with centre at the point $z_0 = 25i$ (Fig. 6). As is seen from the figure, the number with the least argument is represented by the point z_1 which is the point of tangency of the tangent line drawn from the point O to that circle. From the right triangle Oz_1z_0 we find $x_2 = 12$ and $y_1 = 16$. The sought-for number is $z_1 = 12 + 16i$.

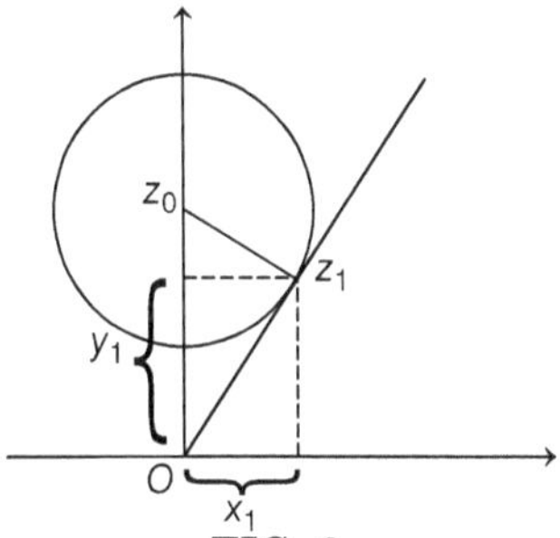

FIG. 6

283. Let us prove that for a complex number $a + bi$ to be representable in the form

$$a + bi = \frac{1 - ix}{1 + ix} \qquad (1)$$

it is necessary and sufficient that $|a + bi| = 1$ and $c + bi \neq -1$.

Necessity. Let equality (1) be fulfilled. Then $|a + bi| = \dfrac{|1 - ix|}{|1 + ix|} = 1$,

since $|1 - ix| = |1 + ix| = \sqrt{1 + x^2}$. Furthermore, $\dfrac{1 - ix}{1 + ix} \neq -1$,

because, if otherwise, we have $1 - ix = -1 - ix$, i. e., $2 = 0$.

Sufficiency. Let $|a + bi| = 1$ and $a + bi \neq -1$. Put $\arg(a + bi) = \alpha$ where $-\pi < \alpha < \pi$. Note that $\alpha \neq \pi$ by virtue of the condition $a + bi \neq -1$. Now we have

$$a + bi = |a + bi|(\cos\alpha + i\sin\alpha) = \cos\alpha + i\sin\alpha \qquad (2)$$

But

$$\cos\alpha = \frac{1 - \tan^2\frac{\alpha}{2}}{1 + \tan^2\frac{\alpha}{2}}, \quad \sin\alpha = \frac{2\tan\frac{\alpha}{2}}{1 + \tan^2\frac{\alpha}{2}}.$$

Substituting these expressions into the right-hand side of formula (2) we get

$$a + bi = \frac{\left(1 + i\tan\frac{\alpha}{2}\right)^2}{\left(1 + i\tan\frac{\alpha}{2}\right)\left(1 - i\tan\frac{\alpha}{2}\right)} = \frac{1 + i\tan\frac{\alpha}{2}}{1 - i\tan\frac{\alpha}{2}} = \frac{1 - ix}{1 + ix},$$

where $x = -\tan\dfrac{\alpha}{2}$.

284. Let $z = r(\cos\varphi + i\sin\varphi)$. Then

$$|z^2+1| = \sqrt{(r^2\cos 2\varphi + 1)^2 + (r^2\sin 2\varphi)^2}$$
$$= \sqrt{r^4 + 2r^2\cos 2\varphi + 1},$$

$$\left|z + \frac{1}{z}\right| = \frac{|z^2+1|}{r} = 1,$$

and $\quad r^4 + r^2(2\cos 2\varphi - 1) + 1 = 0.$

Put $r^2 = t$. The modulus $|z|$ takes on the greatest value when t attains its greatest value. We have

$$t = \frac{1 - 2\cos 2\varphi \pm \sqrt{(1-2\cos 2\varphi)^2 - 4}}{2}.$$

Since we are interested in the greatest value of t, we take the plus sign in front of the radical. It is readily seen that the greatest value of t is attained when

$\cos 2\varphi = -1$ i.e. for $\varphi = \frac{\pi}{2} + k\pi$. This greatest value is equal to $\frac{3+\sqrt{5}}{2}$. Hence, the greatest value of $|z|$ is equal to $\sqrt{\frac{3+\sqrt{5}}{2}} = \frac{1+\sqrt{5}}{2}$.

285. The angle between two neighbouring rays is equal $\frac{2\pi}{n}$. Let $d_1, d_2, \ldots$ be the distances from A and the feet of the perpendiculars which are dropped, in succession, on the rays intersecting at the point A (Fig. 7). We obviously have

$$d_k = d\left(\cos\frac{2\pi}{n}\right)^k \quad (k = 1, 2, \ldots).$$

The length of the k-th perpendicular is

$$L_k = d_{k-1}\sin\frac{2\pi}{n}$$
$$= d\sin\frac{2\pi}{n}\left(\cos\frac{2\pi}{n}\right)^{k-1}.$$

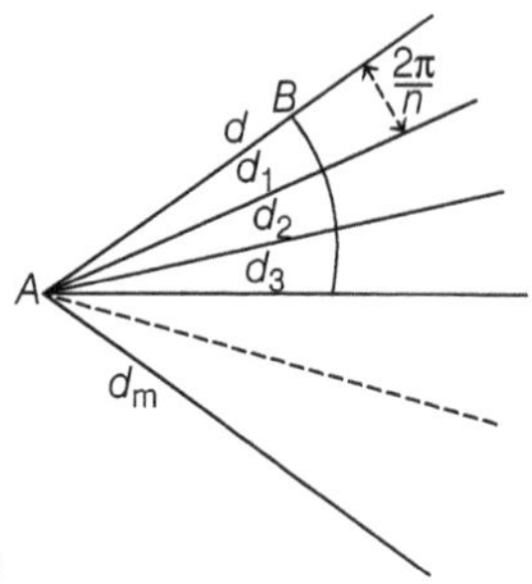

FIG. 7

The total length of the polygonal line consisting of m segments is equal to

$$d\sin\frac{2\pi}{n}\left[1 + \cos\frac{2\pi}{n} + \left(\cos\frac{2\pi}{n}\right)^2 + \ldots + \left(\cos\frac{2\pi}{n}\right)^{m-1}\right].$$

The length L of the whole polygonal line which sweeps out an infinite number of circuits is obtained when m is made to lend to infinity and is expressed as the sum of terms of the geometric progression with common ratio $q = \cos\frac{2\pi}{n}$ $(|\ q\ | < 1)$ and first term $d\sin\frac{2\pi}{n}$:

$$L = d\frac{\sin\frac{2\pi}{n}}{1-\cos\frac{2\pi}{n}} = d\cot\frac{\pi}{n}.$$

When n is increased the length L also increases and approaches infinity as n tends to infinity.

286. *First solution.* Let 1 $abcde$ be the desired number (where the letters a, b, c, d and e denote the digits in the corresponding decimal places). Obviously, $e = 7$, since 1 $abcde \times 3 = abcde$ 1. After 7 has been multiplied by 3, the digit 2 is carried to the next (to the left) decimal place and therefore the product $d \times 3$ has 5 in the unit's place.

Hence, $d = 5$. We thus have 1 $abc57 \times 3 = abc571$. By a similar argument, we find that $c = 8$, $b = 2$ and, finally, $a = 4$. The sought-for number is 142 857.

Second solution. Let again 1 $abcde$ be the number in question. Put $abcde = x$, then the number is equal to $10^5 + x$. By the hypothesis. we have

$$(10^5 + x)3 = 10x + 1,$$

and hence $x = 42\,857$. Consequently, the required number is 142 857.

287. p being divisible by 37, we can write

$$p = 100a + 10b + c = 37k.$$

where k is an integer. It then becomes evident that

$$q = 100b + 10c + a = 10p - 999a = 370k - 37 \times 27a.$$

Consequently, q is also divisible by 37.

A similar reasoning is also applicable to the number r.

288. We have $A = n^3 + (n+1)^3 + (n+2)^3 = 3n^3 + 9n^2 + 15n + 9$. it is obviously sufficient to show that $B = 3n^3 + 15n = 3n(n^2 + 5)$ is divisible by 9. If $n = 3k$ where k is an integer, then B is divisible by 9. For $n = 3k + 1$ we have $n^2 + 5 = 9k^2 + 6k + 6$ and for $n = 3k + 2$ we have $n^2 + 5 = 9k^2 + 12k + 9$. In both cases $n^2 + 5$ is divisible by 3. Hence, B is, divisible by 9 in all cases.

289. *First solution.* The sum S_n can be represented in the following form:

$$S_n = n^3 + 3\,(n^2 + 2n + 1) - n = (n-1)n(n+1) + 3(n+1)^2.$$

The first summand is divisible by 3 because it is the product of three consecutive integers, one of them necessarily being a multiple of three. Hence, S_n is also divisible by 3.

Second solution. We shall prove the assertion by induction. For $n=1$ the number $S_1 = 12$ is divisible by 3.

Suppose that for some n the sum S_n is divisible by 3. We then have
$S_{n+1} = (n+1)^3 + 3(n+1)^2 + 5(n+1) + 3 = S_n + 3(n^2+3n+3)$.
Consequently, S_{n+1} is also divisible by 3.

290. At the base of the pyramid the balls are put in the form of an equilateral triangle. Let the side of this triangle contain n balls. Then, at the base of the pyramid there are $n+(n-1)+(n-2)+\ldots+3+2+1 = \dfrac{n\,(n+1)}{2}$ balls. The second layer of the pyramid contains $(n-1)+(n-2)+\ldots+3+2+1 = \dfrac{(n-1)\,n}{2}$ balls. The third layer contains $\dfrac{(n-2)\,(n-1)}{2}$ balls and so on. The topmost layer contains only one ball. The total number of balls in the pyramid is equal to 120. Hence,

$$120 = \frac{n(n+1)}{2} + \frac{(n-1)n}{2} + \frac{(n-2)\,(n-1)}{2} + \ldots + \frac{3\times4}{2} + \frac{2\times3}{2} + \frac{1\times2}{2}.$$

The right-hand side of the equality is equal to $\dfrac{n\,(n+1)\,(n+2)}{6}$ (see Problem 266, page 192); hence, for defining n we get the equation

$$n(n+1)\,(n+2) = 270. \qquad (1)$$

This equation has an obvious solution $n=8$. To find the other solutions we transpose 720 to the left-hand side and divide the polynomial thus obtained by $n-8$. The quotient is equal to $n^2+11\,n+90$. Since the roots of this latter polynomial are nonreal, equation (1) has no other integral solutions except $n=8$. Thus, the base layer consists of $\dfrac{n(n+1)}{2} = 36$ balls.

291. The number of filled boxes being equal to m we conclude that the number of the inserted boxes is equal to mk. It follows that the total number of the boxes (including the first box) is equal to $mk+1$. Hence, the number of empty boxes is equal to $m\,k+1-m = m(k-1)+1$.

GEOMETRY

A. PLANE GEOMETRY

1. Computation Problems

292. Draw the bisector of the angle A (see Fig. 8). It intersects the side BC at a point D and divides it into parts proportional to b and c. Note then that

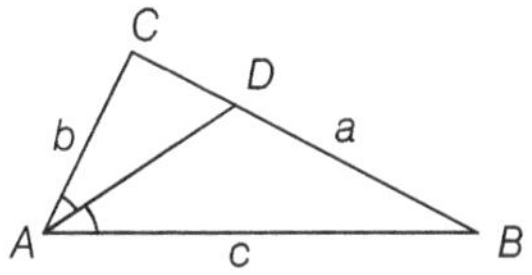

FIG. 8

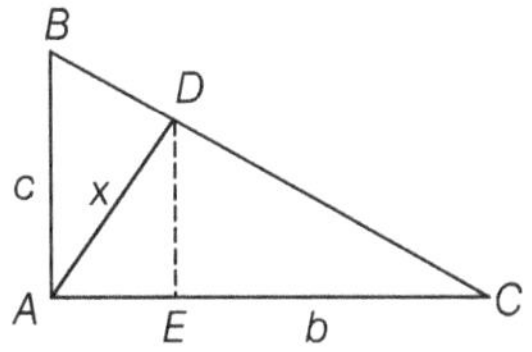

FIG. 9

ΔACD is similar to ΔABC since they have a common angle C, and the angle CAD is equal to the angle B. Hence,

$$\frac{AC}{CD}=\frac{BC}{AC}, \text{ i.e. } \frac{b}{ab/(b+c)}=\frac{a}{b}.$$

Consequently,

$$a=\sqrt{b^2+bc}.$$

293. Let AD be the bisector of the right angle A in ΔABC, and $DE \perp AC$ (Fig. 9). Since

$$\angle DAE=\frac{\pi}{4}, \text{ we have } AE=DE=\frac{x}{\sqrt{2}}$$

where $x = AD$ is the sought-for length. We obviously have

$$\frac{ED}{AB}=\frac{CE}{CA}, \quad \text{i.e. } \frac{\frac{x}{\sqrt{2}}}{c}=\frac{b-\frac{x}{\sqrt{2}}}{b}.$$

Hence, $$x=\frac{bc\sqrt{2}}{b+c}.$$

294. In the triangle ABC (Fig. 10) the medians AD and BE intersect at a point O, $AC = b$ and $BC = a$. Let us find $AB = c$.

Let $OD = x$ and $OE = y$. Taking advantage of the property of medians we find from the triangles AOB, BOD and AOE that

$$4x^2+y^2=\frac{b^2}{4}, 4x^2+4y^2=c^2,$$

$$4x^2+16y^2=a^2.$$

Eliminating x and y we obtain $c^2 = \dfrac{a^2 - b^2}{5}$

The conditions for existence of a triangle with sides a, b and c take the form

$$5(a+b)^2 > a^2 + b^2,\ 5(a-b)^2 < a^2 + b^2.$$

The first inequality is obviously fulfilled for any a and b, and the second one is transformed into the following relation:

$$a^2 - \frac{5}{2}ab + b^2 < 0.$$

Solving this inequality with respect to $\dfrac{a}{b}$ we finally obtain

$$\frac{1}{2} < \frac{a}{b} < 2.$$

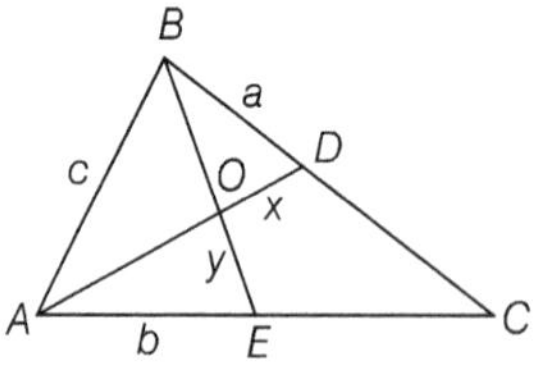

FIG. 10

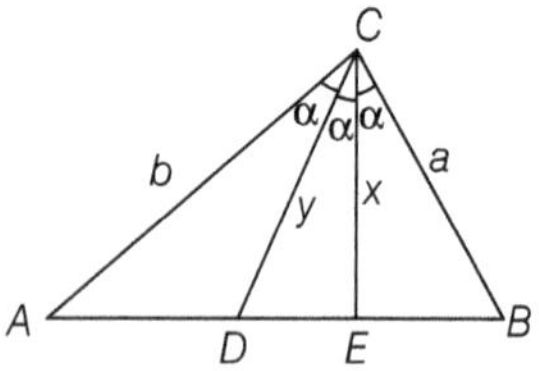

FIG. 11

295. Let $\angle ACD = \angle DCE = \angle ECB = \alpha$ and $CE = x, CD = y$ (Fig. 11). For the area of the triangle ABC we can write the following three expressions:

$$S_{ACD} + S_{DCB} = \frac{1}{2}by\sin\alpha + \frac{1}{2}ay\sin 2\alpha,$$

$$S_{ACE} + S_{ECB} = \frac{1}{2}bx\sin 2\alpha + \frac{1}{2}ax\sin\alpha$$

and

$$S_{ACD} + S_{DCE} + S_{ECB} = \frac{1}{2}by\sin\alpha + \frac{1}{2}xy\sin\alpha + \frac{1}{2}ax\sin\alpha.$$

Equating the left members of these equalities and taking into consideration the condition of the problem we arrive at a system of three equations of the form

$$2a\cos\alpha = x + a\frac{x}{y},$$

$$2b\cos\alpha = y + b\frac{y}{x},\quad \frac{x}{y} = \frac{m}{n}.$$

Solving the system we obtain

$$x = \frac{(n^2 - m^2)ab}{n(bm - an)},\ y = \frac{(n^2 - m^2)ab}{m(bm - an)}.$$

296. Let S be the area of the given triangle ABC (Fig. 12), and put $\frac{AD}{AB} = x$. Then the area of ΔADE is equal to x^2S, and that of ΔABE to xS. By the hypothesis, we get the equation

$$xS - x^2S = k^2.$$

Solving this equation we find

$$x = \frac{1 \pm \sqrt{1 - \frac{4k^2}{S}}}{2}.$$

The problem is solvable if $S \geq 4k^2$. It has two or one solution depending on whether $S > 4k^2$ or $S = 4k^2$, respectively.

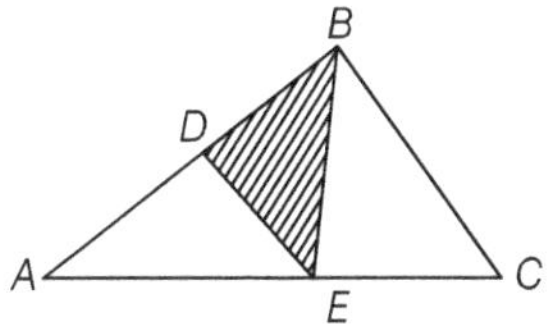

FIG. 12

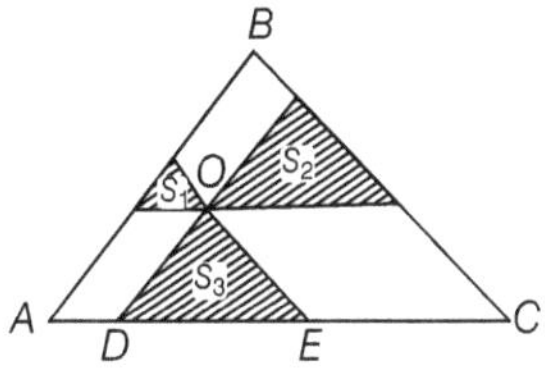

FIG. 13

297. Let S be the area of the given triangle ABC. The constructed triangles with areas S_1, S_2 and S_3 are similar to ΔABC (Fig. 13). Therefore, their areas are in the ratio of the squares of the corresponding sides, whence

$$\sqrt{\frac{S_1}{S}} = \frac{AD}{AC}, \sqrt{\frac{S_2}{S}} = \frac{EC}{AC}, \sqrt{\frac{S_3}{S}} = \frac{DE}{AC}.$$

Adding these equalities termwise we find:

$$S = (\sqrt{S_1} + \sqrt{S_2} + \sqrt{S_3})^2.$$

298. Denote by x the third side of the triangle which is equal to the altitude drawn to it. Using two expressions for the area of the given triangle, we get the equation

$$\frac{1}{2}x^2 = \sqrt{\frac{b+c+x}{2} \cdot \frac{c+x-b}{2} \cdot \frac{x+b-c}{2} \cdot \frac{b+c-x}{2}}.$$

Solving it, we find :

$$x^2 = \frac{1}{5}(b^2 + c^2 \pm 2\sqrt{3b^2c^2 - b^4 - c^4}). \quad (1)$$

The necessary condition for solvability of the problem is

$$3b^2c^2 \geq b^4 + c^4. \quad (2)$$

If it is fulfilled, then both values of x^2 in (1) are positive. It can easily be verified that if (2) is fulfilled, the inequalities $b + c > x \geq |b - c|$ are also fulfilled, the sign of equality appearing only in the case when $x = 0$.

The latter takes place if in (1) we take a minus in front of the radical for $b = c$. Hence, if $b = c$, the problem has a unique solution, namely

$$x = \frac{2}{\sqrt{5}} b.$$

For $b \neq c$ the triangle exists only if inequality (2) is fulfilled. Solving it with respect to $\frac{b}{c}$, we find that it is equivalent to the two inequalities

$$\frac{2}{1+\sqrt{5}} \leq \frac{b}{c} \leq \frac{1+\sqrt{5}}{2}. \qquad (3)$$

Consequently, for $b \neq c$ there exist two triangles if both inequalities (3) are fulfilled with sign <, and only one triangle if at least one of the relations (3) turns in an equality.

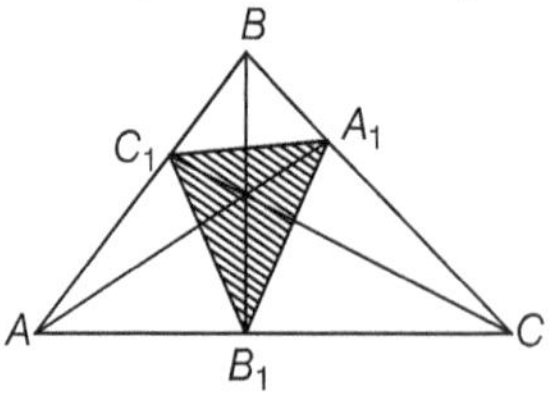

FIG. 14

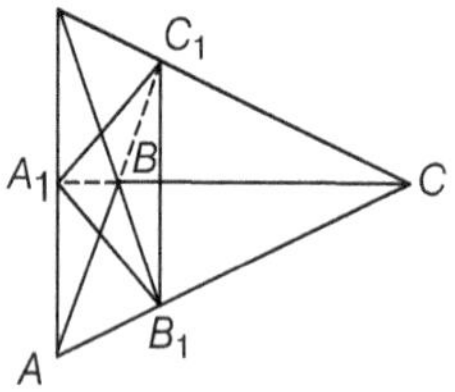

FIG. 15

299. First suppose that ΔABC is acute (Fig. 14). Then

$$S_{ABC} - S_{A_1B_1C_1} = S_{B_1AC_1} + S_{C_1BA_1} + S_{A_1CB_1}. \qquad (1)$$

We have

$$S_{B_1AC_1} = \frac{1}{2} AB_1 . AC_1 \sin A = \frac{1}{2} AB \cos A . AC \cos A \sin A$$

$$= \frac{1}{2} AB . AC \sin A \cos^2 A = S_{ABC} \cos^2 A$$

and, similarly,

$$S_{C_1BA_1} = S_{ABC} \cos^2 B, S_{A_1CB_1} = S_{ABC} \cos^2 C.$$

Substituting these expressions into (1), after some simple transformations we obtain

$$\frac{S_{A_1B_1C_1}}{S_{ABC}} = 1 - \cos^2 A - \cos^2 B - \cos^2 C. \qquad (2)$$

If ΔABC is obtuse (Fig. 15), then, instead of (1), we have

$$S_{ABC} + S_{A_1B_1C_1} = S_{B_1AC_1} + S_{C_1BA_1} + S_{A_1CB_1}$$

and, accordingly, instead of (2),

$$\frac{S_{A_1B_1C_1}}{S_{ABC}} = \cos^2 A + \cos^2 B + \cos^2 C - 1. \qquad (3)$$

Finally, if ΔABC is right, then $S_{A_1B_1C_1} = 0$ which, as is readily seen, also follows from formulas (2) or (3).

300. (1) Let BO and CO be the bisectors of the interior angles of ΔABC (Fig. 16). As is readily seen, the triangles BOM and CON are isosceles. Hence, $MN = BM + CN$.

(2) The relationship $MN = BM + CN$ also holds in the case of the bisectors of exterior angles.

(3) If one of the bisectors divides an interior angle and the other an exterior angle (Fig. 17), then from the interior triangles BMO and CNO we find that $MN = CN - BM$ when $CN > BM$, and $MN = BM - CN$ when $CN < BM$. Thus, in this case

$$MN = |CN - BM|.$$

The points M and N coincide only in the case (3) if ΔABC is isosceles ($AB = AC$).

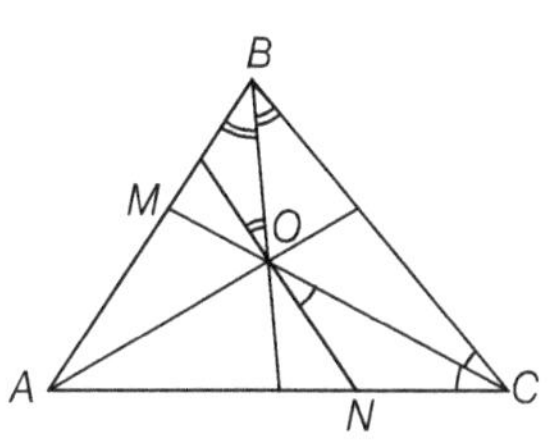

FIG. 16

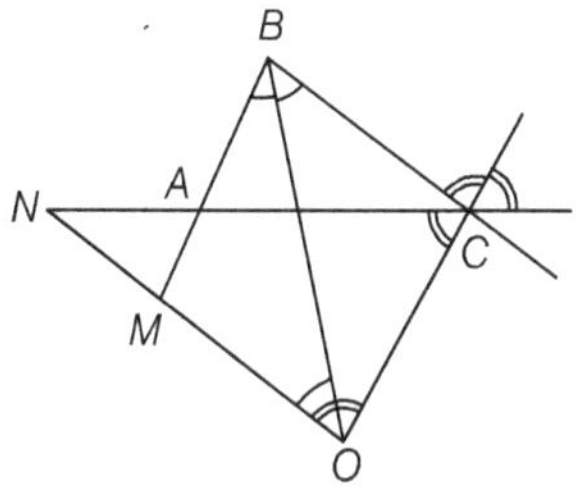

FIG. 17

301. Draw through the point P three straight lines parallel to the sides of the triangle (Fig. 18). The three triangles thus formed (they are shaded in the figure) are also equilateral, and the sum of their sides is equal to the side $AB = a$

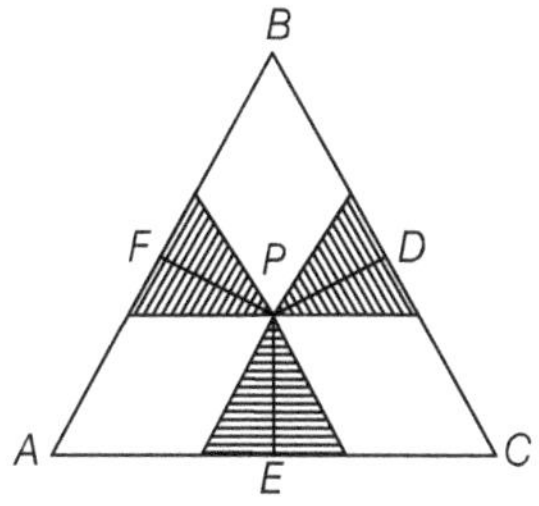

FIG. 18

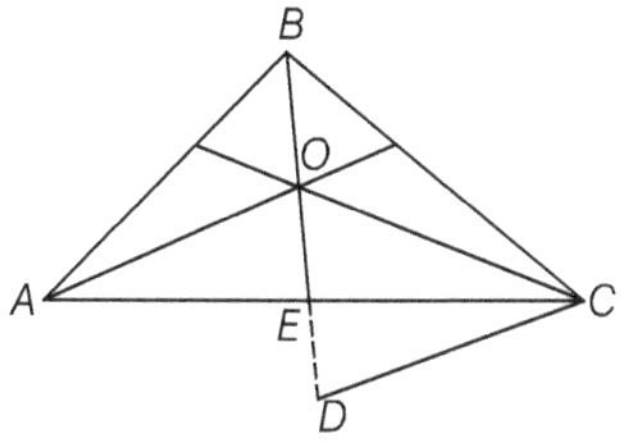

FIG. 19

of the triangle ABC. Consequently, the sum of their altitudes is equal to the altitude of ΔABC and hence

$$PD + PE + PF = \frac{a\sqrt{3}}{2}.$$

The sum $BD + CE + AF$ is equal to the sum of the sides of the shaded triangles added to the sum of the halves of these sides and thus,

$$BD + CE + AF = \frac{3}{2}a.$$

Consequently,

$$\frac{PD + PE + PF}{BD + CE + AF} = \frac{1}{\sqrt{3}}.$$

302. Let O be the point of intersection of the medians in ΔABC (Fig. 19). On the extension of the median BE lay off $ED = OE$. By the property of medians the sides of ΔCDO are $\frac{2}{3}$ the corresponding sides of the triangle formed by the medians. Denoting the area of the latter triangle by S_1, we have $S_1 = \frac{9}{4} S_{CDO}$.

On the other hand, ΔCDO is made up of two, and ΔABC of six triangles whose areas are equal to that of ΔCEO. Therefore, $S_{CDO} = \frac{1}{3} S_{ABC}$. Consequently, $\frac{S_1}{S_{ABC}} = \frac{3}{4}$.

303. Let ABC be the given triangle (Fig. 20). The area of ΔCOB is equal to $\frac{1}{2}ar$, and the area of ΔCOA to $\frac{1}{2}br$. Adding these quantities and expressing the area of ΔABC by Heron's formula, we obtain

$$r = \frac{2}{a+b}\sqrt{p(p-a)(p-b)(p-c)},$$

where $$p = \frac{1}{2}(a + b + c).$$

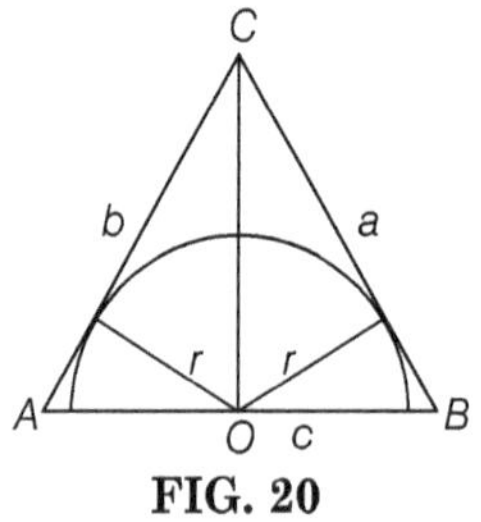

FIG. 20

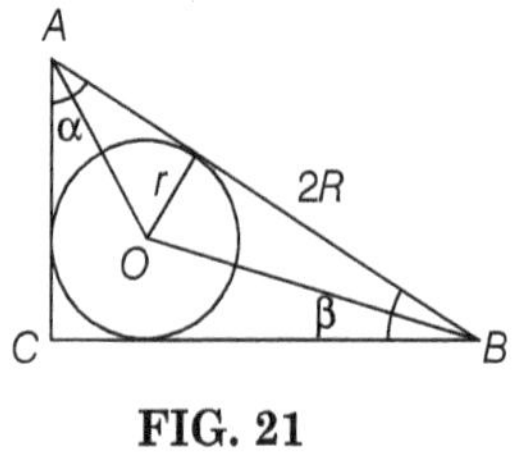

FIG. 21

304. Let R be the radius of the circumscribed circle and r the radius of the inscribed one. Then (Fig. 21) $AB = 2R$, and also

$$AB = r\cot\frac{\alpha}{2} + r\cot\frac{\beta}{2}.$$

Hence

$$\cot\frac{\alpha}{2} + \cot\frac{\beta}{2} = \frac{2R}{r} = 5.$$

Furthermore, $\frac{\alpha}{2} + \frac{\beta}{2} = \frac{\pi}{4}$ and $\cot\left(\frac{\alpha}{2} + \frac{\beta}{2}\right) = 1$, i.e.

$$\frac{\cot\frac{\alpha}{2}\cot\frac{\beta}{2}-1}{\cot\frac{\alpha}{2}+\cot\frac{\beta}{2}}=1,$$

whence $$\cot\frac{\alpha}{2}\cot\frac{\beta}{2}=6.$$

Consequently, $\cot\frac{\alpha}{2}$ and $\cot\frac{\beta}{2}$ are equal to the roots of the quadratic equation $x^2-5x+6=0$.

Finally we obtain

$$\alpha=2\text{ arc tan }\frac{1}{2}, \beta=2\text{ arc tan }\frac{1}{3}.$$

305. Let us denote by a and b the sides of the given rectangle and by φ the angle between the sides of the circumscribed and the given rectangles (Fig. 22). Then the sides of the circumscribed rectangle are equal to

$$a\cos\varphi+b\sin\varphi \text{ and } a\sin\varphi+b\cos\varphi.$$

By the hypothesis, we have

$$(a\cos\varphi+b\sin\varphi)(a\sin\varphi+b\cos\varphi)=m^2,$$

whence we find

$$\sin 2\varphi=\frac{2(m^2-ab)}{a^2+b^2},$$

The condition for solvability of the problem is of the form $0\le\sin 2\varphi\le 1$ which is equivalent to the following two inequalities:

$$\sqrt{ab}\le m\le\frac{a+b}{\sqrt{2}}.$$

306. If $\angle AED=\angle DEC$ (Fig. 23), then also $\angle CDE=\angle DEC$ which imples $CE=CD$. Consequently, E is the point of intersection of the side AB with

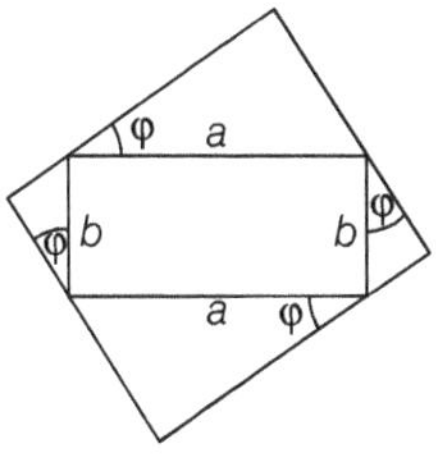

FIG. 22

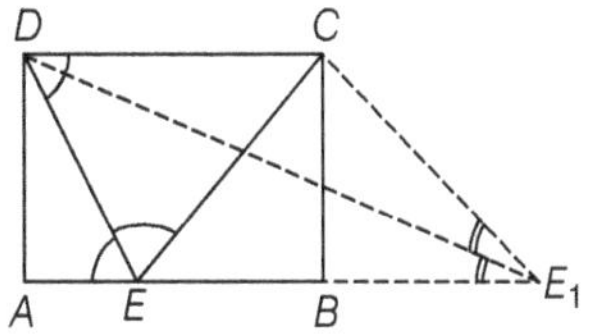

FIG. 23

the circle of radius CD with centre at C. The problem is solvable if $AB\ge BC$, and it has two solutions when $AB>BC$, and only one when $AB=BC$. (The point E_1 in Fig. 23 corresponds to the second solution).

307. Consider one of the nonparallel sides. It is seen from the opposite vertex lying on the lower base at an angle $\frac{\alpha}{2}$ (Fig. 24), and the midline is equal to the line segment joining this vertex to the foot of the altitude drawn from the opposite vertx, i.e. to $h \cot \frac{\alpha}{2}$. Hence, the area of the trapezoid is

$$S = h^2 \cot \frac{\alpha}{2}.$$

308. The midpoints of the diagonals E and F of the trapezoid lie on its midline MN (Fig. 25). But $ME = FN = \frac{a}{2}$, and consequently

$$EF = \frac{b+a}{2} - a = \frac{b-a}{2}.$$

309. The parallelogram is made up of eight triangles of area equal to that of the triangle AOE. The figure (an octagon) obtained by the construction is made

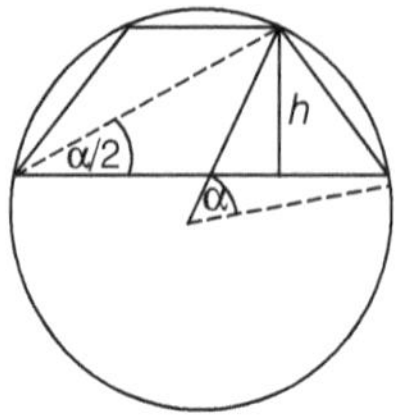

FIG. 24

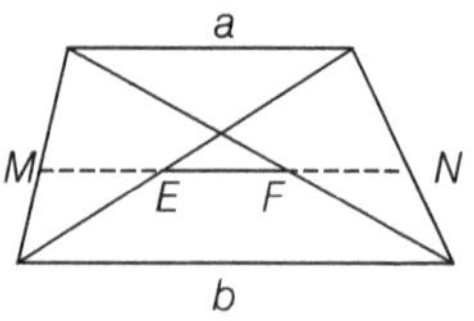

FIG. 25

up of eight triangles whose areas are equal to that of ΔPOQ (Fig. 26). Since $OP = \frac{1}{3}OA$ (by the property of the medians in ΔDAE), and $OQ = \frac{1}{2}OE$, we have

$$S_{POQ} = \frac{1}{6}S_{AOE}.$$

Hence, the sought-for ratio is equal to $\frac{1}{6}$.

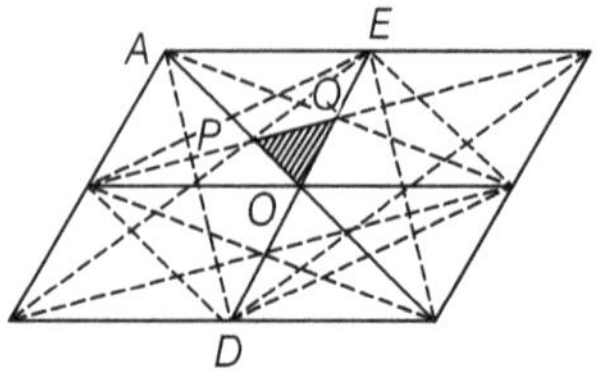

FIG. 26

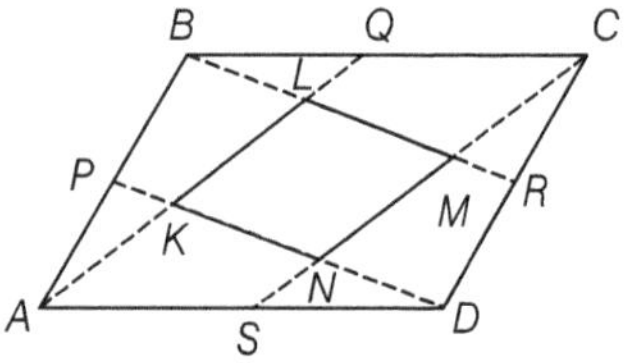

FIG. 27

310. It is obvious that $KLMN$ is a parallelogram (Fig. 27), and $KL = \frac{2}{5} AQ$.

Consequently,

$$S_{KLMN} = \frac{2}{5} S_{AQCS} = \frac{2}{5}\frac{1}{2} a^2 = \frac{1}{5} a^2.$$

311. To the two given chords of length $2a$ and $2b$ there correspond central angles 2α and 2β where

$$\sin\alpha = \frac{a}{R}, \sin\beta = \frac{b}{R}.$$

An arc equal to $2(\alpha \pm \beta)$ is subtended by the chord $2c$ where

$$c = R|\sin(\alpha \pm \beta)| = \left|\frac{a}{R}\sqrt{R^2 - b^2} \pm \frac{b}{R}\sqrt{R^2 - a^2}\right|.$$

312. The sought-for area is equal to the sum of the areas of two sectors with central angles 2α and 2β (Fig. 28) minus twice the area of the triangle with sides R, r, d:

$$S = R^2\alpha + r^2\beta - Rd\sin\alpha.$$

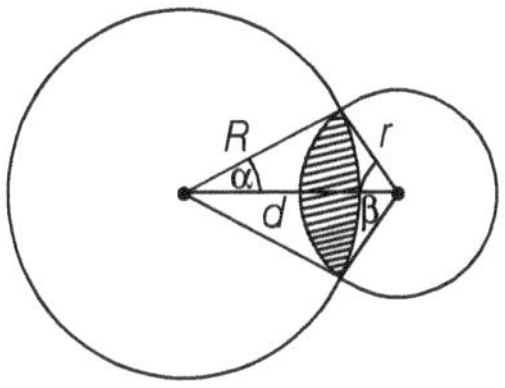

FIG. 28

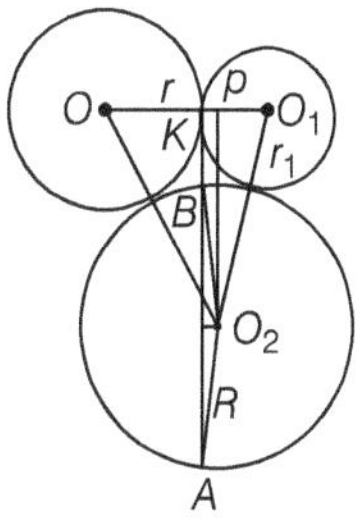

FIG. 29

For determining the angles α and β we have two equations

$$R\sin\alpha = r\sin\beta$$

and

$$R\cos\alpha + r\cos\beta = d.$$

Solving them we find:

$$\cos\alpha = \frac{d^2 + R^2 - r^2}{2Rd}$$

$$\cos\beta = \frac{d^2 + r^2 - R^2}{2rd}.$$

Hence,

$$S = R^2 \operatorname{arc\,cos} \frac{d^2 + R^2 - r^2}{2Rd} + r^2 \operatorname{arc\,cos} \frac{d^2 + r^2 - R^2}{2rd} - Rd\sqrt{1 - \left(\frac{d^2 + R^2 - r^2}{2Rd}\right)^2}.$$

313. Let K be the point of tangency of two circles having radii r and r_1, and P be the foot of the perpendicular dropped from the centre O_2 of the third circle on OO_1 (Fig. 29). Putting $KP = x$, we can write

$$AB = 2\sqrt{R^2 - x^2}. \qquad (1)$$

The quantity x is determined from the equation

$$(R + r)^2 - (r + x)^2 = (R + r_1)^2 - (r_1 - x)^2$$

and is equal to $\frac{r - r_1}{r + r_1} R$. Substituting this value in (1) we obtain

$$AB = \frac{4\sqrt{rr_1}}{r + r_1} R.$$

314. Let O_1 and O_2 be, respectively, the centres of the circles of radii R and r and O_3 be the centre of the third circle. Denote by x the radius of the third circle and by P the point of tangency of this circle and the diameter O_1O_2 (see Fig. 30). Applying the Pythagorean theorem to the triangles O_2O_3P and O_1O_3P we obtain the equality

$$O_2O_3^2 = O_3P^2 + (O_2O_1 + \sqrt{O_1O_3^2 - O_3P^2})^2.$$

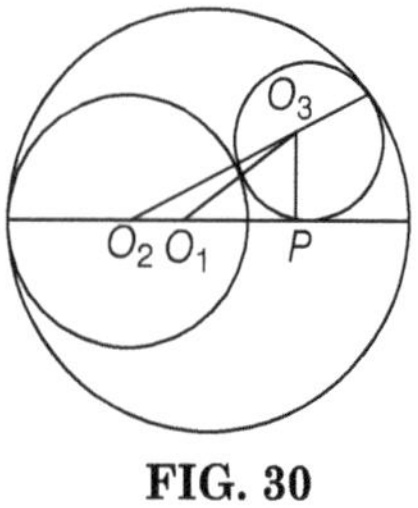

FIG. 30

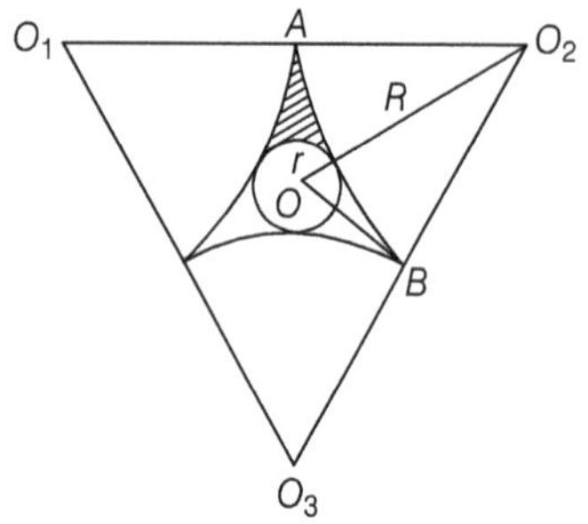

FIG. 31

Substituting the values $O_2O_3 = r + x$, $O_3P = x$, $O_2O_1 = R - r$ and $O_1O_3 = R - x$ into this equality we obtain an equation with x as unknown:

$$(r + x)^2 = x^2 + (R - r + \sqrt{(R - x)^2 - x^2})^2.$$

Solving this equation we find $x = 4Rr\frac{R - r}{(R + r)^2}$.

315. Let O_1, O_2 and O_3 be the centres of the three equal circles and O be the centre of the circle of radius r (Fig. 31). Let us denote be $S_{O_1O_2O_3}$ the area of $\Delta O_1O_2O_3$ and by S_{AO_2B} the area of the sector AO_2B. Then the sought-for area is equal to

$$S = \frac{1}{3}(S_{O_1O_2O_3} - 3S_{AO_2B} - \pi r^2). \qquad (1)$$

If R is the common radius of the three circles, then

$$R = \frac{\sqrt{3}}{2}(R + r),$$

whence, we obtain

$$R = \frac{\sqrt{3}}{2-\sqrt{3}} r = (3 + 2\sqrt{3})r.$$

Then we find

$$S_{O_1O_2O_3} = \frac{1}{2} 2RR\sqrt{3} = \sqrt{3}R^2 = 3(12 + 7\sqrt{3})r^2$$

and $$S_{AO_2B} = \frac{1}{6}\pi R^2 = \frac{\pi}{2}(7 + 4\sqrt{3})r^2.$$

Finally, using formula (1) we obtain

$$S = \left[12 + 7\sqrt{3} - \left(\frac{23}{6} + 2\sqrt{3}\right)\pi\right] r^2.$$

316. Let $O_3D \perp O_1O_2$ (See Fig. 32). We have

$$OO_3^2 = O_1O_3^2 + O_1O^2 - 2O_1O \cdot O_1D = O_2O_3^2 + OO_2^2 - 2OO_2 \cdot DO_2, \qquad (1)$$

where $O_1O_3 = a + r, O_2O_3 = b + r, O_1O = (a + b) - a = b$

and $$OO_2 = (a + b) - b = a.$$

Putting $O_1D = x$ we rewrite the second equality (1) in the form

$$(a + r)^2 + b^2 - 2bx = (b + r)^2 + a^2 - 2a(a + b - x),$$

whence we find $x = a + \frac{a-b}{a+b} r.$

The first equality (1) now takes the form of an equation in one unknown r: $(a + b - r)^2 = (a + r)^2 + b^2 - 2b\left(a + \frac{a-b}{a+b^2} r\right).$

Solving this equation we finally obtain $r = \frac{ab(a+b)}{a^2 + ab + b^2}.$

317. Let us denote by a and b the distance between the given point A an the given straight lines l_1 and l_2, respectively, and by x and y the lengths of the leg s

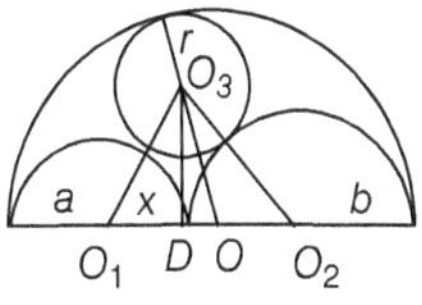

FIG. 32

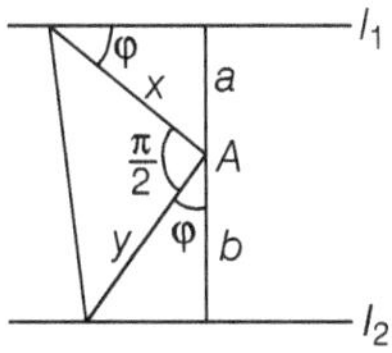

FIG. 33

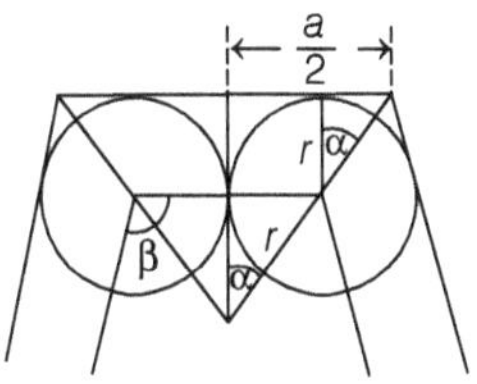

FIG. 34

of the sougth-for triangle (Fig. 33). Noting that $\frac{a}{x} = \sin\varphi, \frac{b}{y} = \cos\varphi$ we obtain two equations $\frac{a^2}{x^2} + \frac{b^2}{y^2} = 1$ and $\frac{1}{2}xy = k^2$.

Transforming these equations we arrive at the system

$$\left.\begin{aligned} xy &= 2k^2, \\ b^2x^2 + a^2y^2 &= 4k^4. \end{aligned}\right\}$$

Solving it, we receive

$$x = \frac{k}{b}\,|\sqrt{k^2+ab} \pm \sqrt{k^2-ab}|,\ y = \frac{k}{a}\,|\sqrt{k^2+ab} \mp \sqrt{k^2-ab}|.$$

The problem is solvable for $k^2 \geq ab$, and has two solutions for $k^2 > ab$ and one solution for $k^2 = ab$.

318. Joining the centres of the circles we obtain a polygon similar to the given one. The centre of the polygon thus constructed coincides with the centre of the given one, and its sides are respectively parallel to the sides of the given polygon (Fig. 34).

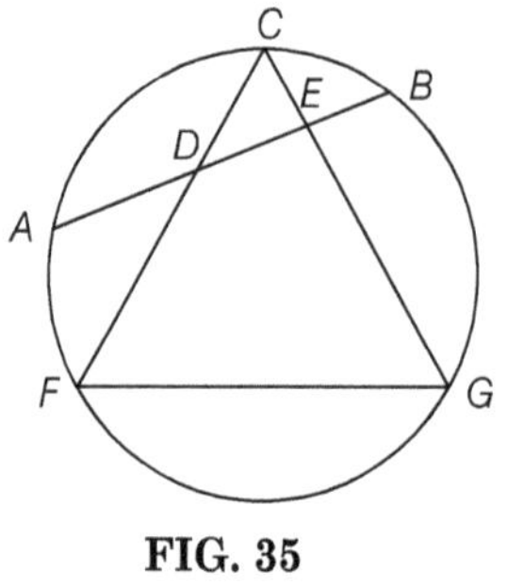

FIG. 35

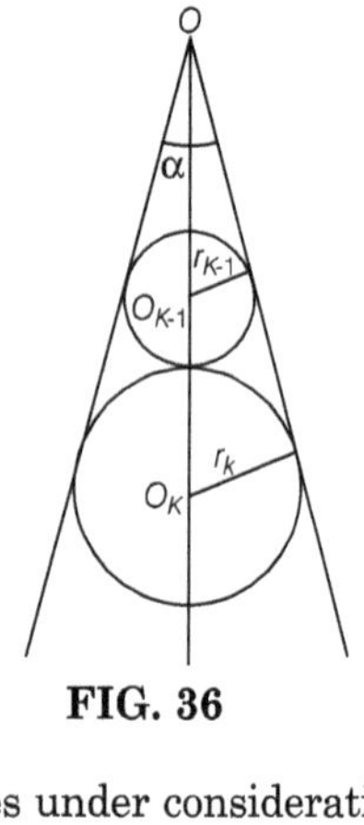

FIG. 36

Let r be the common radius of the circles under consideration. Then the side of the newly constructed polygon is equal to $2r$, and its area is

$$\sigma = nr^2 \cot\frac{\pi}{n}.$$

Furthermore, let $\beta = \frac{\pi(n-2)}{n}$ be the interior angle of the polygon. For the desired area S of the star-shaped figure we obtain the expression

$$S = \sigma - n\frac{r^2}{2}\beta = nr^2\cot\frac{\pi}{n} - n\frac{r^2}{2}\beta.$$

It is obvious (see Fig. 34) that

$$\frac{a}{2} - r = r \tan \frac{\pi}{n},$$

whence we obtain $r = \dfrac{a}{2\left(1 + \tan \dfrac{\pi}{n}\right)}$, and, consequently,

$$S = \frac{a^2}{4} \frac{n \cot \dfrac{\pi}{n} - (n-2)\dfrac{\pi}{2}}{\left(1 + \tan \dfrac{\pi}{n}\right)^2}.$$

319. From Fig. 35 we have

$$\angle CGF = \frac{1}{2}(\breve{FA} + \breve{AC}) \text{ and } \angle CDB = \frac{1}{2}(\breve{FA} + \breve{BC}).$$

The figure $DEGF$ is an inscribed quadrilateral if and only if $\angle CGF = \angle CDB$, i.e. if $\breve{AC} = \breve{BC}$.

320. Let O be the vertex of the acute angle α, and O_k the centre of the kth circle (Fig. 36). Then

$$r_k = O_k, \sin \frac{\alpha}{2}, \; r_{k+1} = (OO_k - r_k - r_{k+1}) \sin \frac{\alpha}{2}$$

and

$$r_{k+2} = r_k - r_k \sin \frac{\alpha}{2} - r_{k+1} \sin \frac{\alpha}{2}.$$

Hence,

$$\frac{r_{k+1}}{r_k} = \frac{1 - \sin \dfrac{\alpha}{2}}{1 + \sin \dfrac{\alpha}{2}},$$

i.e. the radii of the circles form a geometric progression with common ratio

$$\frac{1 - \sin \dfrac{\alpha}{2}}{1 + \sin \dfrac{\alpha}{2}}.$$

321. Let the least angle between the reflected rays and the plane P be equal to α (Fig. 37). Such an angle is formed by the ray passing through the edge C of the mirror after one reflection from the point B. By the hypothesis, we have

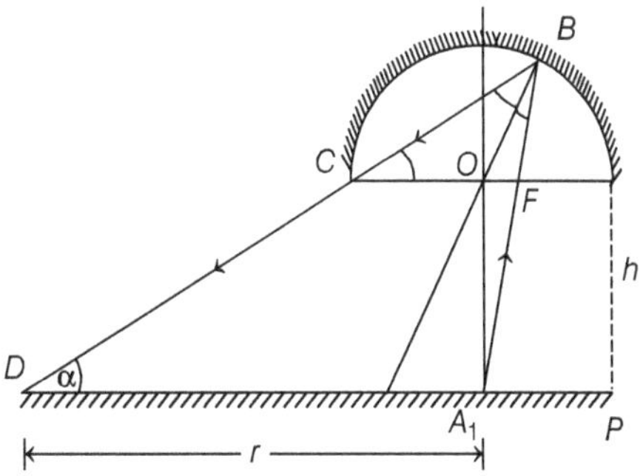

FIG. 37

$CF \| DA$, and hence, $\angle OCB = \angle OBC = \alpha$. From the reflection condition at the point B it follows that $\angle OBF = \alpha$. Therefore, for the triangle OBF we have $\angle BOF = 2\alpha, \angle OFB = 180°-2\alpha-\alpha = 180°-3\alpha$.

Let us denote by h the distance between the mirror and the plane, and by r the radius AD of the illuminated circle. Since the radius of the mirror is equal to 1, we have

$$\frac{h}{r-1} = \tan\alpha. \tag{1}$$

Applying the law of sines to the triangles OBF we find

$$OF = \frac{\sin\alpha}{\sin 3\alpha}.$$

By similarity of the triangles CBF and DBA, their altitudes are proportional to the sides, and thus

$$\frac{AD}{FC} = \frac{h+\sin 2\alpha}{\sin 2\alpha},$$

that is
$$\frac{r}{1+\dfrac{\sin\alpha}{\sin 3\alpha}} = \frac{h+\sin 2\alpha}{\sin 2\alpha}. \tag{2}$$

Solving together (1) and (2) as simultaneous equations we find

$$r = \frac{2\cos 2\alpha}{2\cos 2\alpha - 1}$$

Substituting the given value $\alpha = 15°$ into the latter formula we obtain

$$r = \frac{\sqrt{3}}{\sqrt{3}-1} = \frac{3+\sqrt{3}}{2}.$$

Furthermore, we have $\tan\alpha = \dfrac{\sin 2\alpha}{1+\cos 2\alpha} = \dfrac{\frac{1}{2}}{1+\frac{\sqrt{3}}{2}} = \dfrac{1}{2+\sqrt{3}} = 2-\sqrt{3}$,

and therefore relation (1) yields $h = \frac{1}{2}(1+\sqrt{3})(2-\sqrt{3}) = \dfrac{\sqrt{3}-1}{2}$.

322. We must consider all the different possible cases depending on the value of the ratio $\frac{r}{a}$.

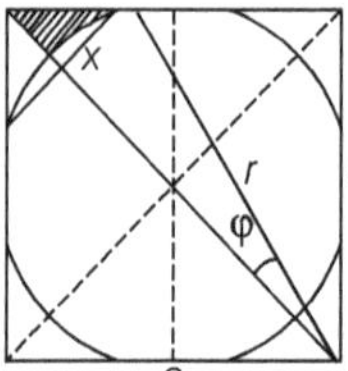

FIG. 38

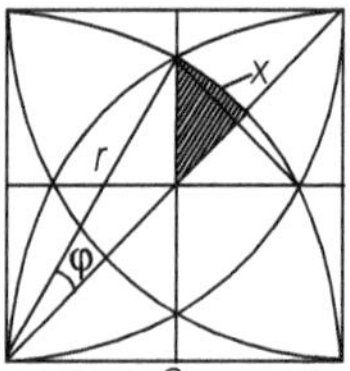

FIG. 39

(1) Let $\frac{r}{a} \geq \sqrt{2}$. The circles do not intersect the square and $S = a^2$.

(2) Let $\frac{\sqrt{5}}{2} \leq \frac{r}{a} < \sqrt{2}$. As is obvious, in this case $S = a^2 - 8\sigma$ where σ is the area of the shaded curvilinear triangles (Fig. 38). We have

$$\sigma = \frac{1}{2} a\sqrt{2}x - \frac{1}{2} r^2\varphi,$$

where $\varphi = \text{arc sin}\,\frac{x}{r}$. To find x we note that

$$x\sqrt{2} + \sqrt{r^2 - a^2} = a$$

which implies $x = \dfrac{a - \sqrt{r^2 - a^2}}{\sqrt{2}}$.

Hence,

$$\sigma = \frac{1}{2} a\,(a - \sqrt{(r^2 - a^2)}) - \frac{1}{2} r^2 \,\text{arc sin}\, \frac{a - \sqrt{r^2 - a^2}}{r\sqrt{2}}.$$

(3) Let $\frac{1}{\sqrt{2}} < \frac{r}{a} < \frac{\sqrt{5}}{2}$. Here $S = 8\sigma$ where σ is the area of the shaded curvilinear triangle (Fig 39). We have

$$\sigma = \frac{1}{2} r^2\varphi - \frac{1}{2}\frac{a}{\sqrt{2}} x,$$

where

$$\varphi = ar\sin\frac{x}{r}.$$

Noting that $\sqrt{r^2 - \left(\frac{a}{2}\right)^2} = \frac{a}{2} + x\sqrt{2}$,

we find $$x = \frac{\sqrt{4r^2 - a^2} - a}{2\sqrt{2}}$$

Consequently, $\sigma = \frac{1}{2} r^2 \,\text{arc sin}\, \frac{\sqrt{4r^2 - a^2} - a}{2\sqrt{2}r} - \frac{a(\sqrt{4r^2 - a^2} - a)}{8}$.

(4) Let $\frac{r}{a} \leq \frac{1}{\sqrt{2}}$. The required area is equal to zero.

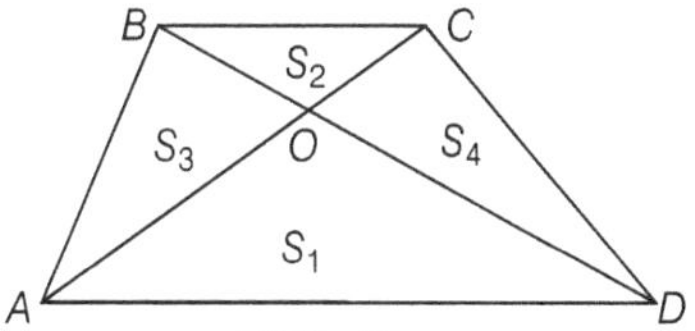

FIG. 40

323. We have (see Fig. 40)

$$S = S_1 + S_2 + S_3 + S_4. \qquad (1)$$

Furthermore,

$$\frac{S_3}{S_2} = \frac{S_1}{S_4} = \frac{AO}{OC},$$

whence $S_3S_4 = S_1S_2$. But we obviously have

$$S_3 + S_1 = S_4 + S_1,$$

which implies $S_3 = S_4$ and $S_3 = S_4 = \sqrt{S_1S_2}$.

Hence, from (1) we obtain

$$S = S_1 + S_2 + 2\sqrt{S_1S_2} = (\sqrt{S_1} + \sqrt{S_2})^2.$$

324. Let us denote by a, b, c and d the lengths of the sides and by m and n the lengths of the diagonals of the quadrilateral (Fig. 41). By the law of cosines we have

$$n^2 = a^2 + d^2 - 2ad\cos\varphi$$

and

$$n^2 = b^2 + c^2 + 2bc\cos\varphi,$$

which yield

$$(bc + ad)n^2 = (a^2 + d^2)\,bc + (b^2 + c^2)ad = (ab + cd)\,(ac + bd).$$

Hence,

$$n^2 = \frac{ab + cd}{bc + ad}(ac + bd).$$

Analogously, we find

$$m^2 = \frac{ad + bc}{ab + cd}(ac + bd).$$

Multiplying these equalities we obtain Ptolemy's theorem:

$$mn = ac + bd.$$

2. Construction Problems

325. Let O_1 and O_2 be the centres of the given circles. Draw the straight line O_1A and another straight line parallel to O_1A passing through the centre O_2 of the second circle. This line intersects the second circle at points M and N

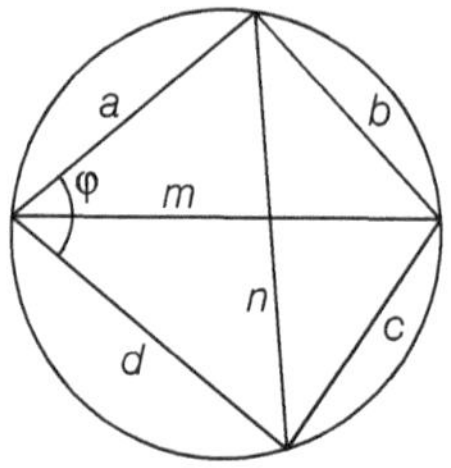

FIG. 41

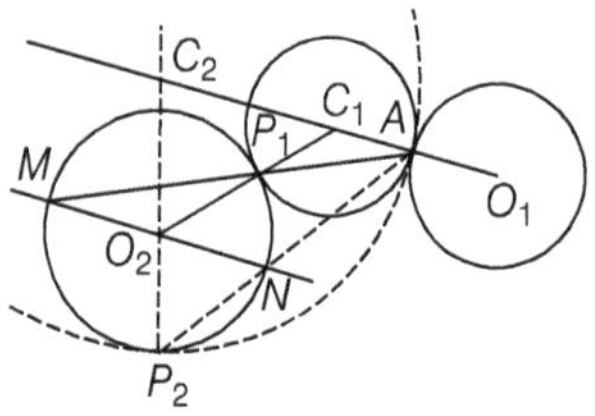

FIG. 42

(Fig. 42). The straight line MA intersects the second circle at a point P_1. The straight line O_2P_1 intersects O_1A at a point C_1. The similarity of the triangles MO_2P_1 and AC_1P_1 implies $C_1A = C_1P_1$.
Hence, the cirlce of radius C_1A with centre at C_1 is the required one. A second solution is obtained with the aid of the point N in just the same way as the first solution with the aid of the point M. If one of the straight lines MA or NA is tangent to the second cirlce, then only one solution remains while the second solution yields this tangent line (which can be interpreted as a cirlce with centre lying at infinity). The latter case takes place if and only if the point A coincides with the point of tangency of one of the the four common tangents to the given circles.

326. Let O be the centre of the given circle, and AB the given line (Fig. 43). The problem is solved by analogy with the preceding one. In the general case it has two solutions. There are three singular cases here: (1) The given line intersects the cirle, and the given point A coincides with one of the points of intersection. Then there are no solutions. (2) The given line is tangent to the circle, and the point A does not coincide with the point of tangency. In this case there is one solution. (3) The given line is tangent to the circle, and the point A coincides with the point of tangency. In this case there is an infinitude of solutions.

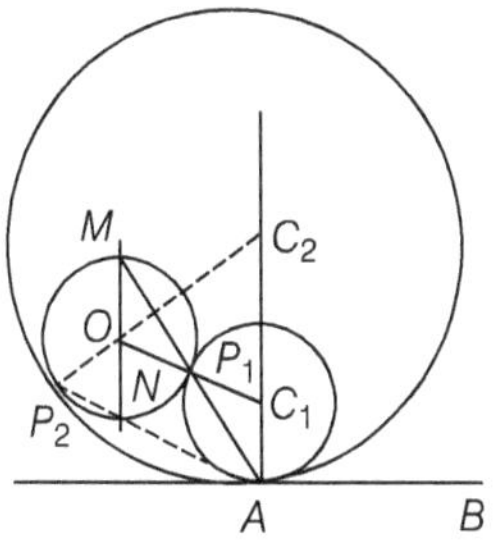

FIG. 43

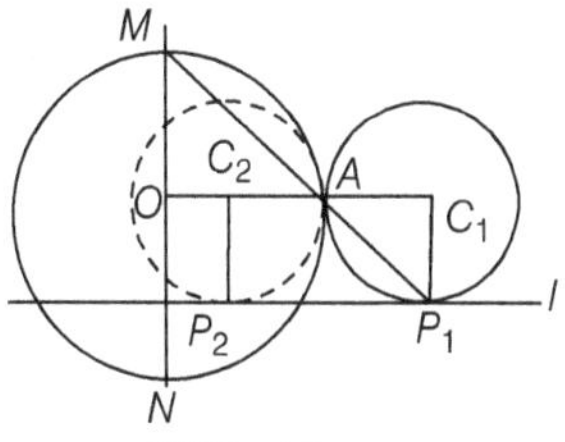

FIG. 44

327. Through the centre O of the given circle draw a straight line perpendicular to the given line l and intersecting the circle at points M and N (Fig. 44). The straight line MA intersects l at a point P_1. C_1 is the point of intersection of the perpendicular erected to l at the point P_1 with the straight line OA. The similarity of the triangles AOM and AC_1P_1 implies that $C_1A = C_1P_1$. Consequently, the circle of radius C_1A with centre at C_1 is the requried one. Another solution is obtained with the aid of the point N in just the same way as the first solution with the aid of the point M. If the straight line l does not pass through one of the points A, M and N, and the point A does not coincide with M or N, the problem always has two solutions.

Suppose that A does not coincide with M or N. If l passes through M or N, the problem has one solution (the second circle coincides with the given one). But if l passes through A, the problem has no solutions.

Let A coincide with M; if l does not pass through M or N, the problem has one solution (the second degenerates into a straight line coinciding with l). If l passes through N the given circle is the solution, and if l passes through M, the problem has an infinite number of solutions.

328. On the given hypotenuse $AB = c$ as on diameter construct a circle with centre at the point O (Fig. 45). Draw $OE \perp AB$ and lay off $OF = h$ on OE. The straight line parallel to AB and passing through F intersects the circle at the sought-for point C. The problem is solvable if $h \le \frac{c}{2}$. The lengths of the legs a and b are found from the system of equations $\left.\begin{aligned} a^2 + b^2 &= c^2, \\ ab &= hc. \end{aligned}\right\}$

Solving this system we obtain

$$a = \frac{1}{2}(\sqrt{c^2 + 2hc} + \sqrt{c^2 - 2hc}), \quad b = \frac{1}{2}(\sqrt{c^2 + 2hc} - \sqrt{c^2 - 2hc}).$$

329. Let us take the line segment AB, and on its extension lay off a segment $AE = AD$ in the direction from A to B (Fig. 46). Construct ΔBCE on BE as base with sides BC and $EC = CD$. On AC as base construct ΔACD with sides AD and CD. The quadrilateral $ABCD$ is the required one because it has the given sides and $\angle DAC = \angle CAE$ (the triangles ACD and ACE are congruent by construction).

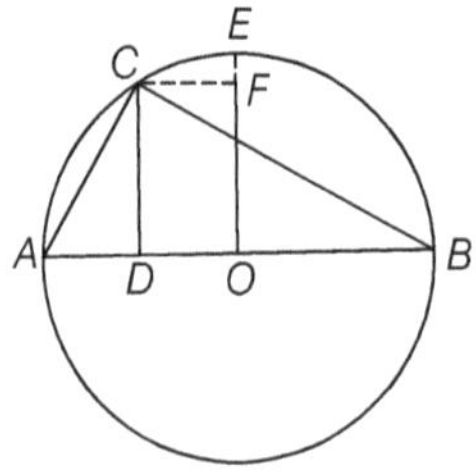

FIG. 45

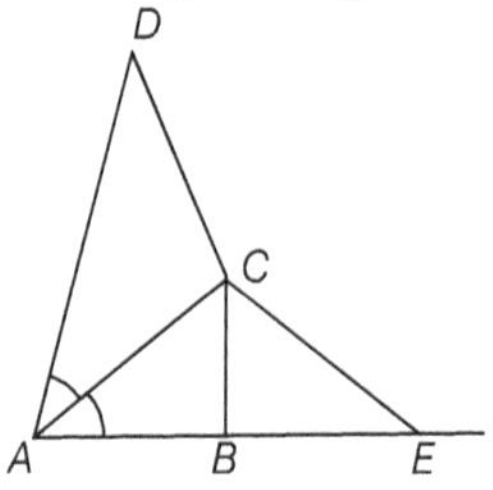

FIG. 46

330. Let H, S and M be, respectively, the points of intersection of the altitude, bisector and median with the circumscribed circle K whose centre is at the point O (Fig. 47). Draw the straight line SO, and through H another straight line parallel to SO whose second point of intersection with K is the point A. Draw the straight line AM intersecting SO at a point P. Through the point P draw a straight line perpendicular to SO which intersects the circle at points B and C.

The triangle ABC is the required one, since $AH \perp BC$, $\breve{BS} = \breve{SC}$ and $BP = PC$. The problem is solvalble if and only if H, S and M do not lie in a straight line, the tangent line to K at the point H is not parallel to SO and the points H and M lie on opposite sides from the straight line SO but not on a diameter of the circle K.

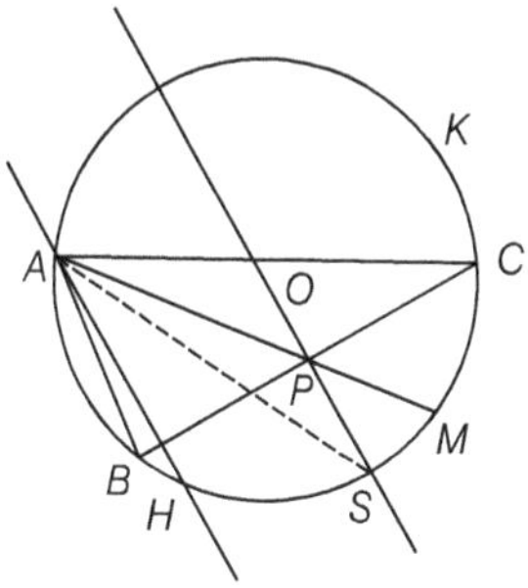

FIG. 47

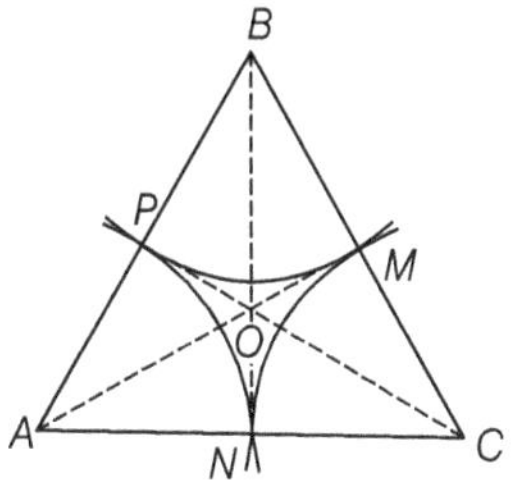

FIG. 48

331. A. Exterior tangency. From the point O of intersection of the bisectors of the interior angles of the triangles ABC drop the perpendiculars OM, ON and OP on the sides of the triangle (Fig. 48). Then $AP = AN$, $BP = BM$ and $CM = CN$. Consequently, the circles of radii AP, BM and CN with centres at A, B and C are tangent to one another at the point P, M, N.

B. Interior tangency. From the point O of intersection of the bisector of angle C and bisectors of the exterior angles A and B draw the perpendiculars OM, ON and OP to the sides of the triangle ABC or to their extensions (Fig. 49). Then

$$AP = AN, BP = BM, CM = CN.$$

Consequently, the circles of radii AP, BM and CN with centres at A, B and C are tangent to one another at the points P, M, N.

Taking the bisectors of the interior angle A and exterior angles B and C, or the interior angle B and exterior angles A and C, we obtain two more solutions.

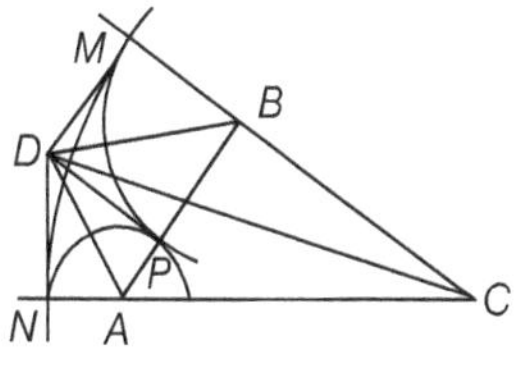

FIG. 49

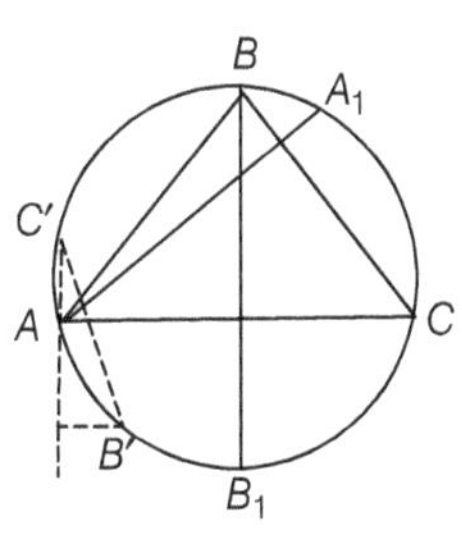

FIG. 50

332. The solution is based on the following property: if the altitudes h_A and h_B of the inscribed triangle ABC intersect the circle at points A_1 and B_1, then the vertex C bisects the arc A_1B_1 (Fig. 50). This is implied by the equality of $\angle A_1AC$ and $\angle B_1BC$, each of which is equal to $\frac{\pi}{2} - \angle ACB$.

Construction. Through A draw a straight line in the given direction to intersect the circle at a point A_1. Let B_1 be the point of intersection of the altitude h_B and the circle. Find the midpoint C of the arc A_1B_1 and draw AC. Then draw $B_1B \perp AC$. The triangle ABC is the sought-for. Taking the midpoint C' of the second of the two arcs A_1B_1, we obtain another solution, namely the triangle $AB'C'$.

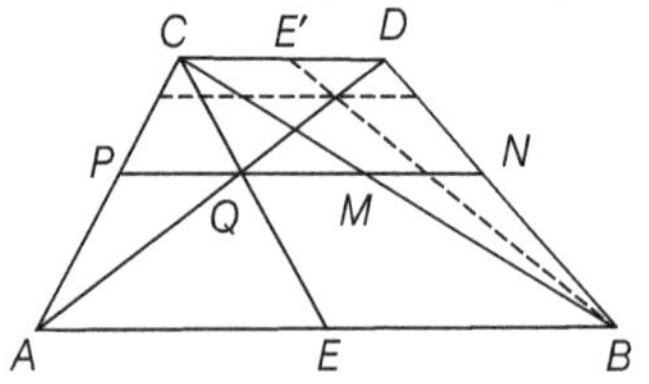

FIG. 51

333. Join the midpoint E of the base AB to the vertex C and find the point Q of intersection of the straight lines EC and AD (Fig. 51). The straight line $PQMN$ parallel to AB is the required one. Indeed,

$$\frac{PQ}{QM} = \frac{AE}{EB} = 1,$$

which results in $PQ = QM$. Furthermore. $\frac{MN}{CD} = \frac{PQ}{CD}$,

whence $MN = PQ$. A second solution is obtained with the aid of the midpoint E' of the base CD like the first soultion with the aid of E.

334. Let B be the given vertex and E, F the given points (Fig. 52). Suppose that the square $ABCD$ has been constructed. The vertex D must lie on the circle constructed on HF as on diameter. Let BD intersect this circle at a point K. Then $\breve{EK} = \breve{KF}$ since $\angle ADB = \angle BDC$.

Construction. On EF as on diameter construct a circle and at its centre erect a perpendicular to EF to intersect the circle at point. K and K' . Join B to K and extend BK to intersect the circle at point D. Draw the straight lines DE and DF and through the point B the straight lines BA and BC perpendicular to them. $ABCD$ is the required square. Using the point K' we obtain another solution. The problem always has two solutions except the case when the point B lies on the circle with diameter EF. In this latter case the problem has no solutions if the point B does not coincide with one of the points K and K'.

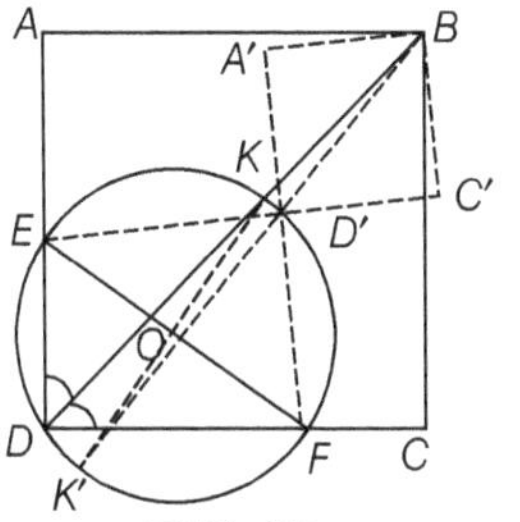

FIG. 52

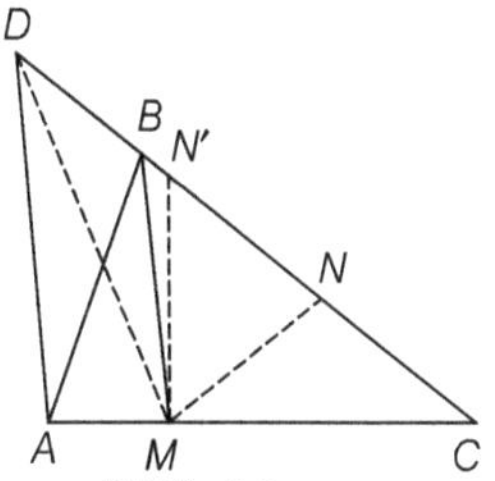

FIG. 53

335. *First solution.* Draw $AD \| MB$ to intersect the extension of BC at a point D (Fig. 53). On the line segment CD find a point N such that

$$\frac{CD}{CN} = k.$$

The straight line MN is the desired one since the area S_{ABM} is equal to the area S_{DBM} and hence $S_{ABC} = S_{DMC}$, and by construction we have $S_{DMC} = k\,S_{NMC}$.

Second solution is obtained by using a point N_1 such that $\dfrac{CD}{N_1D} = k.$

This yields $$\frac{S_{ABC}}{S_{ABN_1M}} = k.$$

Taking into consideration the possibility of an analogous construction based on the vertex C (instead of A), we can easily verify that for $k \neq 2$ the problem always has two solutions and for $k = 2$ only one.

336. To make the construction it is sufficient to determine the altitude $h = KL$ of the rectangle.

Let $KLMN$ be the required rectangle, and KN lie on AC (Fig. 54). If the vertex B is made to move in a straight line parallel to the base AC while the altitude h remains unchanged, the lengths of the base and of the diagonals of the rectangle also remain unchanged (becuase LM and AC are in the ratio of $BH - h$ to BH). Consequently, for determining h the given triangle ABC can be replaced by any other triangle having the same base AC and the same altitude BH.

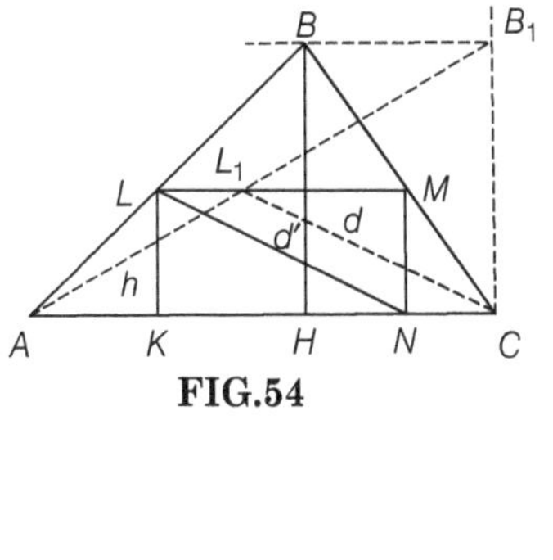

FIG.54

It is most convenient to take a triangle with right base angle. Hence, we perform the following construction. Through B draw a straight line parallel to AC, and through C a straight line perpendicular to AC. Using a compass with opening equal to the length d of the given diagonal, lay off on the hypotenuse AB_1 a line segment AL_1 from the vertex of the right angle C. Through the point L_1 draw a straight line parallel to AC; the point L and M at which it intersects the sides AB and BC are the vertices of the required rectangle. Depending on whether the altitude of the triangle AB_1C draw from C is less than, equal to or greater than the given value of d, the problem has two, one or no solutions.

337. Inscribe the given circle in the given angle. Lay off on the sides of the angle line segments $AC = BD$ of length equal to that of the given side of the triangle from the points of tangency A and B in the direction from the vertex S (Fig. 55). Inscribe in the given angle another circle so that it is tangent to the sides of the angle at points C and D. Draw a common tangent EF to the constructed circles. We shall prove that ΔSEF thus obtained is the required triangle. For

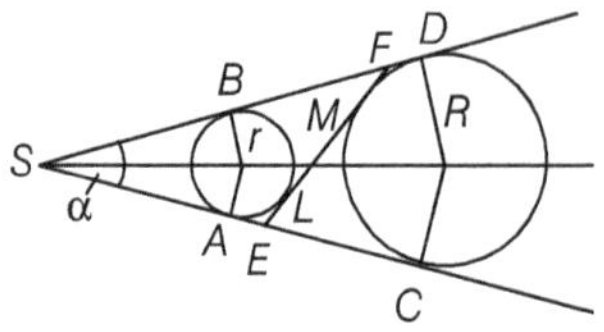

FIG. 55

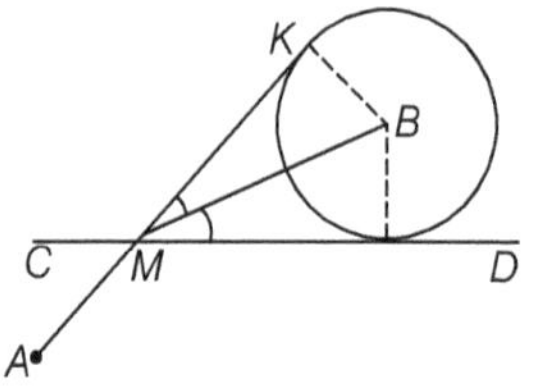

FIG. 56

this purpose it is sufficient to prove that $AC = FE$. It is easily seen that the perimeter of the triangle SEF is equal to $2SC$. On the other hand, it is obviously equal to $2(SA + EL + LF)$. Thus, we have

$$SC = SA + EL + LF, SA + AC = SA + EF, \text{i.e. } AC = EF,$$

which is what we set out to prove.

It is clear, that the problem has two solutions if the circles do not intersect, and only one if they are tangent. The problem has no solution if the circles intersect. Let α be the given angle, r and R the radii of the cirlce and a the given side of the triangle. The distance between the centres of the circles is equal to $\frac{a}{\cos\frac{\alpha}{2}}$. For the problem to be solvable it is necessary that $R + r \geq \frac{a}{\cos\frac{\alpha}{2}}$.

But we have

$$R = r + a \tan\frac{\alpha}{2}$$

and, consequently, there must be

$$2r + a \tan\frac{\alpha}{2} \geq \frac{a}{\cos\frac{\alpha}{2}},$$

that is

$$\frac{2r}{a} \geq \frac{1-\sin\frac{\alpha}{2}}{\cos\frac{\alpha}{2}}.$$

338. Describe a circle with centre at the point B tangent to the straight line CD (Fig. 56). From the point A (if A and B lie on different sides from CD) or from the point A' which is the reflection of A through CD (if A and B are on one side of CD) draw the tangent line AK or $A'K$ to the constructed circle. The point M of intersection of AK (or $A'K$) and CD is the sought-for point. Indeed, we have

$$\angle AMC = \angle KMD = 2\angle BMD.$$

3. Proof Problems

339. Let BO be a median in the triangle ABC. Construct the parallelogram $ABCD$ (Fig. 57). From the triangle BCD we have $2BO < BC + CD$, and since $CD = AB$, we can write

$$BO < \frac{AB+BC}{2}.$$

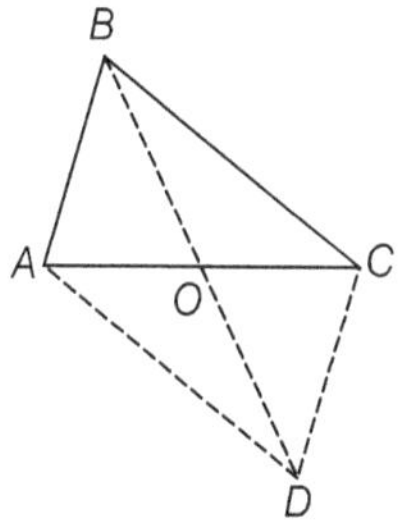

FIG. 57

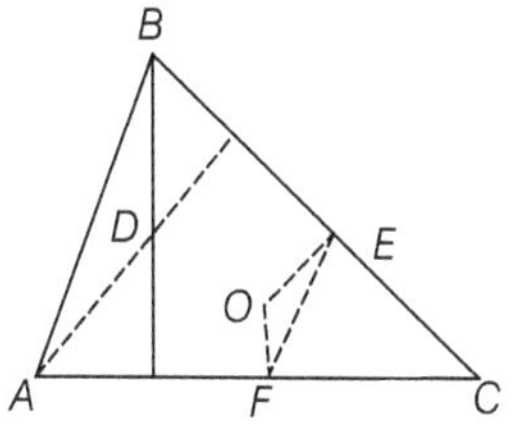

FIG. 58

From ΔAOB and ΔBOC we have

$$BO + \frac{AC}{2} > AB$$

and

$$BO + \frac{AC}{2} > BC.$$

Adding together these inequalities we obtain

$$BO > \frac{AB+BC}{2} - \frac{AC}{2}.$$

340. Let D be the point of intersection of the altitudes (the orthocentre), O be the centre of the circumscribed circle, E and F the midpoints of the sides BC and AC (Fig. 58). The triangles ADB and EOF are similar because $\angle ABD = \angle OFE$ and $\angle BAD = \angle OEF$ (as angles with parallel sides). Hence, $\dfrac{OE}{AD} = \dfrac{EF}{AB} = \dfrac{1}{2}$.

341. See the solution of Problem 301.

342. Let a, b and c be the lengths of the sides of the triangle opposite the angles A, B and C, respectively. We shall prove that the length l_A of the bisector of the angle A is expressed by the formula

$$l_A = \frac{2bc\cos\frac{A}{2}}{b+c} = \frac{2\cos\frac{A}{2}}{\frac{1}{b}+\frac{1}{c}}. \tag{1}$$

Indeed, the area of the triangle ABC is

$$S_{ABC} = \frac{1}{2}bc\sin A = \frac{1}{2}cl_A\sin\frac{A}{2} + \frac{1}{2}bl_A\sin\frac{A}{2},$$

which results in formula (1). Similarly, for the bisector l_B of the angle B we obtain the formula

$$l_B = \frac{2\cos\frac{B}{2}}{\frac{1}{a}+\frac{1}{c}}. \tag{2}$$

Let $a > b$; then $\angle A > \angle B$, and since we have $0 < \dfrac{A}{2} < \dfrac{\pi}{2}$ and $0 < \dfrac{B}{2} < \dfrac{\pi}{2}$ this implies $\cos\dfrac{A}{2} < \cos\dfrac{B}{2}$. Thus, the numerator of fraction (1) is less than that of fraction (2). Furthermore, the denominator $\dfrac{1}{b}+\dfrac{1}{c}$ of fraction (1) is greater than the denominator $\dfrac{1}{a}+\dfrac{1}{c}$ of fraction (2) because $\dfrac{1}{b} > \dfrac{1}{a}$. Consequently, $l_A < l_B$.

343. Let $\angle CPQ = \alpha$ and $\angle PQC = \beta$ (Fig. 59). By the law of sines we have

$$\frac{RB}{\sin\alpha} = \frac{BP}{\sin(\alpha+\beta)}, \frac{PC}{\sin\beta} = \frac{CQ}{\sin\alpha}, \frac{AQ}{\sin(\alpha+\beta)} = \frac{AR}{\sin\beta}.$$

Multiplying these equalities termwise we obtain

$$RB\cdot PC\cdot QA = PB\cdot QC\cdot RA.$$

344. Let $\angle AKB = \alpha$, $\angle AFB = \beta$ and $\angle ACB = \gamma$ (Fig. 60). We have $\alpha = \dfrac{\pi}{4}$, and since $\tan\beta = \dfrac{1}{2}$, $\tan\gamma = \dfrac{1}{3}$ we can write

$$\tan(\beta+\gamma)=\frac{\frac{1}{2}+\frac{1}{3}}{1-\frac{1}{2}\cdot\frac{1}{3}}=1.$$

It follows that $\beta+\gamma=\frac{\pi}{4}$ and $\alpha+\beta+\gamma=\frac{\pi}{4}+\frac{\pi}{4}=\frac{\pi}{2}$.

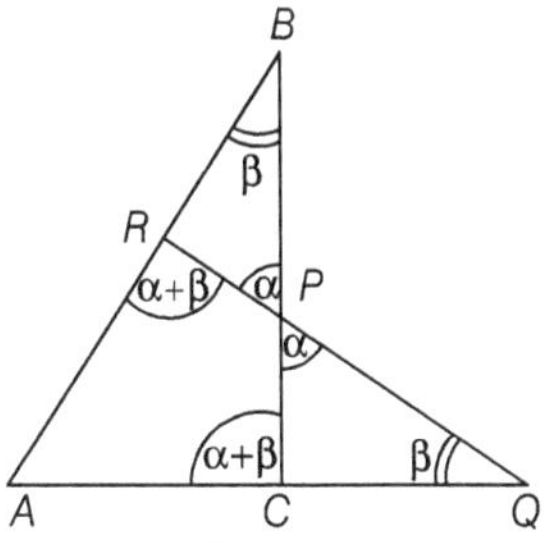

FIG. 59

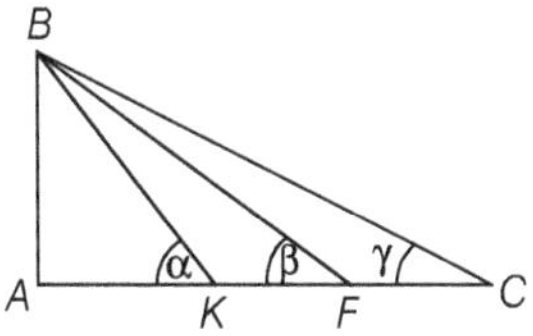

FIG. 60

345. We shall use the converse of the Pythagorean theorem: if the sum of the squares of two sides of a triangle is equal to the square of the third side this triangle is right.

In our case the relationship

$$(a+b)^2+h^2=(c+h)^2$$

is fulfilled because it is equivalent to the obvious equality $ab=ch$.

346. *First solution.* Draw AE so that $\angle EAC=20°$ and $BD\perp AE$ (Fig. 61). Since ΔCAE is similar to ΔABC, we have

$$\frac{CE}{a}=\frac{a}{b}$$

which yields $CE=\frac{a^2}{b}$ and $BE=b-\frac{a^2}{b}$.

On the other hand, $\angle BAD=60°$ and therefore $BD=\frac{\sqrt{3}}{2}b$, $AD=\frac{b}{2}$.

But $AE=a$ and hence $ED=\frac{b}{2}-a$. It follows that

$$BE=\sqrt{\left(\frac{b}{2}-2\right)^2+\frac{3}{4}b^2}.$$

Consequently, $b-\frac{a^2}{b}=\sqrt{\left(\frac{b}{2}-a\right)^2+\frac{3}{4}b^2}$.

Squaring both members and simplifying the obtained equality, we find that this relationship is equivalent to the one to be proved.

Second solution. We have $a = 2b \sin 10°$, and therefore the relationship to be proved is equivalent to
$1 + 8\sin^3 10° = 6\sin 10°$, that is $\sin 30° = 3\sin 10° - 4\sin^3 10°$.

The latter equality holds by virtue of the general formula $\sin 3\alpha = 3\sin\alpha - 4\sin^3\alpha$.

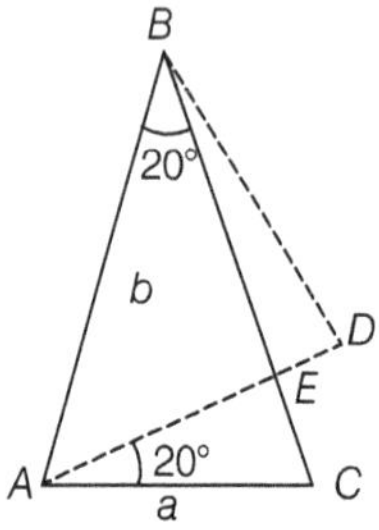

FIG. 61

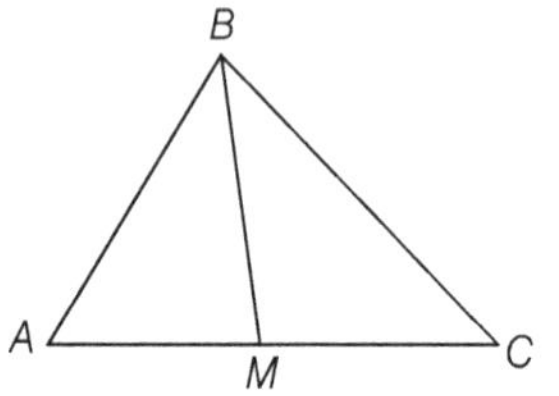

FIG. 62

347. In any triangle, the greater angle is opposite the greater side. Therefore, if

$AC < 2BM$, in ΔABC (Fig. 62)

which is equivalent to the two inequalities

$AM < BM$, $MC < BM$, then

$$\angle ABM < \angle BAM, \angle MBC < \angle BCM.$$

Adding these inequalities we obtain

$\angle ABC < \angle BAM + \angle BCM = \pi - \angle ABC$,

whence $2\angle ABC < \pi$ or $\angle ABC < \frac{\pi}{2}$.

The cases $AC \geq 2BM$ are considered analogously.

348. *First solution.* Let $QQ' \| AC$ and N be the point of intersection of AQ' and QC (Fig. 63). The angles whose values are implied by the conditions of the problem are indicated in the figure by continuous arcs.

Let us show that

$$QP \perp AQ'. \qquad (1)$$

Indeed, we have $NC = AC$; but $AC = PC$ since ACP is an isosceles triangle. Therefore, $NC = PC$ and, consequently, NCP is also an isosceles triangle and hence

$$\angle CNP = \angle NPC = 80°.$$

Now it readily follows that $\angle Q'NP = 180° - 60° - 80° = 40°$, and since $\angle NQ'P = 40°$, the triangles $QQ'P$ and QNP are congruent which implies (1). Now it is clear that $\angle Q'PQ = 50°$ and, consequently,

$$\angle QPA = 180° - 50° - 50 = 80°.$$

Second solution (see Fig. 64). It is easily seen that the angle P is equal to 80° if and only if ΔABP is similar to ΔPCQ (the angles whose values are directly implied by the conditions of the problem are indicated in the figure by continuous arc). Let us prove that these triangles are in fact similar. The angles ABP and PCQ being equal, it is sufficient to establish the relation

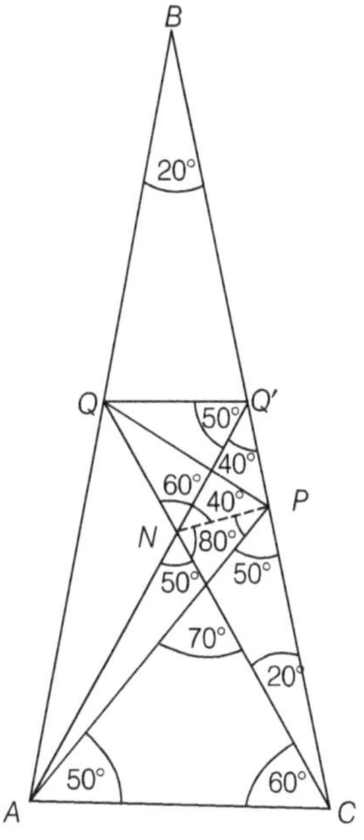

FIG. 63

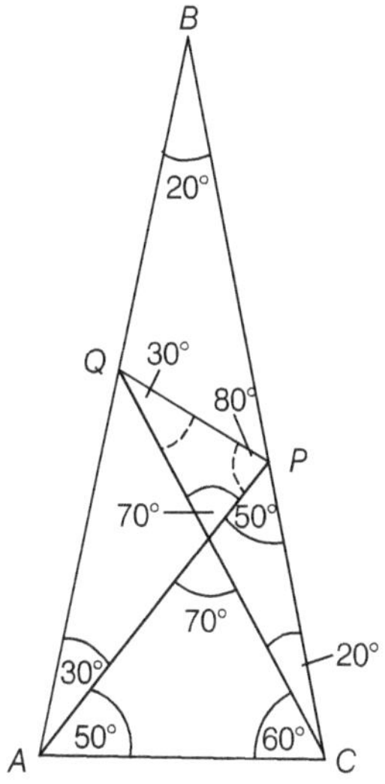

FIG. 64

$$\frac{AB}{CQ}=\frac{PB}{CP}. \tag{1}$$

Put $AB = l$; then from the isosceles triangle CQB we find

$$CQ=\frac{l}{2\cos 20°}.$$

On the other hand, since $PC = AC$, we have

$PC = 2l\sin 10°$, and , besides, $BP = l - 2l\sin 10°$.

Substituting these expressions into (1) we get the equivalent equality

$$4\sin 10°\cos 20° = 1-2\sin 10°. \tag{2}$$

The validity of (2) is readily revealed by noting that

$$\sin 10°\cos 20°=\frac{\sin(10°+20°)+\sin(10°-20°)}{2}=\frac{1}{4}-\frac{1}{2}\sin 10°.$$

349. Let ΔABC be given (Fig. 65). Lay off $AD = c$ on the extension of the side AC. From the equality $a^2 = b^2 + bc$ it follows that $\frac{a}{b} = \frac{b+c}{a}$, which means that the triangles CAB and CBD are similar, and $\angle A = \angle CBD$. Furthermore, $\angle B = \angle BDA = \angle DBA$. Consequently, $\angle A = \angle B + \angle DBA = 2\angle B$.

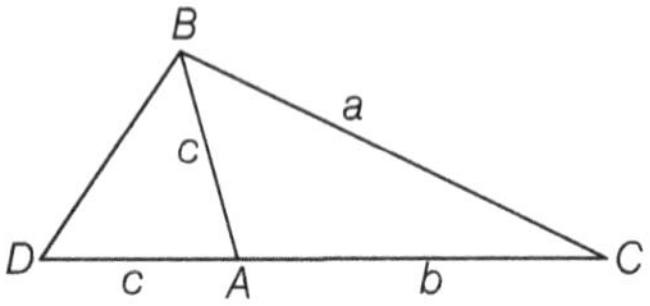

FIG.65

350. Let OC be a median in ΔOAB_1. Let a point D lie on the extension of OC so that $OC = CD$ (see Fig. 66). We shall show that $\Delta AOD = \Delta OA_1B$. Indeed, $AO = OA_1$ by construction. Furthermore, since AOB_1D is a parallelogram, we have $AD = OB_1 = OB$. Lastly, $\angle OAD = \angle A_1OB$ because the sides of these angles are mutually perpendicular: $AO \perp OA_1$ and $OB_1 \perp OB$ by construction, and $AD \| OB_1$. Consequently, $\Delta AOD = \Delta OA_1B$, and two sides of one of them are respectively perpendicular to two sides of the other. Therefore their third sides are also perpendicular, i.e. $OD \perp A_1B$.

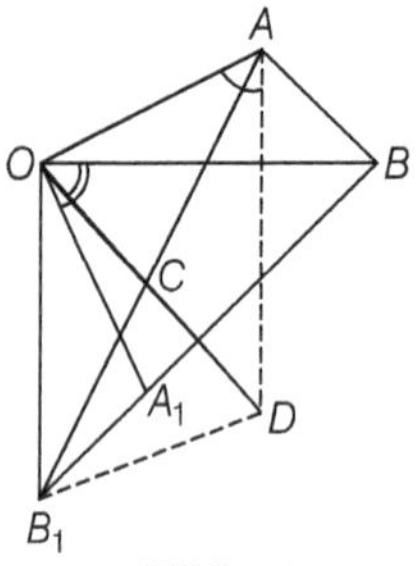

FIG.66

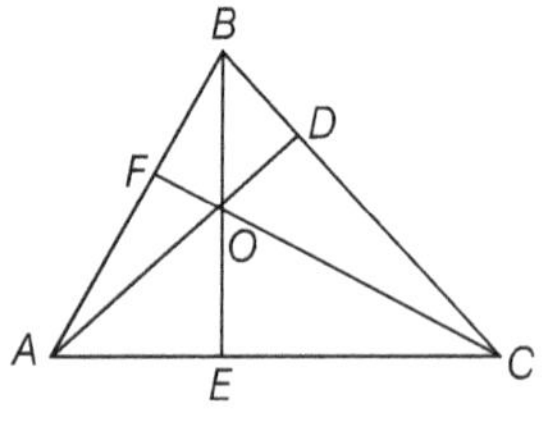

FIG.67

351. Let ABC be an acute triangle, and AD, BE and CF be its altitudes which intersect at a point O (Fig. 67). Each of the quadrilaterals $BDOF$, $CEOD$ and $AFOE$ is inscribed in a circle. According to the theorem on the product of a secant of a circle by its outer portion, we have

$$AD \cdot AO = AB \cdot AF = AC \cdot AE, BE.BO = BC \cdot BD = BA \cdot BF,$$
$$CF \cdot CO = CA \cdot CE = CB \cdot CD.$$

Adding together these equalities we obtain

$$\begin{aligned}2(AD \cdot AO + BE \cdot BO + CF \cdot CO) &= AB \cdot AF + BC \cdot BD \\ &\quad + CA \cdot CE + AC \cdot AE + BA \cdot BF + CB \cdot CD \\ &= AB(AF + BF) + BC(BD + CD) + CA(CE + AE) \\ &= (AB)^2 + (BC)^2 + (CA)^2,\end{aligned}$$

which is what we set out to prove. In the case of an obtuse triangle the product corresponding to the obtuse angle should be taken with the minus sign.

352. By the hypothesis, $b-a=c-b$, i.e. $a+c=2b$. To compute the product Rr let us use the formulas expressing the area S of a triangle in terms of the radii of its circumscribed or inscribed circle and its side. As is known, $S=\frac{1}{2}bc\sin A$, and according to the law of sines we have $\sin A=\frac{a}{2R}$ which implies $S=\frac{abc}{4R}$.

On the other hand, $S=rp$, where $p=\frac{a+b+c}{2}$. Equating both expressions we obtain

$$rR=\frac{abc}{4p}. \qquad (1)$$

Under the conditions of the problem we have $p=\frac{a+b+c}{2}=\frac{3}{2}b$.

Substituting this value in (1) we obtain $6rR=ac$.

353. Let z be the length of the bisector, and m and n the lengths of the line segments into which the base of the triangle is divided by the bisector

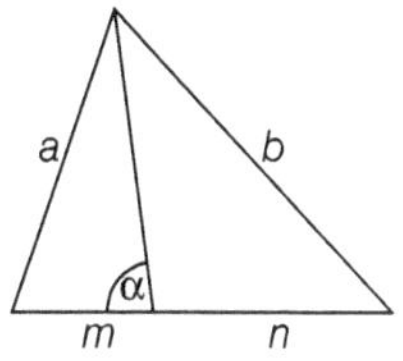

FIG. 68

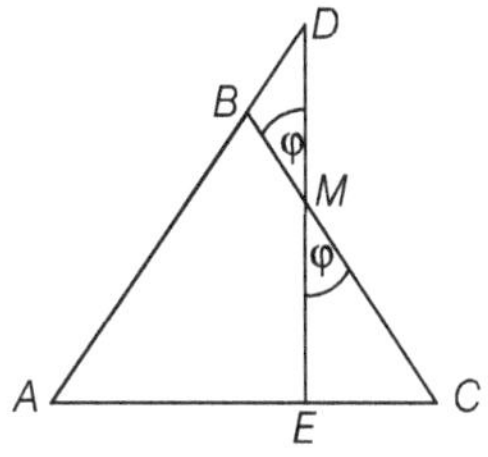

FIG. 69

(Fig. 68). By the law of cosines we have

$$a^2=z^2+m^2-2mz\cos\alpha \text{ and } b^2=z^2+n^2+2nz\cos\alpha.$$

Multiplying the first equality by n and the second one by m, and adding them, we obtain

$$na^2+mb^2=(m+n)(z^2+mn). \qquad (1)$$

By virtue of the relation $\frac{a}{m}=\frac{b}{n}$, we have

$$na^2+mb^2=na\frac{mb}{n}+mb\frac{na}{m}=ab(m+n).$$

Substituting this expression into (1) we obtain the required equation $ab=z^2+mn$.

If $a=b$ and $m=n$, the equality thus proved expresses the Pythagorean theorem: $a^2=z^2+m^2$.

354. By the hypothesis, $BD = EC$ (Fig. 69). If M is the point of intersection of BC and DE, then for the triangles BDM and ECM we obtain

$$\frac{BD}{\sin\varphi} = \frac{DM}{\sin B}, \frac{EC}{\sin\varphi} = \frac{ME}{\sin C},$$

whence it follows that $\dfrac{DM}{ME} = \dfrac{\sin B}{\sin C}$.

But in ΔABC we have $\dfrac{\sin B}{\sin C} = \dfrac{AC}{AB}$.

Consequently, $\dfrac{DM}{ME} = \dfrac{AC}{AB}$.

355. Let BD, BE and BF be, respectively, an altitude, bisector and median in ΔABC. Suppose that $AB < BC$. Then

$$\angle A > \angle C, \angle CBD > \angle ABD,$$

which implies

$$\angle CBD > \frac{1}{2}(\angle ABD + \angle CBD) = \frac{1}{2}\angle B,$$

i.e. $\angle CBD > \angle CBE$. Consequently, the bisector BE passes inside $\angle CBD$, and the point E lies between D and C.

Furthermore, we have $\dfrac{AE}{EC} = \dfrac{AB}{BC} < 1$ and $AE < EC$ whence

$$AE < \frac{1}{2}(AE + EC) = \frac{1}{2}AC,$$

i.e. $AE < AF$. Hence, the point F lies between E and C. Thus, the point E lies between D and F which is what we set out to prove.

356. Consider a triangle ABC. Let BD be a bisector, BM a median and BN the straight line which is the reflection of BM through BD (Fig. 70). If S_{ABN} and S_{MBC} are the areas of the corresponding triangles, then

$$2S_{ABN} = xh_B = nc\sin\angle ABN$$

and

$$2S_{MBC} = \frac{x+y}{2}h_B = ma\sin\angle MBC,$$

where h_B is the altitude dropped from the vertex B onto AC. Since $\angle ABN = \angle MBC$, this implies

$$x = \frac{x+y}{2}\cdot\frac{nc}{ma}. \qquad (1)$$

Similarly, $\quad 2S_{NBC} = yh_B = na\sin\angle NBC$

and $\quad 2S_{ABM} = \dfrac{x+y}{2}h_B = mc\sin\angle ABM.$

Since $\angle NBC = \angle ABM$, it follows that $\quad y = \dfrac{x+y}{2}\cdot\dfrac{na}{mc}. \qquad (2)$

Dividing (1) by (2) termwise we obtain the required proportion

$$\frac{x}{y} = \frac{c^2}{a^2}.$$

357. The straight lines AP, BQ and CR divide the triangle ABC into six triangles: ΔAOR, ΔROB, ΔBOP, ΔPOC, ΔCOQ and ΔQOA (Fig. 71).

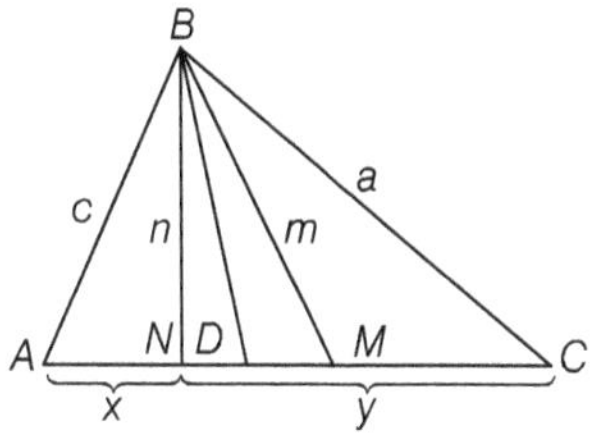

FIG. 70

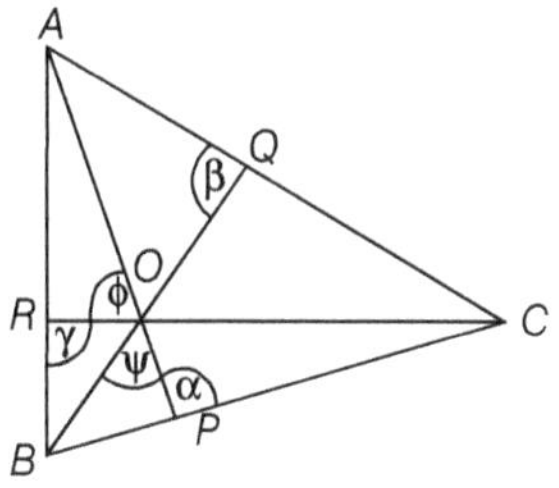

FIG. 71

Applying the law of sines to them we obtain

$$\frac{AR}{\sin\varphi}=\frac{AO}{\sin\gamma},\qquad \frac{AO}{\sin\beta}=\frac{AQ}{\sin\psi},$$

$$\frac{BO}{\sin\gamma}=\frac{BR}{\sin(\varphi+\psi)},\quad \frac{BP}{\sin\psi}=\frac{BO}{\sin\alpha}$$

$$\frac{CQ}{\sin(\varphi+\psi)}=\frac{CO}{\sin\beta},\qquad \frac{CO}{\sin\alpha}=\frac{CP}{\sin\varphi}$$

Multiplying all these equalities termwise we find

$$AR\cdot BP\cdot CQ=BR\cdot AQ\cdot CP.$$

358. Let K and O be, respectively, the centres of the circumscribed and inscribed circles of the triangle ABC, and D the midpoint of the arc AC (see Fig. 72). Each of the angles OAD and AOD is equal to half the sum of the vertex angles at A and B of the triangle ABC. It follows that $OD = AD$.

By the theorem on chords intersecting inside a circle, we have

$$MO\cdot ON=BO\cdot OD.$$

Furthermore, if $OE\perp AB$ and FD is a diameter, the triangles BOE and FDA are similar and therefore $BO{:}OE = FD{:}AD$ which implies $BO\cdot AD = OE\cdot FD$, i.e. $BO\cdot OD = OE\cdot FD$ because $AD = OD$. Hence,

$$MO\cdot ON=OE\cdot FD.$$

Substituting $MO = R + l$, $ON = R - l$, $OE = r$ and $FD = 2R$ in the above equality we arrive at the required result.

$$R^2-l^2=2Rr.$$

359. *First solution.* Let ABC be the given triangle, K_1 the inscribed circle of radius r and K_2 the circumscribed circle of radius R. Let us construct an auxiliary triangle $A_1B_1C_1$ so that its sides are parallel to the sides of ΔABC and pass through its vertices (Fig. 73). Draw tangent lines to the circle K_2 parallel to the sides of $\Delta A_1B_1C_1$ applying the following rule: the tangent line A_2B_2 parallel to the side A_1B_1 is tangent to K_2 at a point belonging to the same arc $\overset{\frown}{AB}$ on which the vertex C lies and so on. The segments of these tangents form a triangle $A_2B_2C_2$.

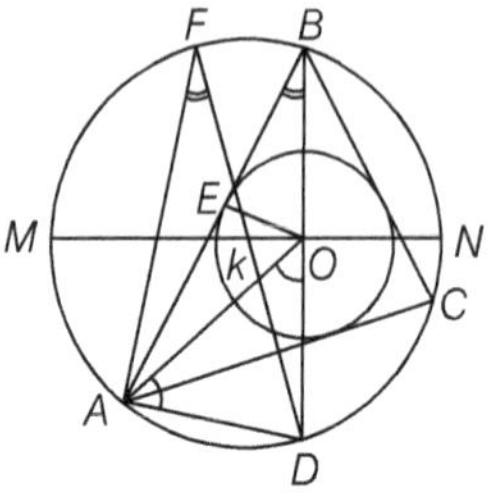

FIG. 72

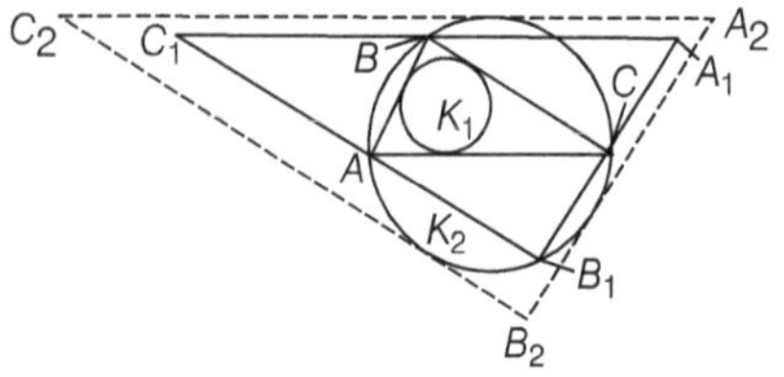

FIG. 73

Then, $\Delta A_1B_1C_1$ lies inside $\Delta A_2B_2C_2$, and the two triangles are similar. Therefore the radius R' of the inscribed circle of $\Delta A_1B_1C_1$ is not greater than the radius R of the inscribed circle K_2 of $\Delta A_2B_2C_2$, i.e. $R' \le R$ On the other hand, the radii of the inscribed circles of the similar triangles $A_1B_1C_1$ and ABC are in the ratio of the corresponding sides of these triangles, i.e. $\frac{A_1B_1}{AB} = 2.$

Thus, $R' = 2r$. Comparing this equality with the inequality $R' \le R$ we finally obtain $2r \le R$.

Second solution Let r and R be the radii of the inscribed and circumscribed circles, S be the area of the given triangle and $p = \frac{a+b+c}{2}$ where a, b and c are the sides. Then

$$\frac{r}{R} = \frac{S}{pR} = \frac{1}{2}\frac{ab\sin C}{pR}$$
$$= \frac{2R\sin A\sin B\sin C}{R(\sin A + \sin B + \sin C)}.$$

But

$$\sin A + \sin B + \sin C = 2\sin\frac{A+B}{2}\cos\frac{A-B}{2} + 2\sin\frac{A+B}{2}\cos\frac{A+B}{2}$$
$$= 4\sin\frac{A+B}{2}\cos\frac{A}{2}\cos\frac{B}{2}$$
$$= 4\cos\frac{A}{2}\cos\frac{B}{2}\cos\frac{C}{2}, \text{ and, consequently,}$$
$$\frac{r}{R} = 4\sin\frac{A}{2}\sin\frac{B}{2}\sin\frac{C}{2}.$$

The problem is thus reduced to proving the inequality

$$\sin\frac{A}{2}\sin\frac{B}{2}\sin\frac{C}{2}\le\frac{1}{8}$$

(see Problem 644).

Third solution. From the formula $l^2 = R^2 - 2Rr$ proved in the foregoing problem it follows that $R^2 - 2Rr \ge 0$ whence we obtain $R \ge 2r$.

360. Let a and b be the lengths of the legs and c the length of the hypotenuse. Comparing two expressions for the area of a triangle, we get

$$S = \frac{1}{2}(a+b+c)\,r = \frac{1}{2}hc,$$

which implies $$\frac{r}{h} = \frac{c}{a+b+c}. \qquad (1)$$

Since $a + b > c$ we have

$$\frac{r}{h} < \frac{c}{c+c} = 0.5.$$

Furthermore, by virtue of the relationship $c^2 = a^2 + b^2$, the inequality $a^2 + b^2 \ge 2ab$ is equivalent to the inequality $2c^2 \ge (a+b)^2$, i.e. $a + b \le c\sqrt{2}$. Therefore,

$$\frac{r}{h} \ge \frac{c}{c\sqrt{2}+c} = \frac{1}{\sqrt{2}+1} = \sqrt{2}-1 > 0.4.$$

361. Let A, B and C be the angles of an acute triangle, and a, b and c be the sides opposite them. Put $P = a + b + c$. The required relationship follows from the equalities

$$ak_a + bk_b + ck_c = Pr, \qquad (1)$$

and $$(b+c)k_a + (c+a)k_b + (a+b)k_c = PR \qquad (2)$$

because, adding them together, we obtain

$$k_a + k_b + k_c = r + R.$$

Equality (1) holds because its left and right members are equal to the doubled area of the triangle. To prove (2) let us note that

$$k_a = R\cos A, k_b = R\cos B, k_c = R\cos C, \qquad (3)$$

and that

$$b\cos C + c\cos B = a,$$
$$c\cos A + a\cos C = b,$$
$$a\cos B + b\cos A = c,$$

whence, by termwise addition, we obtain the equality

$$(b+c)\cos A + (c+a)\cos B + (a+b)\cos C = P.$$

Multiplying the latter relation by R and making use of (3) we obtain the result coinciding with (2).

362. Let A_2B_2, B_2C_2 and C_2A_2 be the midlines in ΔABC and A_3, B_3 and C_3 the midpoints of the segments AA_1, BB_1, CC_1 (Fig. 74). The points A_3, B_3 and C_3 are on the midlines of ΔABC but not at their endpoints because, if otherwise, at least one of the points A_1, B_1 and C_1 coincides with a vertex of ΔABC. Since any straight line not passing through the vertices of the triangle $A_2B_2C_2$ does not intersect all its sides simultaneously, the points A_3, B_3 and C_3 are not in a straight line.

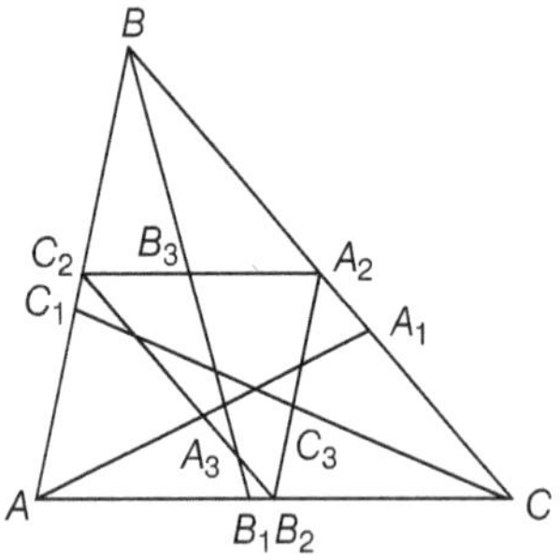

FIG. 74

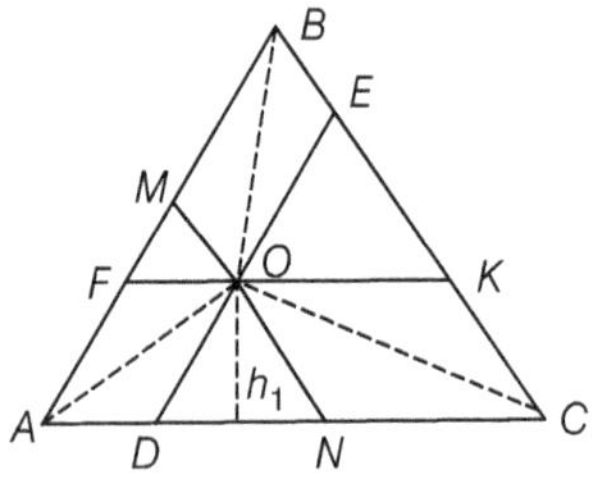

FIG. 75

363. If h_1 is the altitude of ΔDON, h_B the altitude of ΔABC, and S_{AOC} and S_{ABC} are the areas of the corresponding triangles, then (see Fig. 75) we have

$$\frac{S_{AOC}}{S_{ABC}} = \frac{h_1}{h_B} = \frac{OD}{AB} = \frac{AF}{AB},$$

and, similarly, $$\frac{S_{AOB}}{S_{ABC}} = \frac{BE}{BC}, \frac{S_{COB}}{S_{ABC}} = \frac{CN}{CA}.$$

Adding together these equalities we obtain

$$\frac{AF}{AB} + \frac{BE}{BC} + \frac{CN}{CA} = \frac{S_{AOC} + S_{BOC} + S_{AOB}}{S_{ABC}} = \frac{S_{ABC}}{S_{ABC}} = 1.$$

364. (1) Consider the inscribed circle K' of the square. Let its radius be r'. Draw the tangent lines $A'B' \parallel AB$ and $B'C' \parallel BC$ to the circle K' (Fig. 76).

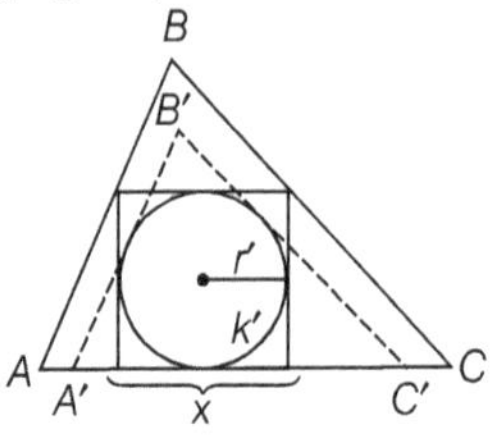

FIG. 76

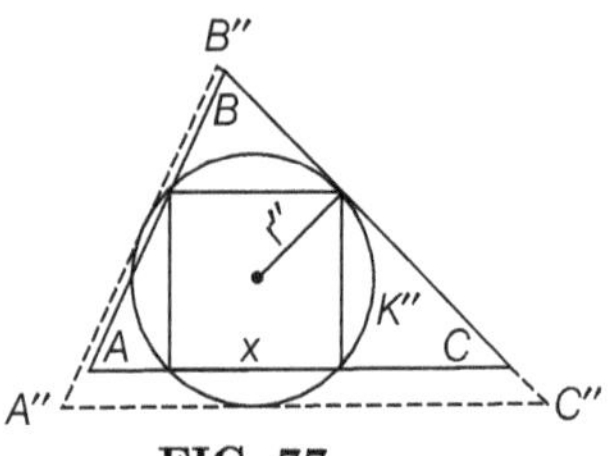

FIG. 77

It is clear that $\Delta A'B'C'$ lies inside ΔABC, and therefore $A'C' < AC$. Since the triangles $A'B'C'$ and ABC are similar, we have $\frac{r'}{r} = \frac{A'C'}{AC} < 1$ which implies $x = 2r' < 2r$.

(2) Consider the circumscribed circle K'' of the square. Let its radius be r''. Draw the tangent lines $A''B'' \parallel AB, B''C'' \parallel BC$ and $A''C'' \parallel AC$ to the circle K'' (Fig. 77). As is clear, ΔABC lies inside $\Delta A''B''C''$ and therefore $A''C'' > AC$. Since $\Delta A''B'C''$ is similar to ABC we have $\frac{r''}{r} = \frac{A''C''}{AC} > 1$, whence it follows that

$$x = \sqrt{2}r'' > \sqrt{2}r.$$

365. Let the point M be the point of intersection of the altitudes AA_1, BB_1 and CC_1 in ΔABC, P be the centre of the circumscribed circle of radius R, C_2, A_2 and B_2 the midpoints of the sides AB, BC and $AC, OM = OP$, $ON \perp AC$ and A_3, B_3 and C_3 the midpoints of AM, BM and CM (Fig. 78). Let

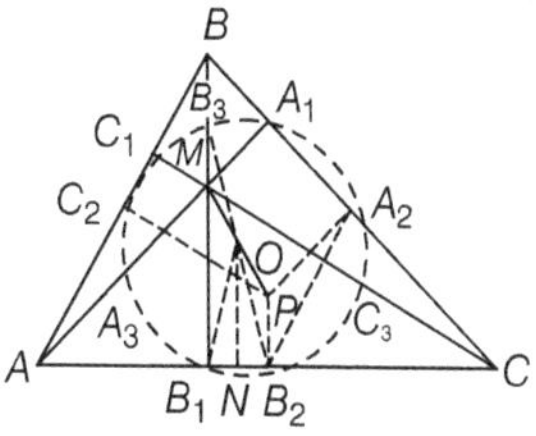

FIG.78

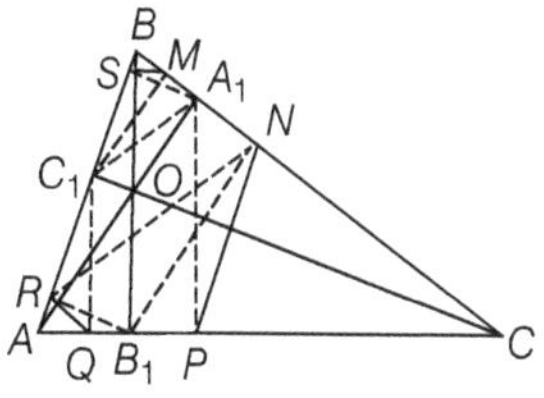

FIG. 79

us prove that the point O is equidistant from A_i, B_i and C_i where $i = 1, 2, 3$. Since ON is the midline of the trapezoid MB_1B_2P, we have $OB_1 = OB_2$. From the similarity of the triangles AMB and PA_2B_2 we conclude that $BM = 2PB_2$, and therefore $B_3M = PB_2$. From the parallelogram MB_3PB_2 we have $OB_3 = OB_2$. But for OB_3 as midline of the triangle PMB we have

$$OB_3 = \frac{1}{2}BP = \frac{R}{2} \text{ and, hence, } OB_3 = OB_2 = OB_1 = \frac{R}{2}.$$

We then prove in just the same way that

$$OA_1 = OA_2 = OA_3 = OC_1 = OC_2 = OC_3 = \frac{R}{2}.$$

366. In ΔABC let AA_1, BB_1 and CC_1 be the altitudes whose point of intersection is $O, C_1M \parallel B_1N \perp BC, A_1P \parallel C_1Q \parallel AC$ and $B_1R \parallel A_1S \perp AB$ (Fig. 79).

(1) Let us prove that $SM \parallel AC$. The triangles BA_1A and BC_1C are similar as right triangles with a common acute angle ABC.

Therefore, $$\frac{BA_1}{BC_1} = \frac{BA}{BC}.$$

Hence, ΔA_1BC_1 is similar to ΔABC and $\angle BA_1C_1 = \angle BAC$. In ΔA_1BC_1 the line segments A_1S and C_1M are altitudes. Therefore, repeating the above argument we can assert that $\angle BSM = \angle BA_1C_1$. Consequently, $\angle BSM = \angle BAC$ and $SM \parallel AC$. We then similarly prove that $PN \parallel AB$ and $RQ \parallel BC$.

(2) To prove that the vertices of the hexagon $MNPQRS$ lie in a circle it is sufficient to show that any four consecutive vertices of the hexagon are in a circle. This follows from the fact that through three points not in a straight line it is possible to draw only one circle. The sets of four consecutive vertices of the hexagon can be classified into the following two types: those in which the intermediate points are on different sides of ΔABC (i.e. $RSMN$, $MNPQ$ and $PQRS$ and those in which the intermediate points are on one side of ΔABC (i.e. $NPQR$, $QRSM$ and $SMNP$).

Consider the quadruples $RSMN$ and $NPQR$ (which belong to different types). From the obvious proportion
$$\frac{BC_1}{BR} = \frac{BO}{BB_1} = \frac{BA_1}{BN}$$
it follows that $NR \parallel A_1C_1$. Therefore,
$$\angle MNR = \angle BA_1C_1 = \angle BAC = BSM.$$
which means that $\angle MNR + \angle MSR = \pi$ and, consequently, the points R, S, M and N lie in one circle. Furthermore, $\angle PNR + \angle PQR = \pi - (\angle PNC + \angle BNR) + \pi - \angle AQR = 2\pi - (\angle ABC + \angle BAC + \angle ACB) = \pi$, whence, it follows that the points N, P, Q and R also lie in a circle. The proof for the rest of the quadruples is carried out in a similar way.

367. Let A_1, B_1 and C_1 be the points of tangency of the inscribed circle and the sides of ΔABC, and D the centre of the inscribed circle (Fig. 80). The segments of the tangent lines drawn from one point to a circle being equal, we have $CA_1 = CB_1, BA_1 = BC_1, AB_1 = AC_1$.

Furthermore, $DB_1 = CA_1$, $B_1C = A_1D$.

Consequently, $AC + BC = CA_1 + A_1B + CB_1 + B_1A$
$= B_1D + A_1D + BC_1 + AC_1 = 2r + 2R$, where r and R are the radii of the inscribed and circumscribed circles.

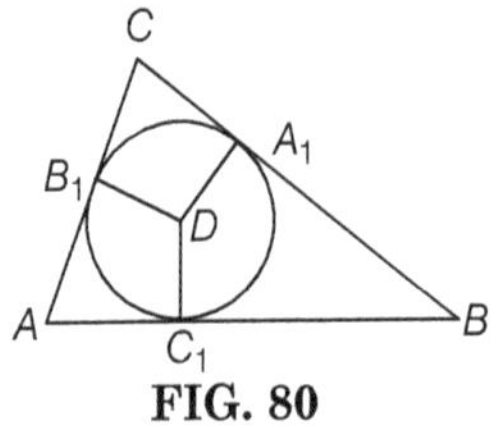

FIG. 80

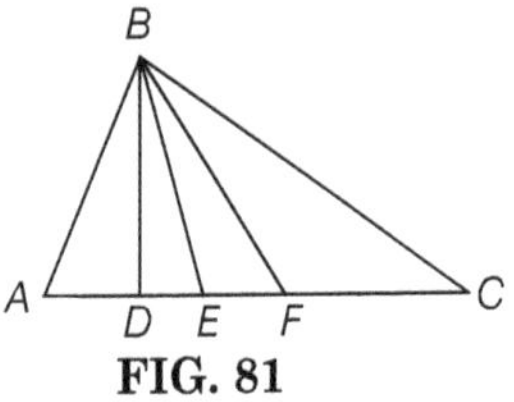

FIG. 81

368. Let in ΔABC the angle ABC be right, BD be the altitude, BE the bisector and BF the median (Fig. 31). Since $BF = FC$, we have $\angle CBF = \angle ACB$.

But $$\angle ABD = \frac{\pi}{2} - \angle BAD = \angle ACB.$$

Hence, $\angle ABD = \angle CBF$ and $\angle DBE = \angle ABE - \angle ABD$ $= \angle CBE - \angle CBF = \angle FBE$, which is what we set out to prove.

369. The symmetry of ABC and $A_1B_1C_1$ about the centre O of the inscribed circle implies that the corresponding points of ΔABC and $\Delta A_1B_1C_1$ lie on a straight line passing through O and are equidistant from this point (Fig. 82). In Particular, $OC = OC_1$, $OB = OB_1$ and BCB_1C_1 is a parallelogram; hence, $BC = B_1C_1$. Analogously, $AC = A_1C_1$ $AB = A_1B_1$ and $ABC = \Delta\, A_1B_1C_1$. Considering the parallelograms ABA_1B_1, BDB_1D_1, ACA_1C_1 and ECE_1C_1 we conclude that $AD = A_1D_1$, $AE = A_1E_1$, and since $\angle A = \angle A_1$, we see that $\Delta ADE = \Delta A_1D_1E_1$. Similarly, $\Delta B_1EK_1 = \Delta BE_1K$ and $\Delta DC_1K = \Delta D_1CK_1$.

Let us denote by S the area of ΔABC, by S_1 the area of ΔADE, by S_2 the area of ΔDC_1K, by S_3 the area of ΔKBE_1. Put $AB = c$, $BC = a$ and $AC = b$, and let h_A, h_B and h_C be the altitudes drawn from the vertices A, B and C, respectively. Then we have

$$S = pr = \frac{ah_A}{2} = \frac{bh_B}{2} = \frac{ch_C}{2},$$

Let AM (AN) be the altitude in ΔADE (in ΔABC). Then

$$S_1 = \frac{DE \cdot AM}{2}.$$

The similarity of the triangles ABC and ADE implies that

$$DE = \frac{a(h_A - 2r)}{h_A}.$$

Hence, $$S_1 = \frac{a(h_A - 2r)^2}{2h_A}$$

$$= \frac{a\left(\frac{2pr}{a} - 2r\right)^2}{2h_A} = \frac{r^2(p-a)^2}{S}.$$

Analogously, $S_2 = \dfrac{r^2(p-c)^2}{S}$, $S_3 = \dfrac{r^2(p-b)^2}{S}$.

Using Heron's formula we obtain

$$S^2S_1^2S_2^2S_3^2 = \frac{r^{12}(p-a)^4(p-b)^4(p-c)^4\,S^2}{S^6} = r^{12}\frac{S^4}{p^4} = r^{16}.$$

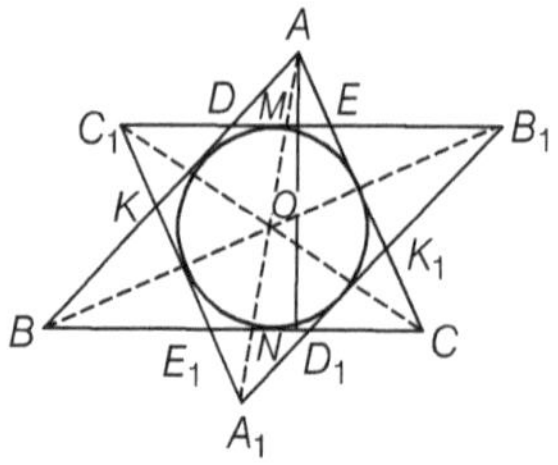

FIG. 82

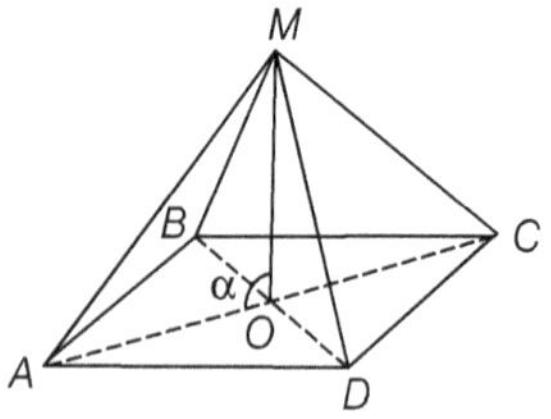

FIG. 83

370. From Fig. 83 we see that

$$MA^2 = MO^2 + AO^2 - 2MO \cdot AO \cos\alpha$$

and $$MC^2 = MO^2 + CO^2 + 2MO \cdot CO \cos\alpha.$$

We have $AO = CO$, and therefore adding these equalities we get

$$MA^2 + MC^2 = 2MO^2 + 2AO^2. \qquad (1)$$

Similarly, $$MB^2 + MD^2 = 2MO^2 + 2BO^2.$$

Consequently, the difference

$$(MA^2 + MC^2) - (MB^2 + MD^2) = 2(AO^2 - BO^2)$$

is independent of the position of point M.

371. Let O be the point of intersection of the straight lines AA_1 and CC_1 (see Fig. 84). The problem reduces to proving that

$$\angle AOB + \angle AOB_1 = 180°. \qquad (1)$$

Note that $\Delta C_1BC = \Delta ABA_1$ because $C_1B = AB, BC = BA_1$ and $\angle C_1BC = 60° + \angle ABC = \angle ABA_1$. Therefore $\angle OC_1B = \angle OAB$, and OAC_1B is an inscribed quadrilateral of a circle. Hence, $\angle AOB = 120°$. We then analogously show that $BOC = 120°$. But this implies that $\angle AOC = 120°$, and it follows that $AOCB_1$ is an inscribed quadrilateral of a circle. Hence it follows that $\angle AOB_1 = \angle ACB_1 = 60°$. Therefore equality (1) holds.

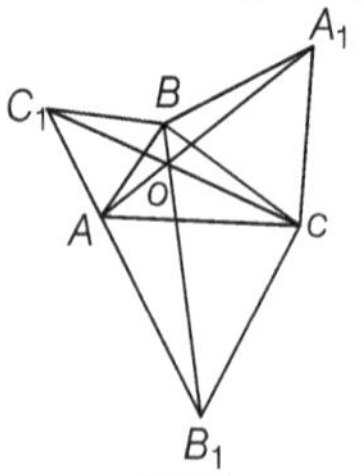

FIG. 84

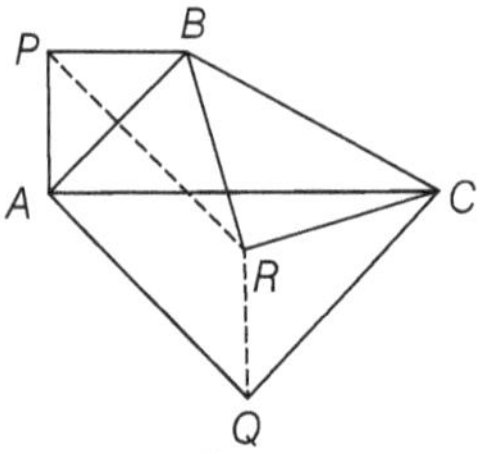

FIG. 85

372. From Fig. 85 we have

$$\angle PBR = \angle ABC$$

and $$\frac{PB}{AB} = \frac{BR}{BC}.$$

Therefore ΔPBR is similar to ΔABC and, analogously, ΔQRC is similar to ΔABC. Hence we obtain

$$\angle APR = \angle APB - \angle BPR = \angle APB - \angle BAC,$$

and thus, $\angle APR + \angle PAQ = \angle APB + 2\angle PAB = \pi$,

that is $PR \parallel AQ$. We similarly prove that $QR \parallel AP$.

373. Let h_B, h_C and h_D be respectively, the distances from the vertices B, C and D of the parallelogram to the straight line AO (Fig. 86). Then the following property takes place: the greatest of the three distances is equal to the sum of the other two. For instance, if AO intersects the side BC (as in Fig. 86), then, drawing $BE \parallel AO$ and $CE \perp AO$ we conclude, by the congruence of the triangles BEC and $AD'D$, that $h_D = h_B + h_C$.

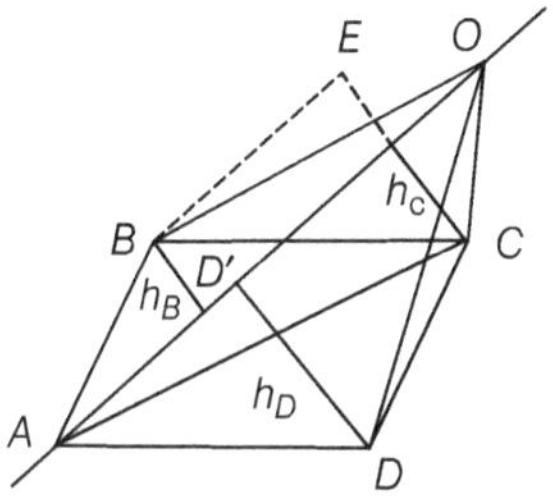

FIG. 86

Analogously, if AO intersects the side CD, then $h_B = h_C + h_D$ and if AO does not intersect the sides BC and CD, then $h_C = h_B + h_D$. From this property, for the case shown in Fig. 86, we immediately receive the equality of the areas of the triangles:

$$S_{AOC} = S_{AOD} - S_{AOB}.$$

Generally, it is obvious, that we can write the formula

$$S_{AOC} = |S_{AOD} \pm S_{AOB}|.$$

where the plus sign is taken if the points B and D lie on one side of AO and the minus sign if these points are on opposite sides of the line.

The same argument can be repeated for the straight line CO which leads to the formula $S_{AOC} = |S_{COD} \pm S_{COB}|$.

where the rule of choosing the sign is obtained from the above by replacing the straight line AO by CO.

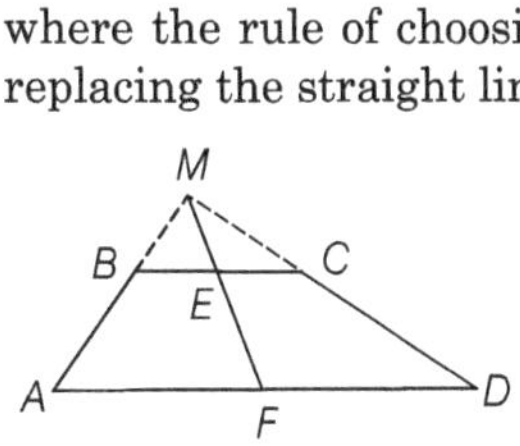

FIG. 87

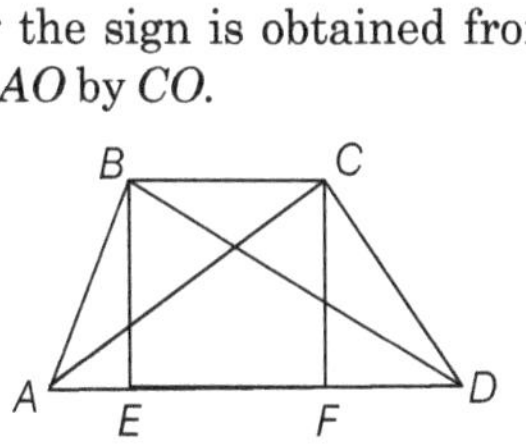

FIG. 88

374. Extend the sides AB and CD of the trapezoid $ABCD$ to obtain the triangle AMB and join M with the midpoint F of the base AD (Fig. 87). Then

$$ME = \frac{BC}{2}, MF = \frac{AD}{2}.$$

Consequently,
$$EF = \frac{AD - BC}{2}.$$

375. Let $ABCD$ be the given trapezoid with bases AD and BC and let $BE \perp AD, CF \perp AD$ (Fig. 88). We have

$$AC^2 - AF^2 = CD^2 - FD^2,$$
$$BD^2 - ED^2 = AB^2 - AE^2.$$

Adding these equalities we get

$$AC^2 + BD^2 = AB^2 + CD^2 + AF^2 - FD^2 + ED^2 - AE^2$$
$$= AB^2 + CD^2 + AD(AF - FD + ED - AE)$$
$$= AB^2 + CD^2 + AD \cdot 2EF = AB^2 + CD^2 + 2AD \cdot BC.$$

376. Let $ABCD$ be the given trapezoid with parallel sides AD and BC, E being the midpoint of BC and F the midpoint of AD. Denote by O the

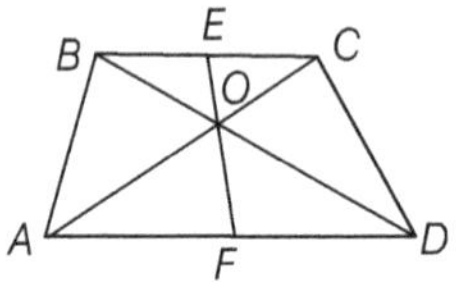

FIG. 89

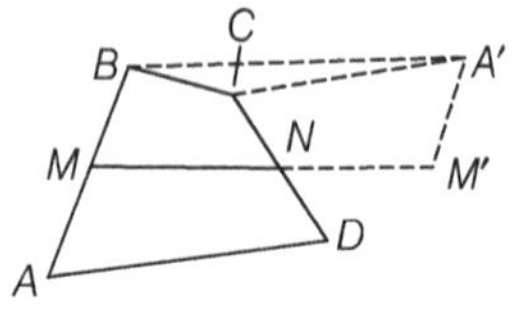

FIG. 90

point of intersection of the diagonals (Fig. 89). The triangles AOF and COE are similar (this is implied by the similarity of the triangles AOD and COB). Therefore $\angle AOF = \angle COE$, i.e. EOF is a straight line.

377. Let $ABCD$ be the given quadrilateral, M and N being the midpoints of the sides AB and CD, respectively (see Fig. 90). Turn the quadrilateral $AMND$ through 180° in its plane about the vertex N. Then the vertex D coincides with C and the vertices M and A occupy the positions M' and A' respectively. Furthermore, the points M, N and M' lie in a straight line and besides, $M'A' \parallel MB$ and $M'A' = MB$. Therefore $MBA'M'$ is a parallelogram, and $A'B = M'M = 2MN$. By the hypothesis we have $BC + AD = 2MN$, and therefore $BC + CA' = A'B$. Consequently, the point C lies on the line segment $A'B$ because, if otherwise, $BC + CA' > A'B$ in the $\Delta BCA'$. It follows that $BC \parallel MN \parallel AD$, i.e. $ABCD$ is a trapezoid.

378. Let us express the area of a quadrilateral in terms of the diagonals and the angle between them. Let O be the point of intersection of the diagonals of a quadrilateral $ABCD$ shown in Fig. 91, and $\angle BOA = \alpha$.

Then

$$S_{ABCD} = S_{AOB} + S_{COD} + S_{AOD} + S_{BOC} =$$
$$= \frac{1}{2} AO \cdot OB \cdot \sin\alpha + \frac{1}{2} OC \cdot OD \cdot \sin\alpha + \frac{1}{2}$$
$$BO \cdot OC \cdot \sin\alpha + \frac{1}{2} AO \cdot OD \sin\alpha$$
$$= \frac{1}{2} BD \cdot AC \cdot \sin\alpha.$$

This formula implies validity of the assertion to be proved.

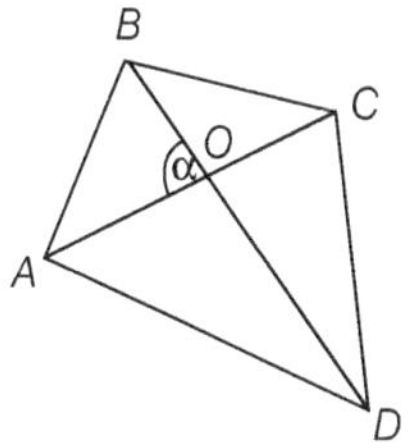

FIG. 91

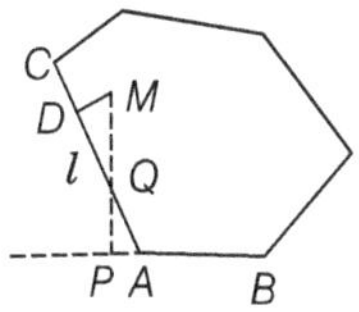

FIG. 92

379. Let M be an interior point of a convex polygon, and AB its side whose distance from M is the least. We shall prove that the foot P of the perpendicular drawn from M to AB lies on AB but not on its extension (Fig. 92). Indeed, if P lies on the extension of AB, then MP intersects a side l of the polygon at a point Q, and, since the polygon is convex, $MQ < MP$. But the distance DM from M to l is less than MQ, and, consequently, less than MP which contradicts the choice of the side AB.

380. Let AA_1, BB_1, CC_1 and DD_1 be the bisectors of the interior angles of a parallelogram $ABCD$, and let $PQRS$ be the quadrilateral formed by their intersection (Fig. 93). Obviously, $BB_1 \parallel DD_1$ and $AA_1 \parallel CC_1$. Furthermore, $\angle APB = \pi - (\angle BAP + \angle ABP)$
$= \pi - \frac{1}{2}(\angle BAD + \angle ABC) = \pi - \frac{1}{2}\pi = \frac{1}{2}\pi$, which means that $PQRS$ is a rectangle. The triangles BAB_1 and CDC_1 are isosceles because the bisectors of their vertex angles are perpendicular to their bases. Therefore $BP = PB_1$ and $D_1R = RD$, and hence $PR \parallel AD$. Thus, $PRDB_1$ is a parallelogram, and we have

$$PR = B_1D = AD - AB_1 = AD - AB.$$

381. Let O_1, O_2, O_3 and O_4 be the centres of the squares constructed on the sides of a parallelogram $ABCD$ (Fig. 94). We have $\Delta O_1BO_2 = \Delta O_3CO_2$ since $O_1B = O_3C, BO_2 = CO_2$ and $\angle O_1BO_2 = \angle MBN + \frac{\pi}{2} = \angle DCB + \frac{\pi}{2}$
$= \angle O_3CO_2$.

Hence, $O_1O_2 = O_3O_2$ and $\angle O_1O_2O_3 = \angle O_1O_2B + \angle BO_2C - \angle O_3O_2C$

$$= \angle BO_2C = \frac{\pi}{2}.$$

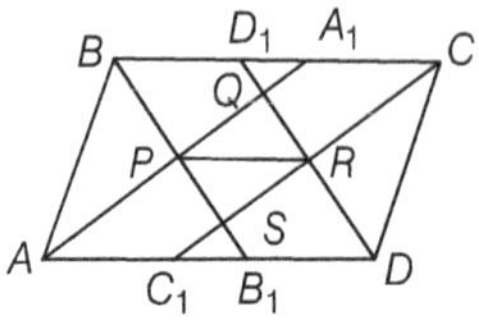

FIG. 93

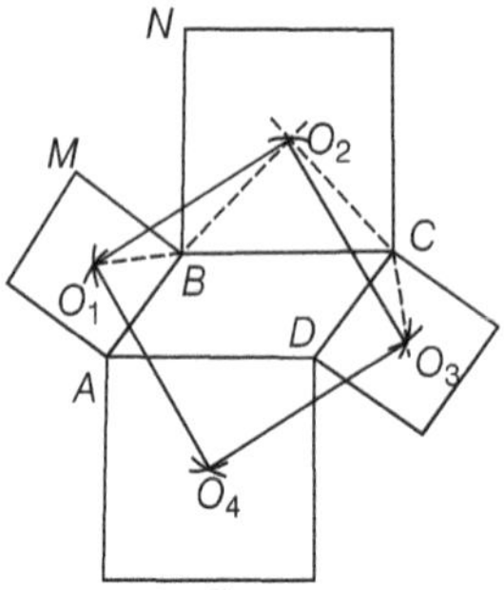

FIG. 94

We similarly prove that $O_2O_3 = O_3O_4 = O_4O_1$ and

$$\angle O_2O_3O_4 = \angle O_3O_4O_1 = \angle O_4O_1O_2 = \frac{\pi}{2}.$$

Consequently, $O_1O_2O_3O_4$ is a square.

382. Let AP, BQ, CR and DS be the bisectors of the interior angles of the quadrilateral $ABCD$ (Fig. 95). Let A, B, C and D be the magnitudes of these angles. Then

$$\angle ASD = \pi - \frac{1}{2}A - \frac{1}{2}D, \angle BQC = \pi - \frac{1}{2}B - \frac{1}{2}C.$$

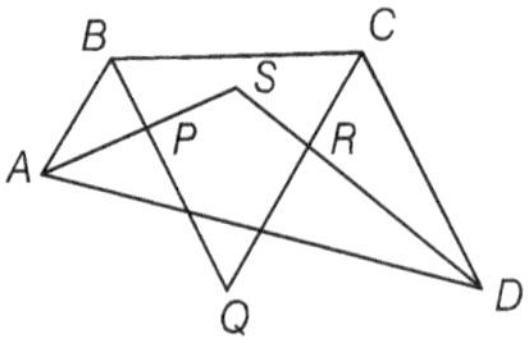

FIG. 95

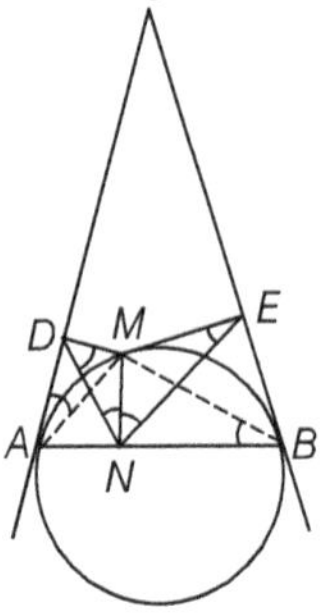

FIG. 96

Adding together these equalities we obtain

$$\angle ASD + \angle BQC = 2\pi - \frac{1}{2}(A + B + C + D)$$

$$= 2\pi - \frac{1}{2}2\pi = \pi.$$

Hence, the points P, Q, R and S lie in a circle.

383. Let A and B be the points of tangency, M an arbitrary point of the circle, and $MN \perp AB$, $MD \perp AC$, $ME \perp BC$ (see Fig. 96). Let us prove that the triangles DMN and NME are similar. To this end we note that about the quadrilaterals $ADMN$ and $NMEB$ it is possible to circumscribe circles because $\angle MNA + \angle ADM = \frac{\pi}{2} + \frac{\pi}{2} = \pi$

and $$\angle MEB + \angle BNM = \frac{\pi}{2} + \frac{\pi}{2} = \pi.$$

Therefore, $\angle MND = \angle MAD$ and $\angle MEN = \angle MBN$. But $\angle MAD = \angle MBN$, because each of these angles is measured by half the arc AM. Thus, $\angle MND = \angle MEN$. We similarly establish the equality $\angle NDM = \angle ENM$. The similarity of the triangles DMN and NME implies the required relationship $\frac{DM}{MN} = \frac{MN}{ME}$.

384. Let ABC be an inscribed triangle of a circle. Denote by D an arbitrary point of the circle, and by L, M and N the feet of the perpendiculars (Fig. 97). Join the point M to N and the point N to L. We shall prove that the angles ANM and LNC are equal.

First note that

$$\angle ANM = \angle ADM, \qquad (1)$$

because about the quadrilateral $MAND$ it is possible to circumscribe a circle.

By the same reason,

$$\angle LNC = \angle LDC. \qquad (2)$$

On the other hand, we have $\angle ADC = \angle MDL$. (3)

Indeed, $\angle ADC + \angle B = 180°$ because the sum of these angles can be thought of as an angle inscribed in the circle subtended by the whole circumference of the circle. At the same time $\angle MDL + \angle B = 180°$ because about the quadrilateral $MBLD$ it is possible to circumscribe a circle. Consequently, equality (3) holds true. As is clear from the figure, in this case we have $\angle LDC = \angle ADM$, and then (1) and (2) imply the validity of the required equality $\angle ANM = \angle LNC$.

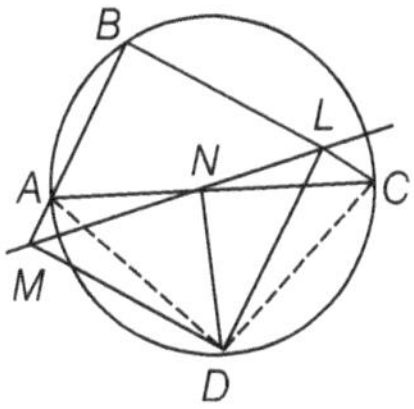

FIG. 97

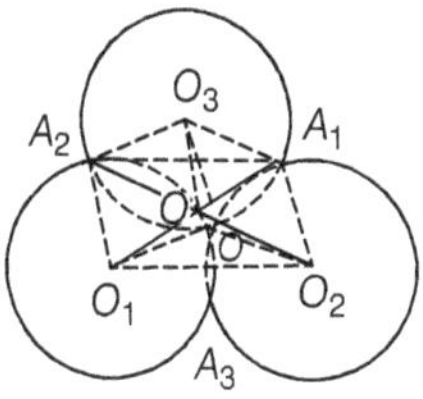

FIG. 98

385. Let us prove that every two of the three line segments O_1A_1, O_2A_2 and O_3A_3 shown in Fig. 98 intersect at their midpoints. This will imply that all the three line segments intersect in one point. For example, we shall prove that the line segments O_1A_1 and O_2A_2 are bisected by the point B of their intersection. Since the circles are equal, we conclude that $O_2A_1O_3O$ and $O_1A_2O_3O$ are rhombuses. It follows that the line segments O_2A_1, OO_3 and O_1A_2 are parallel and equal. Therefore, $O_1A_2A_1O_2$ is a parallelogram and its diagonals O_1A_1 and O_2A_2 are bisected at the point of intersection.

386. Let O be the centre of the smaller circle (Fig. 99). Then $AK \parallel OC$ since $AK \perp BK$, and $OC \perp BK$. Furthermore, $OA = OC$. Hence,

$$\angle KAC = \angle ACO = \angle CAO.$$

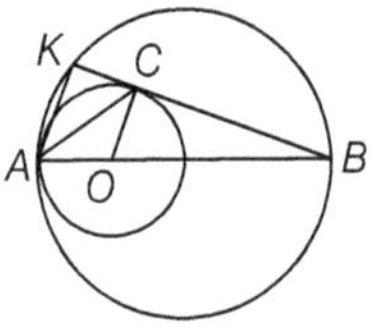

FIG. 99

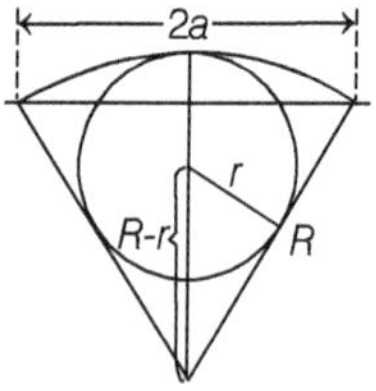

FIG. 100

387. As is clear from Fig. 100, $\dfrac{R-r}{r} = \dfrac{R}{a}$, which is equivalent to the equality $\dfrac{1}{r} = \dfrac{1}{R} + \dfrac{1}{a}$.

388. There are three possible cases here. They are shown in Fig. 101, a, b and c. In the first case the fixed tangents are parallel, $\angle COD = \alpha + \beta = \dfrac{\pi}{2}$, and therefore, $CE \cdot ED = OE^2$, i.e. $AC \cdot BD = r^2$ where r is radius of the circle. In the second and third cases, using the notation indicated in the figure, we find that $\alpha + \beta \pm \gamma = \dfrac{\pi}{2}$, i.e. $\alpha \pm \gamma = \dfrac{\pi}{2} - \beta$, whence it follows that ΔAOC is similar to ΔBDO and

therefore $$\frac{AC}{AO} = \frac{OB}{BD}.$$

Consequently, $$AC \cdot BD = AO^2 = r^2.$$

389. Let M be the point of intersection of mutually perpendicular chords AB and CD (Fig. 102). Draw $AK \parallel CD$, then BK is a diameter, $AK < CD$ and $BK^2 = AB^2 + AK^2 < AB^2 + CD^2$.

Furthermore, $KD = AC$ and hence $KB^2 = BD^2 + KD^2$

$$= BM^2 + DM^2 + AM^2 + CM^2.$$

390. Let $AC = CD = DB$ (Fig. 103). Draw $OE \perp AB$. Then OE is an altitude and OC is a median in ΔAOD. The bisector of $\angle AOD$ lying between the median and altitude (see Problem 355), we have $\angle AOC < \angle COD$.

391. Let AB be a diameter of a circle and E the point of intersection of its chords AD and EC (Fig. 104). We have
$AE \cdot AD = AE^2 + AE \cdot ED = AC^2 + EC^2 + AE \cdot ED.$

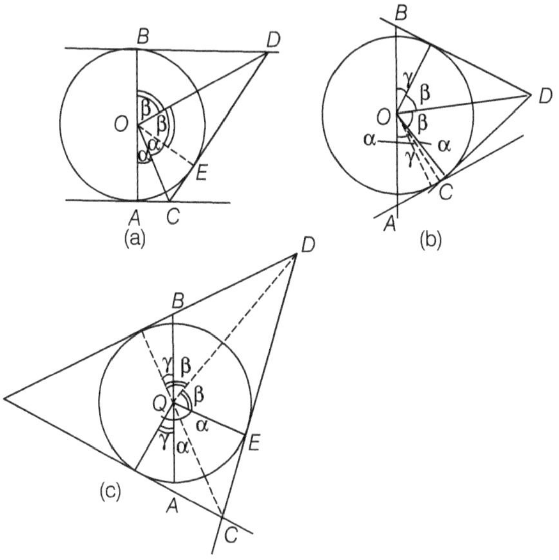

FIG. 101

By the property of intersecting chords, we can write

$$AE \cdot ED = BE \cdot EC.$$

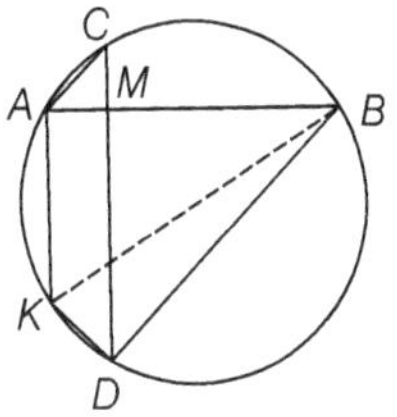

FIG. 102

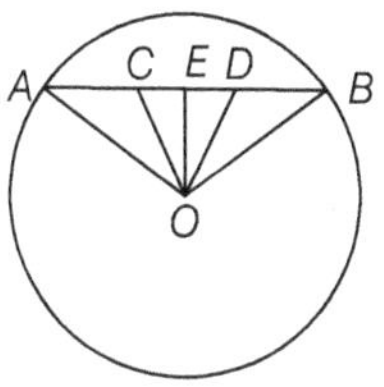

FIG. 103

Therefore,

$$AE \cdot AD = AC^2 + EC^2 + BE \cdot EC$$
$$= AC^2 + EC \cdot BC = AC^2 + (BC - BE)BC$$

$= AC^2 + BC^2 - BE \cdot BC$ and thus, finally, $AE \cdot AD + BE \cdot BC = AB^2$.

392. Let A and B be the given points, O be the centre of the given circle, R its radius and r the common radius of the inscribed circles with centres at

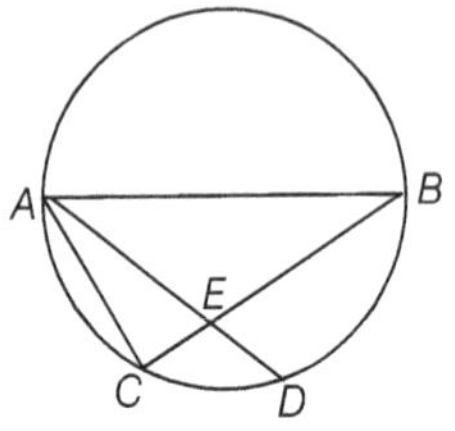

FIG. 104

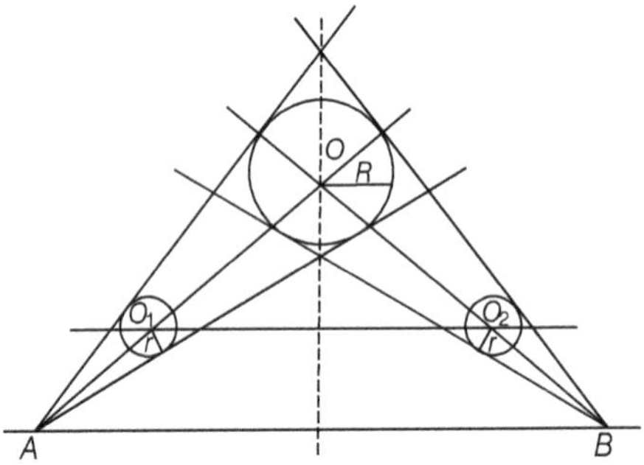

FIG.105

O_1 and O_2 (Fig. 105). Then

$$\frac{R}{r} = \frac{OA}{O_1A} = \frac{OB}{O_2B}.$$

Taking the proportion derived from the above by inversion and addition we obtain $\frac{OA}{OO_1} = \frac{OB}{OO_2}$.

Consequently, $O_1O_2 \parallel AB$.

393. Let r_1 and r_2 be the radii of the semicircles inscribed in a given semicircle of radius R shown in Fig. 106. Since $R = r_1 + r_2$, the shaded area is expressed as $S = \frac{1}{2}\pi R^2 - \frac{1}{2}\pi r_1^2 - \frac{1}{2}\pi r_2^2$

$$= \frac{1}{2}\pi[(r_1 + r_2)^2 - r_1^2 - r_2^2] = \pi r_1 r_2.$$

But $h^2 = 2r_1 \cdot 2r_2 = 4r_1r_2$ and, consequently, $S = \frac{1}{4}\pi h^2$.

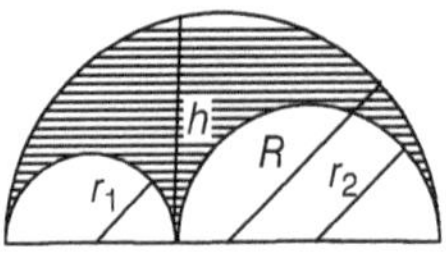

FIG. 106

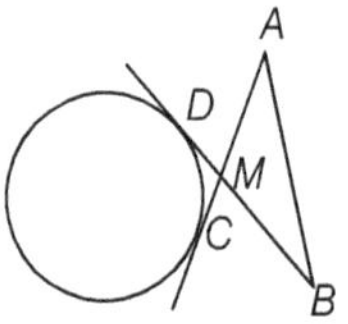

FIG. 107

394. If the straight line joining the points A and B (Fig. 107) *does not intersect* the given circle, then the tangent lines AC and BD can be drawn so that the point M of their intersection lies on the line segments AC and BD. In ΔAMB we have

$AM + BM > AB > |AM - BM|$,

and, since $AC > AM, BD > BM, MC = MD$,
we obtain $AC + BD > AB > |AC - BD|$.

If the straight line AB *does intersect* the circle, then there are two possible cases, namely:(a) the chord cut off by the circle on the straight line AB lies on the line segment AB; (b) the chord is not on AB.

In the case (a) shown in Fig. 108 we have $AB > AE + BF > AC + BD$, because the hypotenuses AE and BF in the right triangles AEC and BFD are greater than the legs AC and BD.

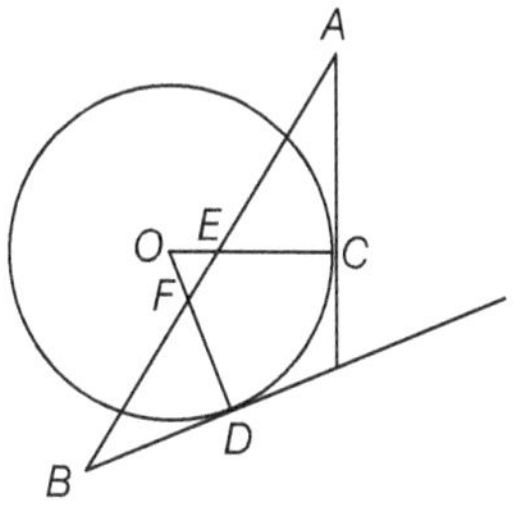

FIG. 108

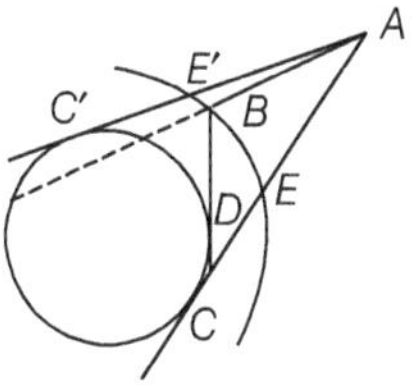

FIG. 109

In the case (b) the line segment AB lies inside the angle CAC' (Fig. 109). Draw through B a circle concentric with the given one. Let it intersect AC and AC' at points E and E'. Then $EC = BD$ and $AE > AB$. Hence, $AB < AE = AC - EC = AC - BD$.

395. Let us introduce the following notation (see Fig. 110):
$\angle PCM = \angle QCN = \alpha$, $\angle NML = \angle NKL = \gamma$, $\angle LCP = \angle QCK = \beta$, $QC = x, PC = y, AC = CB = a$,

By the theorem on intersecting chords of a circle, we have $NQ \cdot QK = AQ \cdot QB = a^2 - x^2$.

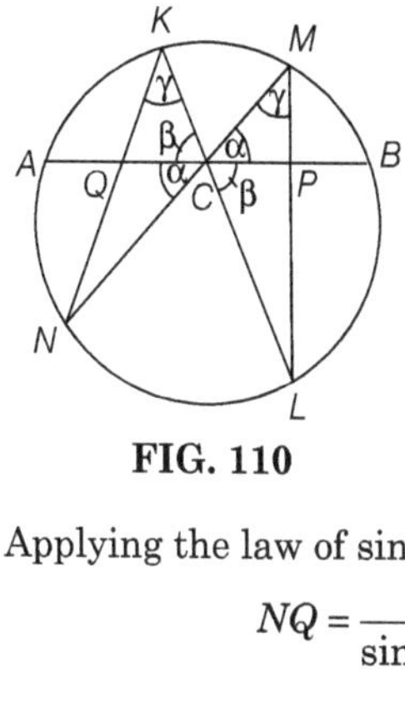

FIG. 110

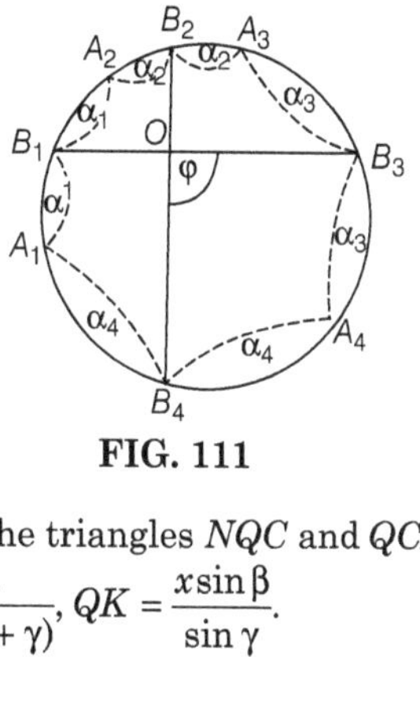

FIG. 111

Applying the law of sines to the triangles NQC and QCK we get

$$NQ = \frac{x \sin\alpha}{\sin(\alpha + \beta + \gamma)}, \quad QK = \frac{x \sin\beta}{\sin\gamma}.$$

Hence, $NQ \cdot QK = \dfrac{x^2 \sin\alpha \cdot \sin\beta}{\sin\gamma \sin(\alpha+\beta+\gamma)} = a^2 - x^2,$

which results in

$$x^2 = \frac{a^2 \sin\gamma \sin(\alpha+\beta+\gamma)}{\sin\alpha \sin\beta + \sin\gamma \sin(\alpha+\beta+\gamma)}.$$

We similarly find $$y^2 = \frac{a^2 \sin\gamma \sin(\alpha+\beta+\gamma)}{\sin\alpha \sin\beta + \sin\gamma \sin(\alpha+\beta+\gamma)}.$$

Thus, $x = y$.

396. Let B_1, B_2, B_3 and B_4 be the midpoints of the arcs A_1A_2, A_2A_3, A_3A_4 and A_4A_1 (Fig. 111). Let α_i be the central angle corresponding to the arc $A_iB_i (i = 1, 2, 3, 4)$. Denote by φ the angle formed by the line segments B_1B_3 and B_2B_4. Then

$$\varphi = \frac{\alpha_1 + \alpha_2 + \alpha_3 + \alpha_4}{2}.$$

But we have $2\alpha_1 + 2\alpha_2 + 2\alpha_3 + 2\alpha_4 = 2\pi$,

and therefore, $$\varphi = \frac{\pi}{2}.$$

397. Consider a closed polygonal line without self-intersection and take two points A and B on it in such a way that the perimeter is divided into two equal parts. Let O be the midpoint of the line segment AB. Draw a circle of radius $\frac{p}{4}$ with centre at O where p is the perimeter of the whole polygonal line.

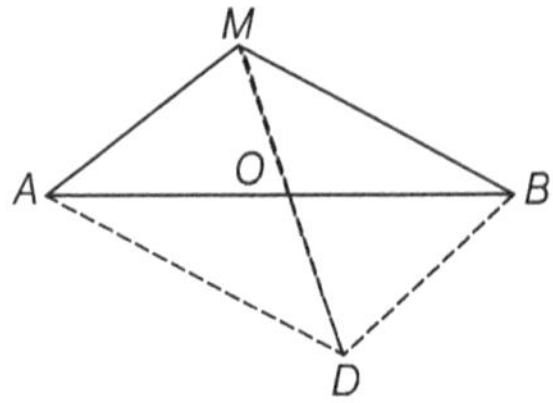

FIG. 112

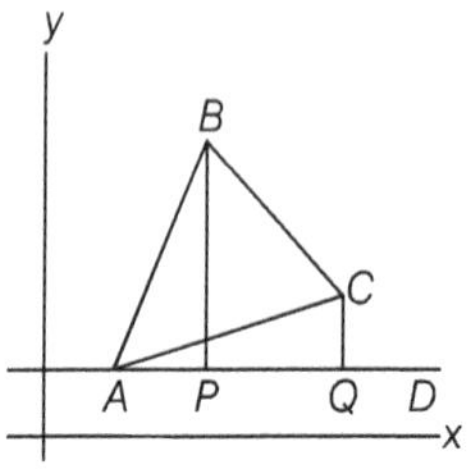

FIG. 113

We shall prove that this circle is a required one. Indeed, if otherwise, then there exists a point M belonging to the polygonal line and lying outside this circle. The length of the portion of the polygonal line containing the point M is not less than $AM + BM$ and hence, $AM + BM \le \frac{p}{2}$. But at the same time $AM + BM \ge 2MO$. Indeed, from the parallelogram $AMBD$ (Fig. 112) we have $DM = 2MO < BM + BD$

$$= AM + BM.$$

Since $MO > \frac{p}{4}$, it follows from the inequality $AM + BM \geq 2MO$ that $AM + BM > \frac{p}{2}$. Thus we arrive at a contradiction.

398. Through the vertex A of a given ΔABC draw a straight line AD parallel to one of the given straight lines x and y and not intersecting the triangle. Drop the perpendiculars BP and CQ to AD from the points B and C (Fig. 113).

Suppose that the distances from the vertices of the triangle ABC to the straight lines x and y are expressed by integers. Then the lengths of the line segments AP, AQ, BP and CQ are also expressed by integers. It follows that

$$\tan \angle BAP = \frac{BP}{AP} \text{ and } \tan \angle CAQ = \frac{CQ}{AQ}$$

are rational numbers, and, hence, the number

$$\tan \angle BAC = \frac{\tan \angle BAP - \tan \angle CAQ}{1 + \tan \angle BAP \tan \angle CAQ} = \frac{\frac{BP}{AP} - \frac{CQ}{AQ}}{1 + \frac{BP}{AP}\frac{CQ}{AQ}}$$

is also rational. Therefore, it is impossible that $\angle BAC = 60°$ because $\tan 60° = \sqrt{3}$ is an irrational number. Consequently, ABC is not an equilateral triangle.

399. Let the straight lines A_1B and AB_1 intersect at a point O, and $OD \perp AB$ (Fig. 114). Since ΔABA_1 is similar to ΔDBO, and ΔBAB_1 to ΔDAO, we have

$$\frac{OD}{a} = \frac{BD}{AB}, \frac{OD}{b} = \frac{AD}{AB}, \text{ which yields } OD\left(\frac{1}{a} + \frac{1}{b}\right) = \frac{AD + BD}{AB} = 1.$$

Hence, the distance $OD = \frac{ab}{a+b}$ is independent of the positions of the point A and B (provided the distances a and b remain unchanged).

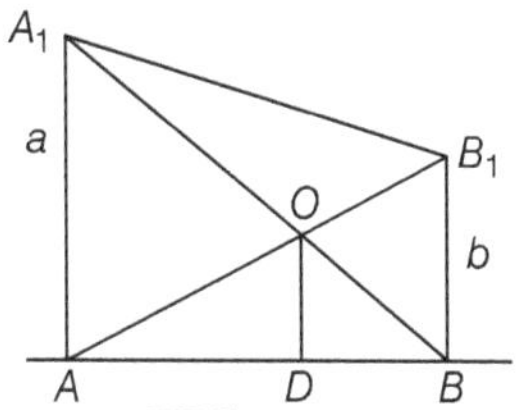

FIG. 114

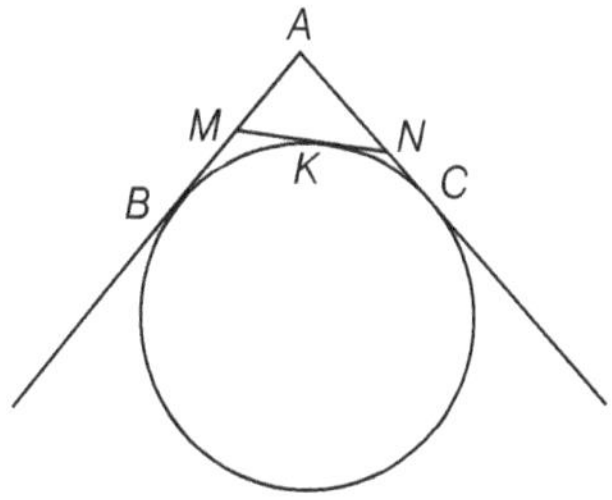

FIG. 115

400. If K is the point of tangency of the line segment MN with the circle (Fig. 115), then $BM = MK$ and $KN = NC$ and consequently

$$MN = BM + CN. \quad (1)$$

But $MN < AM + AN$.

Therefore, $2MN < BM + AM + CN + AN = AB + AC$,

when it follows that $MN < \frac{AB + AC}{2}$.

On the other hand, $MN > AN$ and $MN > AM$ because MN is the hypotenuse in the triangle AMN. Therefore, $2MN > AN + AM$ and, by virtue of (1).

$3MN > AN + NC + AM + MB = AB + AC$.

Hence, $MN > \frac{AB + AC}{3}$.

401. Let ABC be the given triangle (see Fig. 116). $AB = BC$, $BO \parallel AC$ and O be the centre of a circle tangent to AC. Denote by D and E the points of intersection of this circle with AB and BC. Extend AB to intersect the circle a second time at a point F. Let us prove that $FE \perp BO$. Note that $\angle OBF = \angle OBE$, since these angles are equal to the base angles A and C in the triangle ABC. Furthermore, $BF = BE$. Indeed, if $BF > BE$, then laying off on BF the line segment $BE' = BE$ we obtain the congruent triangles OBE and OBE', and $OE' = OE$ which is impossible because the point E' lies inside the

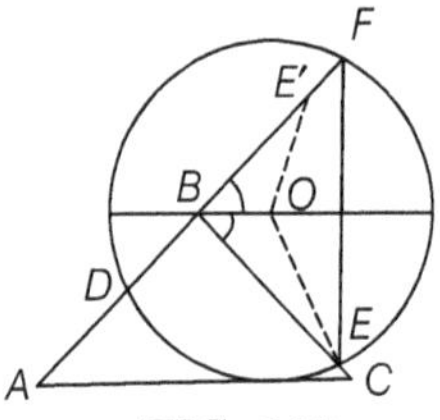

FIG. 116

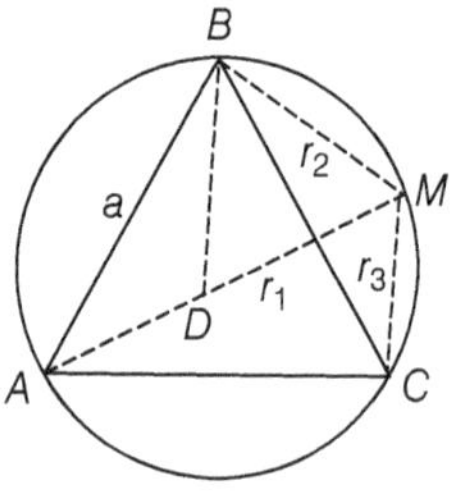

FIG. 117

circle of radius OE. It is similarly proved that the inequality $BF < BE$ is also impossible. Hence, BO is the bisector of the vertex angle in the isosceles triangle FBE and therefore, it is the altitude to its base which implies that $FE \perp BO$.

Therefore, $\angle DFE = \frac{1}{2} \angle ABC$ is independent of the position of the point O on the straight line BO. Consequently, the magnitude of the arc DE subtending the inscribed angle DFE (whose measure is half the arc DE) remains constant as the circle rolls upon AC.

402. Using the notation introduced in the solution of Problem 324 we find
$$n^2 = \frac{ab+cd}{bc+ad}(ac+bd),\ m^2 = \frac{bc+ad}{ab+cd}(ac+bd).$$
dividing these inequalities termwise we get $\frac{n}{m} = \frac{ab+cd}{bc+ad}$.

403. Let ABC be an equilateral triangle with side a. Denote by r_1, r_2 and r_3 the distances from a point M on the circumscribed circle to the vertices of the triangle (Fig. 117). Note first that for the position of the point M indicated in Fig. 117 we have $r_1 = r_2 + r_3$.

Indeed, laying off $DM = r_2$ we obtain an equilateral triangle BMD and hence it follows that $\angle ABD = \angle CBM$ which implies that $\Delta ABD = \Delta CBM$, and $AD = r_3$. Now, applying the law of cosines to ΔBMC we obtain $a^2 = r_2^2 + r_3^2 - 2r_2r_3 \cos 120° = r_2^2 + r_3^2 + r_2r_3$. Consequently, $r_1^2 + r_2^2 + r_3^2 = (r_2 + r_3)^2 + r_2^2 + r_3^2 = 2(r_2^2 + r_3^2 + r_2r_3) = 2a^2$.

404. Let the side AB of a quadrilateral $ABCD$ intersect a circle, and the sides BC, CD and DA be tangent to it at points E, F and G (Fig. 118). Since $CE = CF$ and $DF = DG$, the inequality $AB + CD > BC + DA$ is equivalent to the inequality $AE > BE + AG$, which was proved in Problem 394.

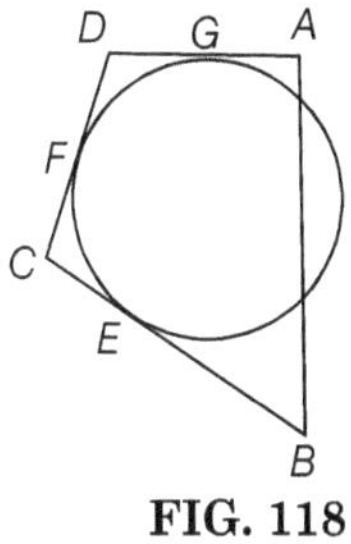

FIG. 118

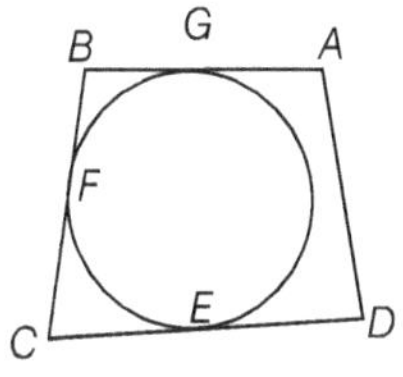

FIG. 119

405. Let the side AD of a quadrilateral $ABCD$ not intersect a circle, and the sides BC, CD and BA be tangent to it at points F, E, G (Fig. 119). The inequality $AD + CB < DC + BA$ is equivalent to the inequality $AD < DE + AG$, which was proved in Problem 394.

406. Let R be the radius of the given semicircles. If $r_1, r_2, \ldots\ldots, r_n$ are the radii of the inscribed circles and $d_1, d_2, \ldots. d_n$, are their diameters (Fig. 120),

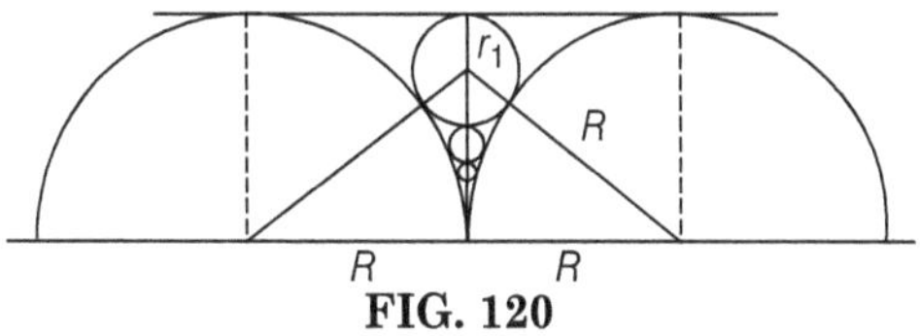

FIG. 120

then it is clear that the sum $d_1 + d_2 + \ldots + d_n$ tends to R when n increases unlimitedly, i.e.

$$d_1 + d_2 + \ldots + d_n + \ldots = R. \tag{1}$$

Besides, we have

$$(R + r_1)^2 = R^2 + (R - r_1)^2, 2r_1 = d_1 = \frac{R}{1 \times 2}$$

and $$(R + r_2)^2 = R^2 + (R - d_1 - r_2)^2, \; 2r_2 = d_2 = \frac{R}{2 \times 3}.$$

Let $d_n = \dfrac{R}{n(n+1)}$. Let us prove that

$$d_{n+1} = \frac{R}{(n+1)\,(n+2)}.$$

We have

$$(R + r_{n+1})^2 = R^2 + (R - d_1 - d_2 - \ldots - d_n - r_{n+1})^2. \tag{2}$$

But $$d_1 + d_2 + \ldots + d_n = R\left(\frac{1}{1 \times 2} + \frac{1}{2 \times 3} + \ldots + \frac{1}{n(n+1)}\right)$$

$$= R\left(1 - \frac{1}{2} + \frac{1}{2} - \frac{1}{3} + \ldots + \frac{1}{n} - \frac{1}{n+1}\right) = R\frac{n}{n+1}$$

Substituting this expression into (2) we find

$$d_{n+1} = 2r_{n+1} = \frac{R}{(n+1)(n+2)}.$$

Putting $R = 1$ in equality (1) we get

$$\frac{1}{1 \times 2} + \frac{1}{2 \times 3} + \ldots + \frac{1}{n(n+1)} + \ldots = 1.$$

407. Let O be the centre of the billiards. Denote by B the first point of reflection and by C the second point of reflection. Let us prove that if $\angle ABC \neq 0$, then ΔABC is isosceles (Fig. 121). Indeed, ΔBOC is isosceles and, hence, $\angle OBC = \angle OCB$. According to the law of reflection, the angle of incidence is equal to the angle of reflection and therefore $\angle OBC = \angle OBA$ and $\angle OCB = \angle OCA$. Consequently, $\angle ABC = \angle ACB$. It follows that the centre O lies on the altitude AD drawn to the side BC. The position of the point B to which the ball should be directed so that it passes through the point A after it has been reflected at B and C, can be specified by the magnitude of the angle $\angle BOD = \alpha$. We have $OD = R\cos\alpha, BD = R\sin\alpha$,

$$BA = \frac{BD}{\cos 2\left(\dfrac{\pi}{2} - \alpha\right)} = \frac{BD}{\cos 2\alpha}.$$

Since BO is the bisector of the angle B in ΔABD, if follows that

$$\frac{BD}{BA}=\frac{OD}{OA}.$$

This implies $-\cos 2\alpha = \dfrac{R\cos\alpha}{a}$,

whence we obtain the equation

$$\cos^2\alpha + \frac{R}{2a}\cos\alpha - \frac{1}{2} = 0.$$

Finding $\cos\alpha$ from this equation we obtain

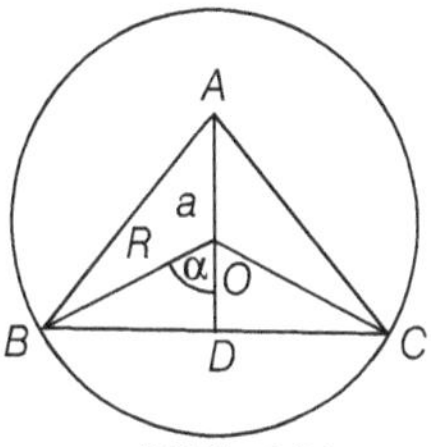

FIG. 121

$$\cos\alpha = -\frac{R}{4a} + \sqrt{\left(\frac{R}{4a}\right)^2 + \frac{1}{2}}.$$

The second root is discarded since, by virtue of the inequality $R > a$, it gives a value of $\cos\alpha$ less than -1.

If now we suppose that $\angle ABC = 0$, then a second solution of the problem appears in which the points B and C are the two extremities of the diameter passing through the point A.

408. Let S be the vertex of the given angle α, A_1 the first point of reflection of the ray, SB_1 the side of the angle on which the point A_1 lies, and SB_0 its other side. We shall denote the consecutive points of reflection of the ray from the sides of the angle by $A_2, A_3, \ldots\ldots$, the path of the ray inside the angle being the polygonal line $AA_1A_2A_3\ldots$ (Fig. 122).

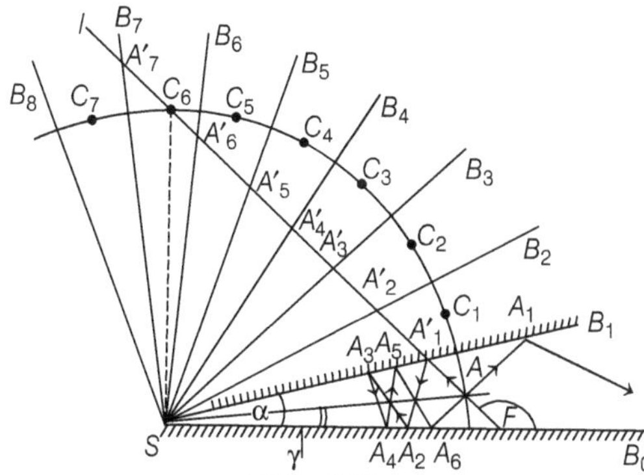

FIG. 122

Let us construct, in succession, the angles B_1SB_3, $B_2SB_3, \ldots$, equal to the angle $\alpha = \angle B_0SB_1$ setting them off in the direction of rotation from SB_0 to SB_1. Lay off the line segment $S'A_m = SA_m$, $m = 2, 3, 4, \ldots$ (the points A'_1 and A_1 are coincident) on the side SB_m. We shall prove that the points $A'_1, A'_2 \ldots$ lie on a straight line l. To this end, it is sufficient to prove that every three consecutive points A'_m, A'_{m+1} and A'_{m+2} (here we put $m = 0, 1, 2, \ldots$) are in a straight line. For this purpose , we note that $\Delta A'_m SA'_{m+1} = \Delta A_m SA_{m+1}$, which implies $\angle A'_m A'_{m+1} S = \angle A_m A_{m+1} S$.

Analogously, $\Delta A'_{m+1} SA'_{m+2} = \Delta A_{m+1} SA_{m+2}$

and, consequently, $\angle SA'_{m+1} A'_{m+2} = \angle SA_{m+1} A_{m+2}$

But accoeding to the law of reflection the angle of incidence is equal to the angle of replection, and hence $\angle SA_{m+1} A_{m+2} = \angle A_m A_{m+1} B$.

Therefore, $\angle A_m A_{m+1} S + \angle SA'_{m+1} A'_{m+2} = \angle A_m A_{m+1} S + \angle A_m A_{m+1} B = \pi$.

We see that path of the ray, that is the polygonal line $AA_1 A_2 \ldots$, is thus developed on the straight line l. Since this straight line can intersect only a finite number of sides SB_m, we conclude that the number of reflections of the ray is finite.

It is clear that if SB_n is the last of the sides intersected by l, then $n\alpha < \beta$ and $(n+1)\alpha \geq \beta$. Thus, the number of reflections is equal to an integer n such that the inequalities $n < \frac{\beta}{\alpha} \leq n+1$ are satisfied.

To find out the conditions for the ray returning to the point A after it has been reflected several times let us construct a sequence of points $C_1, C_2, \ldots$ so that the point C_1 is the reflection of the point A through SB_1, the point C_2 is the reflection of the point C_1 through SB_2, etc. (generally, the point C_m is the reflection of the point C_{m-1} through SB_m). It is clear that the condition that the ray again passes through the point A is equivalent to the condition that the straight line l passes through one of the points $C_m (m = 1, 2, \ldots\ldots)$.

To formulate this condition analytically, let us introduce the angle $\gamma = \angle ASB_0$ and consider the following two possible cases:

(a) if C_k is the point through which the straight line l passes, then k is an even number

(b) the point C_k corresponds to an odd number k.

In the case (a) (which is shown in Fig. 122 for $k = 6$) we have $\angle ASC_k = k\alpha$. ΔASC_k is isosceles and therefore

$$\angle SAC_k = \frac{\pi}{2} - \frac{k\alpha}{2}.$$

On the other hand, $\angle SAC_k$ is equal to $\gamma + \pi - \beta$, and consequently

$$\frac{\pi}{2} - \frac{k\alpha}{2} = \gamma + \pi - \beta,$$

which yields

$$k = \frac{2\beta - 2\gamma - \pi}{\alpha}. \quad (1)$$

In the case (b) we have

$$\angle ASC_k = (k+1)\alpha - 2\gamma$$

and, as above, we come to the relationship

$$\frac{\pi}{2}-\frac{(k+1)\alpha-2\gamma}{\alpha}=\gamma+\pi-\beta,$$

whence $$k+1=\frac{2\beta-\pi}{\alpha}. \qquad (2)$$

Reversing the argument we can easily show that if one of the relationships (1) or (2) is fulfilled for an integer k, the straight line l passes through the point C_k. Consequently, the ray passes through the point A once again if and only if one of the numbers (1) or (2) is an even integer.

4. Loci of Points

409. The required locus of points consists of two circular arcs: the arc BE with centre at the midpoint C of the arc AB of the given circle and the arc BF with centre at the midpoint of the second arc AB of the circle, EAF being the tangent line to the given circle at the point A (Fig. 123).

Proof. Let N be a point of the sought-for locus obtained with the aid of a point M on the lower arc AB. By the construction, the triangle NMB is isosceles, and thus $\angle BNA=\frac{1}{2}\angle BMA=\frac{1}{2}\angle BCA.$

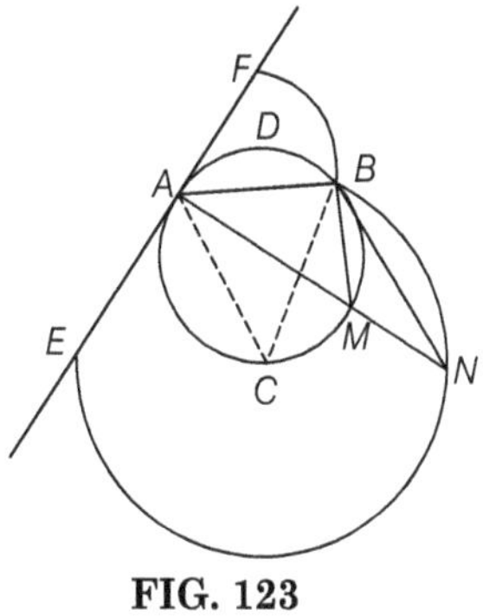

FIG. 123

Consequently, the point N lies on the circle with centre at C passing through the points A and B. Furthermore, the point N must be inside $\angle BAE$, i.e. it lies on the arc BE of the circle with centre at the point C. Conversely, if N lies on this arc, then $\angle BNA=\frac{1}{2}\angle BCA=\frac{1}{2}\angle BMA,$ whence it follows that $\angle BNA=\angle NBM$ and ΔNMB is isosceles. Hence, the point N is obtained by the above construction. When the point M is on the upper arc AB, the proof is carried out in an analogous way.

410. The desired locus of points consists of two straight lines l and k symmetric with respect to the perpendicular BB' to the given parallel lines drawn through the point O. The straight line l passes through the point C perpendicularly to OC, and $B'C = OB$ (Fig. 124).

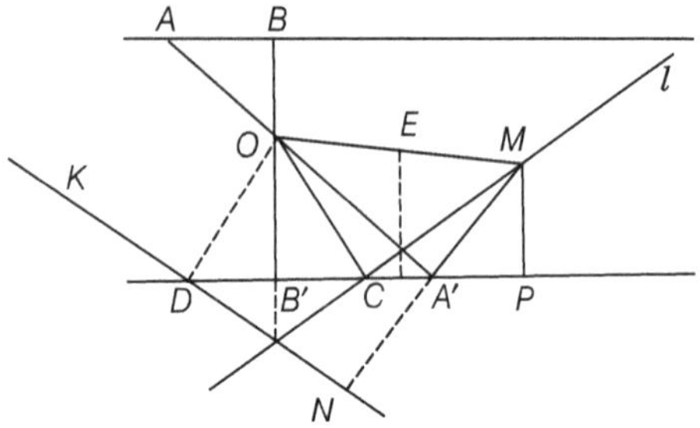

FIG. 124

Proof. Let M and N be two points constructed with aid of a secant AA'. We shall only carry out the proof for the point M (for N it is quite analogous). Let $MP \perp B'C$ then the angles OAB and $A'MP$ are equal as angles with perpendicular sides. Therefore, the right triangles OAB and $A'MP$ with equal hypotenuses OA and $A'M$ are congruent. Hence, $A'P = OB = B'C$. It follows that if E is the midpoint of OM, then the points M, A', C and O lie in a circle with centre at E and consequently, $MC \perp OC$, i.e. the point M lies on the straight line l. Conversely, if M is a point on the straight l and the angle $MA'O$ is right then $A'P = B'C = OB$ which implies the congruence of the triangles OAB and $A'MP$, and finally, the equality $OA = A'M$. Consequently, the point M is obtained by the above construction.

411. In the case of intersecting straight lines the required locus of points consists of four line segments forming a rectangle $ABCD$ whose two vertices are on the given straight lines l and m and the other two vertices are at the given distance a from them (Fig. 125).

Proof. Let M be a point such that $MK \perp l$, $ML \perp m$ and $MK + ML = a$ where a is the length of the given line segment. Through M draw a straight line AB so that $OA = OB$ and $MN \parallel OB$. Let $AP \perp OB$ and Q be the point of intersection of AP and MN. The equality $AN = MN$ shows that $MK = AQ$ and, hence, $AP = AQ + QP = MK + ML = a$.

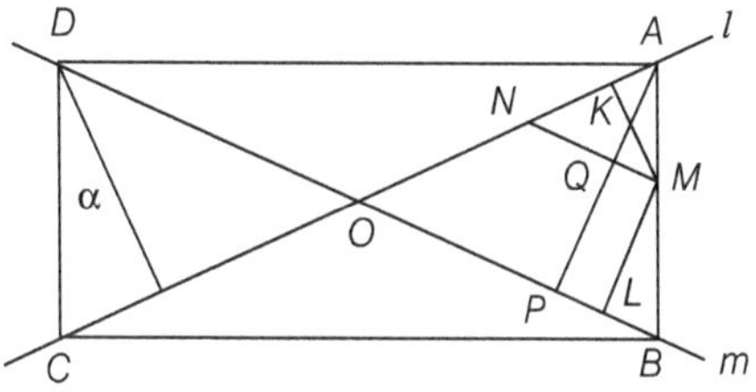

FIG. 125

Consequently, the point A is a vertex of the above rectangle. The same is true for the point B, and hence the point M lies on a side of this rectangle. Conversely, if M lies on a side of this rectangle, then reversing the argument we see that $MK + ML = AP = a$.

If the given straight lines l and m are *parallel* and the distance between them is equal to h, then the desired locus of points exists only if a $\geq h$ and is a pair of straight lines parallel to the given ones for $a > h$ or the whole strip contained between l and m for $a = h$.

412. In the case of *intersecting* straight lines the required locus consists of eight half-lines which are the extensions of the sides of the rectangle $ABCD$ indicated in the solution of Problem 411 (Fig. 126). The proof is then analogous to the one given there.

If the given lines l and m are *parallel* and the distance between them is equal to h, then the sought-for locus exists only if $a \leq h$ and is a pair of straight lines parallel to the given ones for $a < h$ or the portion of the plane which is the exterior of the strip contained between l and m for $a = h$.

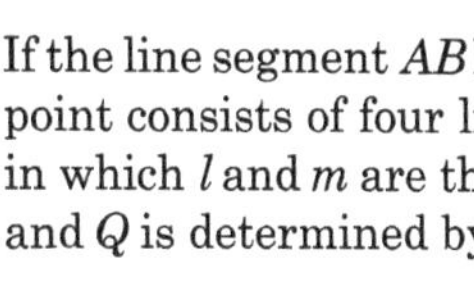

FIG. 126

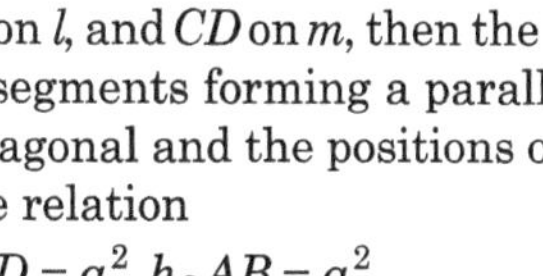

FIG. 127

413. If the line segment AB lies on l, and CD on m, then the desired locus of point consists of four line segments forming a parallelogram $PQRS$ in which l and m are the diagonal and the positions of the vertices P and Q is determined by the relation

$$h_P CD = a^2, h_Q AB = a^2, \tag{1}$$

where h_P and h_Q are the distance from the points P and Q to the straight lines m and l (Fig. 127).

Proof. Note that for fixed l and m the required locus of points is completely specified by the lengths of the given line segments AB and CD and the constant a and is independent of the position of these line segments on the straight lines l and m. Indeed, if this position is varied, the areas of the triangles AMB and CMD remain constant. Therefore it is sufficient to consider the particular case when the line segments AB and CD have a common endpoint coincident with the point of intersection of the straight lines l and m. In this case the segments AB and CD are two sides of a triangles whose third side lies in one of the four angles formed by the intersecting lines l and m. For example, in FIG. 127 the endpoints A and C are made to coincide, BD being the third side. Let M be a point of the required locus lying inside the angle BAD. Then the area of ΔBMD is equal to

$$S_{BMD} = |S_{AMB} + S_{CMD} - S_{ABD}| = |a^2 - S_{ABD}|$$

It follows that the distance between the point M and the straight line BD is independent of its position on the straight line $PQ \parallel BD$. For the points P and Q relationships (1) are fulfilled.

Conversely, let M be a point on the straight line PQ with the points P and Q constructed according to (1). From the relation

$$\frac{AP}{AB} = \frac{S_{APD}}{S_{ABD}} = \frac{a^2}{S_{ABD}}, \frac{CQ}{CD} = \frac{S_{CQB}}{S_{CDB}} = \frac{a^2}{S_{ABD}}$$

it follows that $\frac{AP}{AB} = \frac{CQ}{CD}$, i.e. $PQ \parallel BD$. Therefore

$$S_{AMB} + S_{CMD} = S_{ABD} + S_{BMD} = S_{ABD} + S_{BPD} = S_{APD} = a^2.$$

Consequently, the point M belongs to the required locus. The other sides of the parallelogram $PQRS$ are obtained analogously by making the other end points of the line segments coincide, namely QR is obtained when B coincides with C, RS when B coincides with D and SP when A coincides with D.

414. The required locus is a circle which is the reflection of the given circle K through the given chord AB (Fig. 128).

Proof. Construct a chord $AD \perp AB$ in the circle K. Let ΔABC be inscribed in K, and M be the point of intersection of its altitudes (i.e. its orthocentre). As is easily seen, $AMCD$ is a parallelogram because DA and CM are parallel as perpendiculars to BC ($DC \perp BC$ because BD is a diameter in K). Therefore, the point M lies on the circle K' obtained from K by shifting the latter by the distance AD in the direction of the chord DA. It is clear that K' is the reflection of K through AB. Conversely, let M be a point on K' and

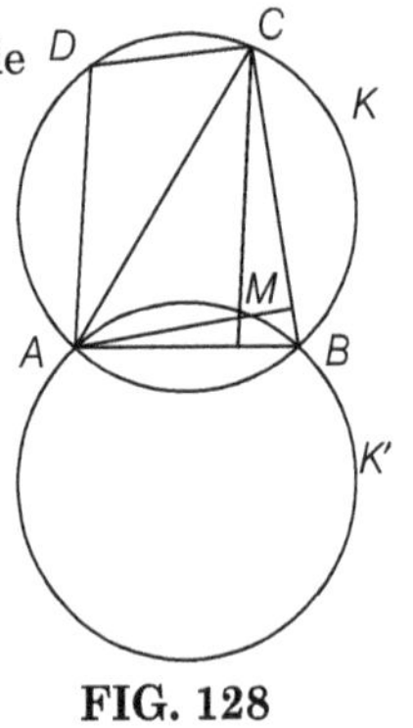

FIG. 128

$MC \perp AB$. Since $MC = AD$, the figure $AMCD$ is a parallelogram, and therefore $AM \parallel DC$.

But $DC \perp BC$ because $ABCD$ is insribed in K and the angle BAD is right. Therefore $AM \perp BC$, and M is the point of intersection of the altitudes in ΔABC. Consequently, M belongs to the required locus.

415. Let O be the centre and R the radius of the given circle (Fig. 129). The required locus of points is a straight line l perpendicular to the straight line OA and intersecting it at a point B such that

$$OB = \frac{R^2}{OA}. \qquad (1)$$

Proof. Through the point M draw a straight line $l \perp OA$ to intersect the straight line OA at the point B. Let C be the point of intersection of the line segment OM and the chord KL. The similarity of the triangles OAC and OMB implies that

$$\frac{OB}{OC} = \frac{OM}{OA},$$

whence

$$OB = \frac{OM \cdot OC}{OA}. \qquad (2)$$

By the construction, KC is an altitude in the right triangle OKM, and hence $OM \cdot OC = R^2$.

Substituting this expression in (2) we obtain the equality (1).

Conversely, let M be a point on the straight line l perpendicular to OA and such that OB is determined by equality (1). Draw the tangent line MK and

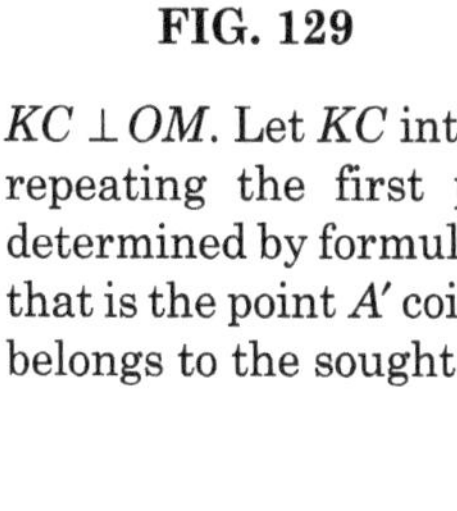

FIG. 129

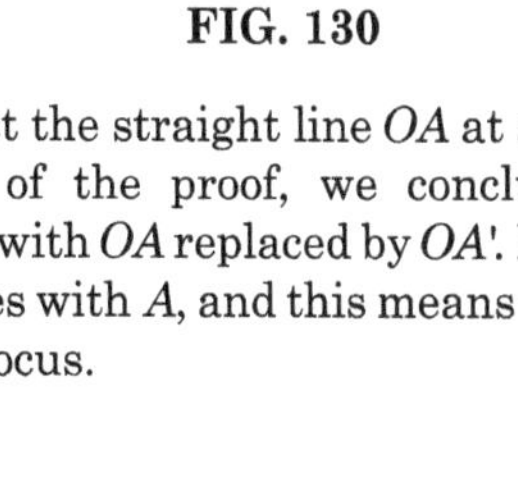

FIG. 130

$KC \perp OM$. Let KC intersect the straight line OA at a point A'. Then, repeating the first part of the proof, we conclude that OB is determined by formula (1) with OA replaced by OA'. Hence, $OA' = OA$, that is the point A' coincides with A, and this means that the point M belongs to the sought-for locus.

416. Let $\dfrac{AM}{BM} = \dfrac{p}{q} > 1.$

Draw the bisectors MP and MQ of the two adjacent angles with vertex M and sides MA and MB (Fig. 130). Then, by the property of bisectors, we have $\dfrac{AP}{BP} = \dfrac{p}{q}$ and $\dfrac{AQ}{BQ} = \dfrac{p}{q}$. (1)

It follows that the position of the points P and Q is independent of the position of the point M. Besides, $\angle PMQ = \dfrac{\pi}{2}$ and therefore the point M lies on the circle K with diameter PQ. Conversely, let the points P and Q be constructed according to (1), and K be a circle with diameter PQ. If a point M lies on this circle, then $\angle PMQ = \dfrac{\pi}{2}$. Through the point B draw $RS \parallel AM$, then $\dfrac{AM}{BR} = \dfrac{AQ}{BQ} = \dfrac{p}{q}, \dfrac{AM}{BS} = \dfrac{AP}{BP} = \dfrac{p}{q}$ (2)

whence $BR = BS$ and hence BM is a median in ΔRMS. Since $\angle RMS$ is right, we have $BM = BR$ and by virtue of (2), $\dfrac{AM}{BM} = \dfrac{p}{q}$.

Therefore the point M belongs to the locus question, 9 – 323

To express the diameter PQ in terms of the length a of the line segment AB we find from the relations

$$PB = AB - AP = a - \frac{p}{q} PB \quad \text{and} \quad BQ = AQ - AB = \frac{p}{q} BQ - a$$

the expressions $PB = a\dfrac{q}{p+q}$ and $BQ = a\dfrac{q}{p-q}$, and, hence,

$$PQ = \frac{2a}{\dfrac{p}{q} - \dfrac{q}{p}}.$$

If $p = q$, the required locus is obviously the perpendicular to the line segment AB drawn through its midpoint.

417. The sought-for locus of points is the perpendicular to the line segment AB drawn through its midpoint E with the point E deleted.

Proof. The triangle ADB is isosceles since $\angle CAD = \angle CBD$ because these angles are subtended on equal arcs CD of two congruent circles (Fig. 131). Therefore, the point D lies on the perpendicular to the line segment AB drawn through its midpoint E, and vice versa, if we take an arbitrary point D on this perpendicular which does not coincide with the point E, then the circles passing through the point A, C and D and through B, C and D are congruent. Indeed, for instance, this can be deduced from the equalities $R_1 = \dfrac{CD}{2\sin\alpha} = \dfrac{CD}{2\sin\beta} = R_2$,

where $\alpha = \angle BAD$ and $\beta = \angle CBD$.

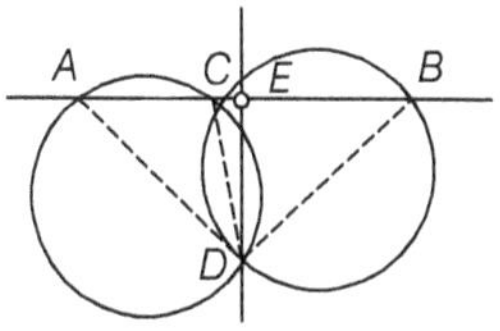

FIG. 131

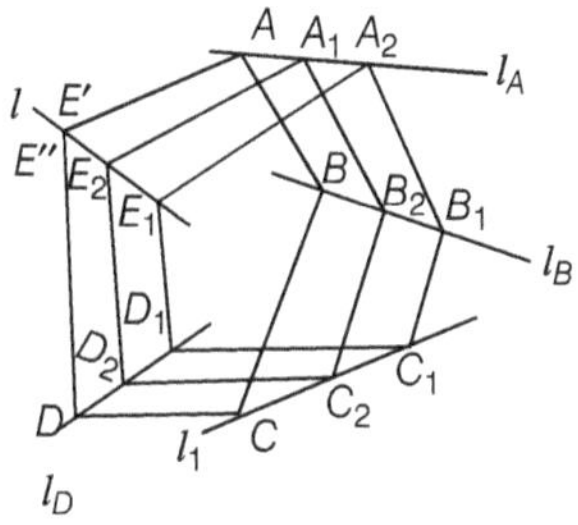

FIG. 132

418. The required locus of points is the straight line drawn through two different positions of the last vertex.

Proof. Let for example, $A_1B_1C_1D_1E_1$ and $A_2B_2C_2D_2E_2$ be two different configurations of the deformed polygon the vertices A, B, C and D sliding, respectively, along straight lines l_A, l_B, l_C and l_D (Fig. 132). Consider the straight line l passing through the positions E_1 and E_2 of the last vertex. Let now the vertex on the line l_A occupy the position A, and on l_D the corresponding position D. The side parallel to A_2E_2 intersects l at a point E', and the side parallel to D_2E_2 at a point E''.

By the construction, we have $\dfrac{E'E_2}{E_2E_1} = \dfrac{AA_2}{A_2A_1} = \dfrac{BB_2}{B_2B_1}$

$$= \frac{CC_2}{C_2C_1} = \frac{DD_2}{D_2D_1} = \frac{E''E_2}{E_2E_1},$$

which shows that $E'E_2 = E''E_2$,

i.e. the points E' and E'' coincide. This means that the last vertex E lies on the line l at the point E' coincident with E''.

The converse is obvious because the configuration of the deformed polygon can be reconstructed beginning with any point E on l.

419. The required locus is a circle passing through the endpoints of the chord AB and a point M_1 obtained by the indicated construction.

Proof. Let us introduce the necessary notation.There is one and only one position C_1D_1 of the chord CD parallel to AB and such that on the given circle K it is possible to choose a direction v of describing K such that when the chord CD moves in this direction starting from the position C_1D_1 the endpoints of the chords AB and CD coincide, in succession, at the points A, B, C_1 and D_1 (such a direction v may only become indeterminate when AC and BD are parallel).

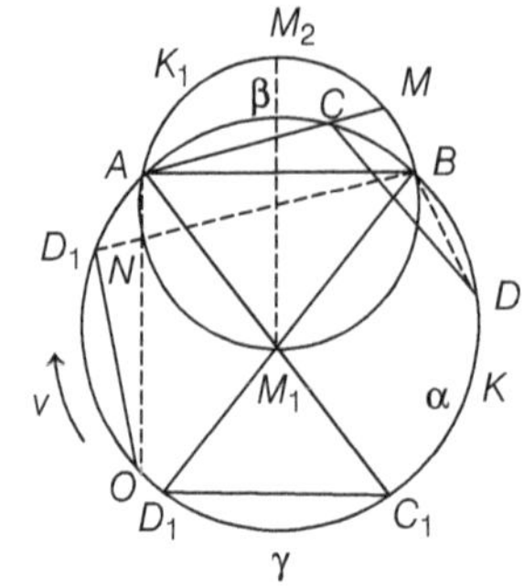

FIG.133

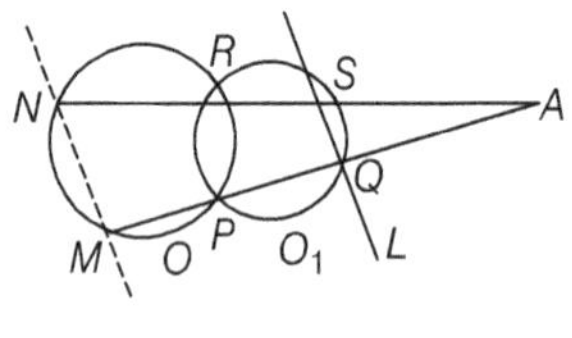

FIG.134

Let us denote by α the arc AB of the given circle K to which the points C_1 and D_1 belong and by β the other arc AB. (Let γ be the arc C_1D_1 which does not obtain the points A and B). Furthermore, let M_1 be the point of intersection of the straight lines AC_1 and BD_1.. The point M_1 lies inside K. Consider the circumscribed circle K_1 of ΔABM_1 (Fig. 133). We shall prove that for any position of the chord CD other than C_1D_1 the point of intersection of AC and BD remains on K_1.

As long as both points C and D lie on the arc α, the point M is inside K, and then $\angle AMB = \frac{1}{2}(\beta + \gamma)$. (1)

But if atleast one of these points is on the arc β, the point M lies outside K, and then $\angle AMB = \frac{1}{2}(\alpha - \gamma)$. (2)

In the former case M lies on the arc AM_1B of the circle K_1 because according to (1), the angle AMB is independent of the position of CD, and, hence, is equal to $\angle AM_1B$. In the latter case, since the sum of the right-hand members of (1) and (2) is equal to $\frac{1}{2}(\alpha + \beta) = \frac{1}{2} \cdot 2\pi = \pi$, the point M is on the arc AB of the circle K_1 lying outside K.

It is obvious that the converse is also true, i.e. any point M of the circle K_1 can be obtained by the above construction for an appropriate choice of the position of the chord CD.

420. Let us designate the given circle by O and the given straight line by L (Fig. 134). Denote by M the second point of intersection of PQ and O. Take any circle O_1 passing through the points P and Q and intersecting the circle O for the second time at a point R and the straight line L at a point S. Let N be the second point of intersection of the line RS with the circle O. (We shall prove that $MN \| L$. To this end, let us take advantage of the following well-known theorem proved in plane geometry: given a circle and a point A, then for any straight line passing through A and intersecting this circle at points

A_1 and A_2 the product of the line segments AA_1 and AA_2 is a constant independent of the choice of the straight line. Denote by A the point of the intersection of the straight lines PQ and RS. We first apply the above theorem to the circle O.) The point A and the straight lines AP and AR. (Since AP intersects the circle O for the second time at the point M, and AR at the point N, we have $AM \cdot AP = AN \cdot AR$. (1)

Now we apply this theorem to the circle O_1, the point A and the same straight lines. Since AP intersects O_1 for the second time at the point Q, and AR at the point S, we can write

$$AQ \cdot AP = AS \cdot AR.$$

From (1) and (2) we derive the equality $\dfrac{AM}{AN} = \dfrac{AQ}{AS}$. (3)

Equality (3), by virtue of the converse of the theorem on proportionality of line segments cut off by parallel straight lines on the sides of an angle, implies that $MN \parallel QS$ which is what we set out to prove.

Thus, for any circle of the type O_1 the point N can be specified as the second point of intersection of the straight line passing through M and parallel to L with the circle O_1. Consequently, all the possible straight lines RS obtained for various circles O_1 intersect the circle O at the point N.

The singular cases in which (1) and (2) do not imply (3), namely, when the points R and P or Q and S coincide, or when $PQ \parallel RS$, may be considered as limiting cases. For these cases the validity of the above argument can be established on the basis of the continuity properties.

5. The Greatest and Least Values

421. If A is the vertex of the right angle in ΔABC, and C and B lie on the given parallel lines l_1 and l_2 (Fig. 135), then

$$AB = \frac{a}{\sin\varphi},\ AC = \frac{b}{\cos\varphi}.$$

Hence, the area of the triangle ABC is equal to

$$S_{ABC} = \frac{1}{2} AB \cdot AC = \frac{ab}{\sin 2\varphi}.$$

It follows that S_{ABC} attains the least value (equal to ab) for $\varphi = \dfrac{\pi}{4}$.

422. If R and r are the radii of the circumscribed and inscribed circles (Fig. 136), then

$$2R = r\cot\frac{\alpha}{2} + r\cot\left(\frac{\pi}{4}-\frac{\alpha}{2}\right).$$

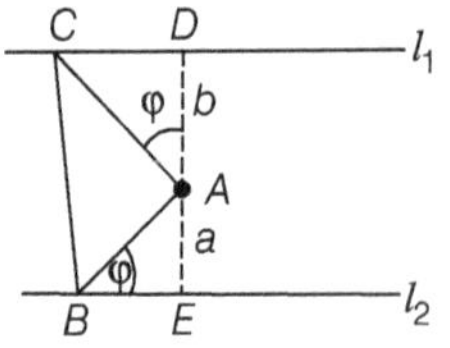

FIG. 135

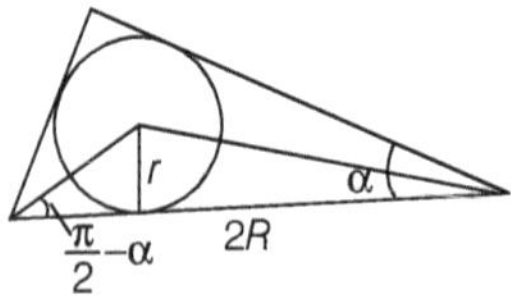

FIG. 136

Noting that

$$\cot\frac{\alpha}{2} + \cot\left(\frac{\pi}{4}-\frac{\alpha}{2}\right) = \frac{\cos\frac{\alpha}{2}\sin\left(\frac{\pi}{4}-\frac{\alpha}{2}\right) + \cos\left(\frac{\pi}{4}-\frac{\alpha}{2}\right)\sin\frac{\alpha}{2}}{\sin\frac{\alpha}{2}\sin\left(\frac{\pi}{4}-\frac{\alpha}{2}\right)} =$$

$$= \frac{2\sin\frac{\pi}{4}}{\cos\left(\alpha-\frac{\pi}{4}\right)-\cos\frac{\pi}{4}} = \frac{2}{\sqrt{2}\cos\left(\alpha-\frac{\pi}{4}\right)-1},$$

we obtain
$$\frac{R}{r} = \frac{1}{\sqrt{2}\cos\left(\alpha-\frac{\pi}{4}\right)-1}.$$

The ratio $\frac{R}{r}$ attains the least value when $\cos\left(\alpha-\frac{\pi}{4}\right)=1$, i.e. when $\alpha=\frac{\pi}{4}$ because we consider the interval $0<\alpha<\frac{\pi}{2}$. In this case $\frac{R}{r}=\frac{1}{\sqrt{2}-1}=\sqrt{2}+1.$

423. Let a right triangle with vertex C and legs a_1 and b_1 be cut off from a rectangle $ABCD$ with sides a and b. Consider the pentagon $ABEFD$ thus obtained (Fig. 137). It is clear, that one of the vertices (say C_1) of the sought-for rectangle $AB_1C_1D_1$ must lie on the line segment EF. The problem is thus reduced to finding the position of this vertex.

To find the point C_1 extend the sides AB and AD of the rectangle to intersect the extension of the line segment EF. This results in a triangle AMN. Let $AM=m$, $AN=n$ and $B_1C_1=AD_1=x$.

The similarity of the triangles AMN and D_1C_1N implies that $\frac{C_1D_1}{m}=\frac{n-x}{n}$, whence we find $C_1D_1=\frac{m}{n}(n-x)$.

Hence, for the area S of the rectangle $AB_1C_1D_1$ which is equal to $AD_1 \cdot C_1D_1$ we get the expression

$$S = \frac{m}{n}(n-x)x.$$

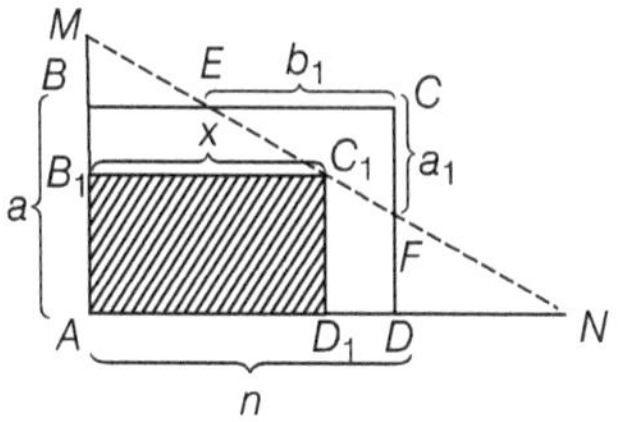

FIG. 137

Transforming this expression to the form $S = \frac{m}{n}\left[\frac{n^2}{4} - \left(\frac{n}{2} - x\right)^2\right]$ (1)

we conclude that the greatest value of S is attained when $\frac{n}{2} - x = 0$, i.e. for $x = \frac{n}{2}$.

Let C_0 be the position of the vertex C_1 corresponding to $x = \frac{n}{2}$.

Noting that expression (1) for S decreases when $\left|\frac{n}{2} - x\right|$ increases, i.e. when the point C_1 moves from the point C_0 to the vertex M or F, we find that there are three possible cases here, namely:

(1) The point C_0 lies on the line segment EF; then the vertex C_1 of the required rectangle coincides with C_0.

(2) The point C_0 lies on the line segment ME; then C_1 must coincide with E.

(3) The point C_0 lies on the segment FN; then C_1 must coincide with F.

We now must establish a criterion for distinguishing between these cases with the aid of the magnitudes of the quantities a, a_1, b and b_1 given in the formulation of the problem.

Let us first find the quantity n. The similarity of the triangle ECF and NDF implies that

$$\frac{n-b}{a-a_1} = \frac{b_1}{a_1}$$

Whence we find $n = b + \frac{b_1}{a_1}(a - a_1)$. (2)

Now note that the point C_0 is within the line segment EF if the inequalities $b - b_1 < x < b$ are fulfilled.

Substituting $x = \frac{n}{2}$ with the known value of n into the above we obtain

$$b - b_1 < \frac{b}{2} + \frac{b_1}{2a_1}(a - a_1) < b.$$

The latter inequalities are readily transformed to the form

$$-1 < \frac{a}{a_1} - \frac{b}{b_1} < 1. \qquad (3)$$

If the inequality $-1 < \frac{a}{a_1} - \frac{b}{b_1}$ is violated, the point C_0 falls on the line segment ME, and if the inequality $\frac{a}{a_1} - \frac{b}{b_1} < 1$ does not hold C_0 falls on FN.

Thus, we arrive at the following final results: if for given a, b, a_1 and b_1 both inequalities (3) are fulfilled, then the vertex C_1 of the rectangle of the greatest area lies within the line segment EF, and the side x of this rectangle is computed by the formula

$$x = \frac{b}{2} + \frac{b}{2a_1}(a - a_1);$$

if the left inequality in (3) does not hold true, the vertex C_1 coincides with the point E, and if the right inequality is not fulfilled, then C_1 coincides with F.

424. Draw a circle passing through the points A and B and tangent to the other side of the angle (Fig. 138). The point of tangency is then the required point, since for any point C' on that side the angle $AC'B$ is measured by half the difference between the arcs AB and A_1B_1, whereas $\angle ACB$ is measured by half the arc AB.

Furthermore, we have $(OC)^2 = OB \cdot OA$. Consequently, the problem is reduced to the well-known construction of the geometric mean of the lengths of two. given line segments (OA and OB).

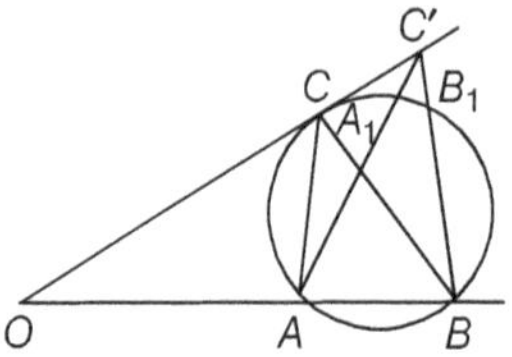

FIG. 138

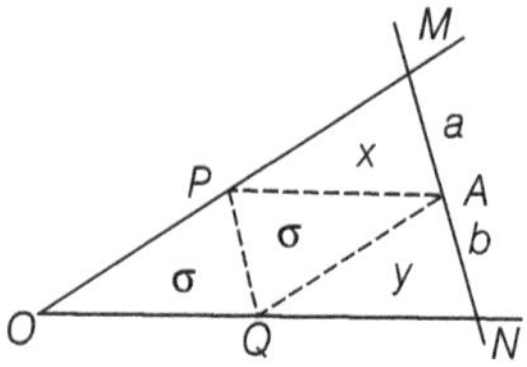

FIG. 139

425. Consider the following three possible configurations of the line segment AB and the straight line l.

(a) $AB \parallel l$. For any point M of the straight line l we have $|AM - BM| \geq 0$, and there exists a point M_0 for which $|AM_0 - BM_0| = 0$. This point is the foot of the perpendicular dropped from the midpoint of AB onto l. There is no point M for which the quantity $|AM - BM|$ attains the greatest value. This is implied by the inequality $|AM - BM| \leq AB$ in which the sign of equality only appears when A, B and M lie in a straight line.

(b) $AB \perp l$. Since $|AM - BM| \leq AB$, the quantity $|AM - BM|$ for the point of intersection of the straight lines l and AB takes on the greatest value equal to the length of AB. There is no point M for which the quantity $|AM - BM|$ attains the least value.

(c) The straight line AB is neither parallel nor perpendicular to l. It is clear that $|AM - BM|$ attains the least value if M is the point of intersection of l and the perpendicular to the line segment AB erected at its midpoint. The greatest value is attained by $|AM - BM|$ when the point M is the point of intersection of AB with l.

426. Let MN be a position of the secant, $AP \parallel ON$ and $AQ \parallel OM$ (Fig. 139). Let us introduce the following notation:

$x =$ the area of ΔAPM,

$y =$ the area of ΔAQN,

$\sigma =$ the area of ΔAPQ,

$S =$ the area of ΔOMN,

$\alpha = AM$,

$b = AN$.

We have:

$$S = 2\sigma + x + y.$$

It is clear that
$$\frac{x}{\sigma} = \frac{a}{b}, \quad \frac{y}{\sigma} = \frac{b}{a}.$$

Consequently, $S = \sigma\left(2 + \dfrac{a}{b} + \dfrac{b}{a}\right) = 4\sigma + \sigma\dfrac{(a-b)^2}{ab}$.

The least value $S = 4\sigma$ is attained for $a = b$ which is what we set out to prove.

427. Let $a + b = q$ (Fig. 140). By the law of cosines, we have

$$c^2 = a^2 + b^2 - 2ab\cos\varphi = a^2 + (q-a)^2 - 2a(q-a)\cos\varphi =$$
$$= q^2 + 2a^2(1+\cos\varphi) - 2aq(1+\cos\varphi) =$$
$$= q^2\frac{1-\cos\varphi}{2} + 2(1+\cos\varphi)\left(a - \frac{q}{2}\right)^2.$$

Since q and φ remain unchanged, the least value c is attained for

$$a = \frac{q}{2} = \frac{a+b}{2}, \text{ i.e. for } a = b.$$

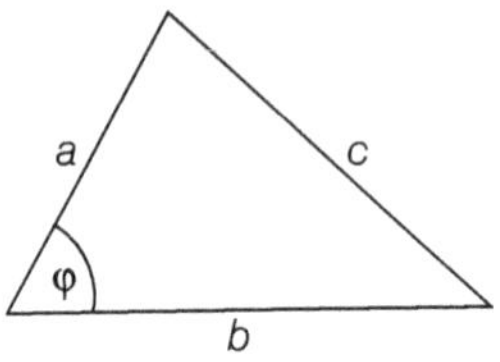

FIG. 140

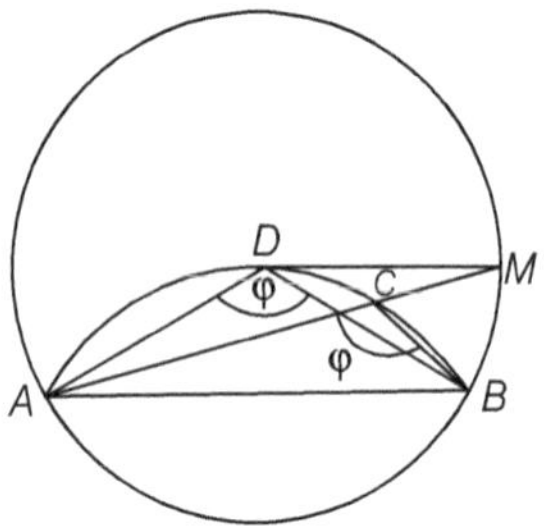

FIG. 141

428. *First solution.* Consider ΔABC with base AC and designate by a, b and c the lengths of the sides opposite the angles A, B and C, respectively; put $a + b + c = p$.

From the relations

$$\frac{a}{\sin A} = \frac{c}{\sin (A+B)} = \frac{b}{\sin B}$$

we find

$$p = b + b\frac{\sin A}{\sin B} + b\frac{\sin (A+B)}{\sin B} = b + \frac{b}{\sin \frac{B}{2}} \sin \left(A + \frac{B}{2}\right).$$

Since $b > 0$ and $\sin \frac{B}{2} > 0$, the quantity p attains the greatest value when $A + \frac{B}{2} = \frac{\pi}{2}$. In this case $A = C$ and ΔABC is isosceles.

Second solution. On the given line segment AB as chord construct a segment of a circle so that the chord AB subtends an angle of the given magnitude φ inscribed in that circle (Fig. 141). Consider the isosceles triangle ADB and a scalene triangle ACB inscribed in the segment of the circle. Draw the circle of radius $AD = DB$ with centre at the point D, extend AC to intersect this circle at a point M and join M and B.

We obtain $AD + DB = AD + DM > AM = AC + CM$.

But in ΔBCM we have

$\angle CBM = \angle ACB - \angle CMB = \angle CMB,$

because the angle ACB is equal to the angle ADB and is measured by the arc AB, and $\angle AMB$ is measured by half the arc AB. Hence, $CM = CB$ and $AD + DB > AC + CB$.

429. Let us designate the radii of the circumscribed circles of the triangles ACD and BCD by R_1 and R_2, respectively. Put $\angle ADC = \varphi$, $AC = b$ and $BC = a$ (Fig. 142). Then we have

$$2R_1 = \frac{b}{\sin\varphi}$$

and

$$2R_2 = \frac{a}{\sin(\pi - \varphi)} = \frac{a}{\sin\varphi}$$

and hence $\frac{R_1}{R_2} = \frac{b}{a}$. The radii R_1 and R_2 attain the least values when $\varphi = \frac{\pi}{2}$; in this case D is the foot of the altitude CD.

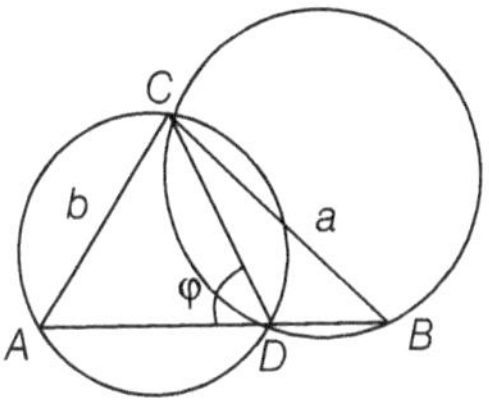

FIG. 142

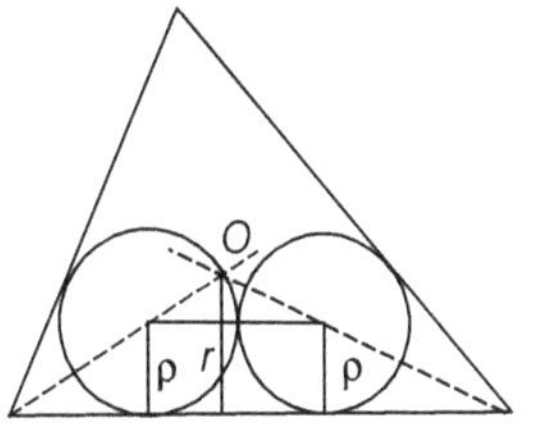

FIG. 143

430. Each of the cut-off circles must be tangent to two sides of ΔABC (see Fig. 143). Furthermore, the circles must be tangent to each other. Indeed, if otherwise, the radius can be increased. Therefore, the centres of the circles lie on two bisectors of interior angles, for example, OA and CO where O is the centre of the inscribed circle of ΔABC. If r is the radius of the inscribed circle of ΔABC and ρ the radius of the cut-off circles, then from ΔAOC we have $\frac{r-\rho}{2\rho} = \frac{r}{b}$, whence, we find $\frac{\rho}{r} = \frac{b}{b+2r} = 1 - \frac{2r}{b+2r}$.

This formula shows that ρ assumes the greatest value when the longest side is taken as b.

B. SOLID GEOMETRY

1. Computation Problems

431. Let a be the side of the base, d the length of the diagonals of the lateral faces of the prism and l the lateral edge (Fig. 144). We have

$$V = \frac{a^2\sqrt{3}}{4}\, l.$$

From ΔA_1BC_1 we obtain that $\frac{1}{2}a = d\sin\frac{\alpha}{2}$. Therefore,

$$l = \sqrt{d^2 - a^2} = \frac{a}{2\sin\frac{\alpha}{2}}\sqrt{1 - 4\sin^2\frac{\alpha}{2}}$$

and, consequently,

$$V = \frac{a^3\sqrt{3}}{8\sin\frac{\alpha}{2}}\sqrt{1 - 4\sin^2\frac{\alpha}{2}}.$$

It follows that $a = \sqrt[3]{\dfrac{8V\sin\frac{\alpha}{2}}{\sqrt{3 - 12\sin^2\frac{\alpha}{2}}}}$.

432. Let H be the altitude of the pyramid, and a the length of the side of the base.

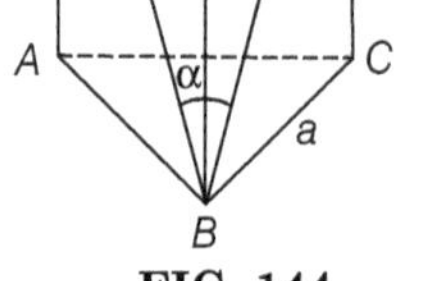

FIG. 144

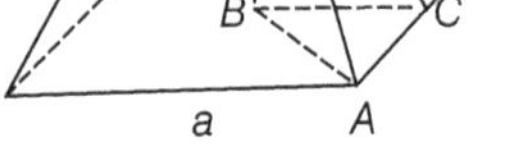

FIG. 145

The similarity of the triangles OMS and ABS (Fig. 145) implies that

$$\frac{h}{\frac{a}{\sqrt{2}}} = \frac{\sqrt{\frac{1}{4}H^2 - h^2}}{H}. \tag{1}$$

Analogously, from the triangles OKS and CBS we obtain

$$\frac{b}{\frac{a}{2}} = \frac{\sqrt{\frac{1}{4}H^2 - b^2}}{H}. \quad (2)$$

Dividing equality (1) by (2) termwise we obtain

$$\sqrt{\frac{H^2-4h^2}{H^2-4b^2}} = \frac{h}{b\sqrt{2}}, \text{ whence } H = \frac{2bh}{\sqrt{2b^2-h^2}}.$$

Substituting this expression into (1) we easily find

$$a^2 = \frac{8b^2h^2}{h^2-b^2}.$$

Finally, for the volume V we receive the expression

$$V = \frac{16}{3}\frac{b^3h^3}{(h^2-b^2)\sqrt{2b^2-h^2}}.$$

433. Let H be the altitude of the pyramid, x the slant height, R the radius of the inscribed circle of the base, r the radius of the circumscribed circle of the base and a the side of the base. From the similarity of the triangles CA_1B_1 and CAB (Fig. 146) we get

$$\frac{H-h}{H} = \frac{R}{r}, \text{ whence } H = \frac{hr}{r-R}.$$

But from ΔADB we have $r = \dfrac{R}{\cos\frac{\pi}{n}}$, and

therefore $H = \dfrac{h}{1-\cos\frac{\pi}{n}}$.

Furthermore, for the area of the base and for the volume we have the formulas $S_{base} = n\frac{1}{2}r^2\sin\frac{2\pi}{n}$ and $V = \frac{1}{3}S_{base}H$, and therefore

$$r^2 = \frac{6V}{Hn\sin\frac{2\pi}{n}}.$$

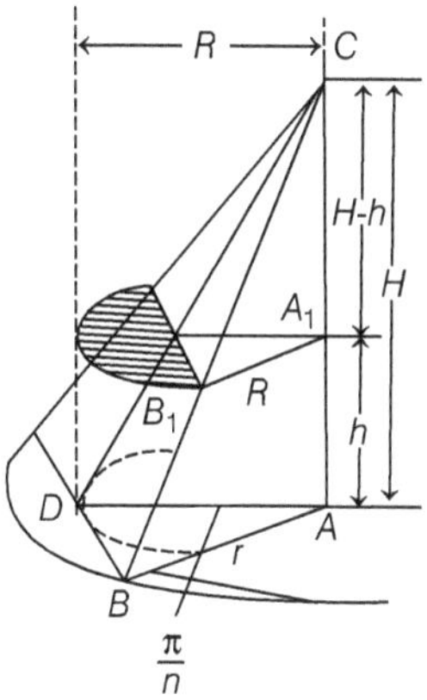

FIG. 146

Substituting in the latter relation the above expression for H, we find

$$r = \sqrt{\frac{6V\left(1-\cos\frac{\pi}{n}\right)}{nh\sin\frac{2\pi}{n}}}.$$

Since $x = \sqrt{R^2 + H^2}$ and $\frac{a}{2} = r \sin \frac{\pi}{n}$, the lateral surface area is equal to $n \frac{1}{2} xa = nr \sin \frac{\pi}{n} \sqrt{R^2 + H^2}$, and, finally,

$$S_{lat} = n \sin \frac{\pi}{n} \sqrt{\frac{6V\left(1 - \cos \frac{\pi}{2}\right)}{nh \sin \frac{2\pi}{n}} \left[\frac{3V\left(1 - \cos \frac{\pi}{n}\right)}{nh \tan \frac{\pi}{n}} + \frac{h^2}{\left(1 - \cos \frac{\pi}{n}\right)^2}\right]}.$$

434. Let M and N be the midpoints of the edges ES and DS (Fig. 147). It is easily seen that $AMNC$ is a trapezoid because $MN \parallel ED$ and $ED \parallel AC$. It is also obvious that $MN = \frac{1}{2} q$.

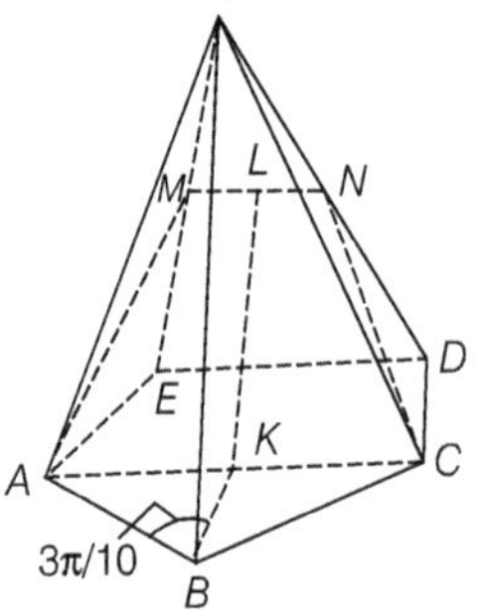

FIG. 147

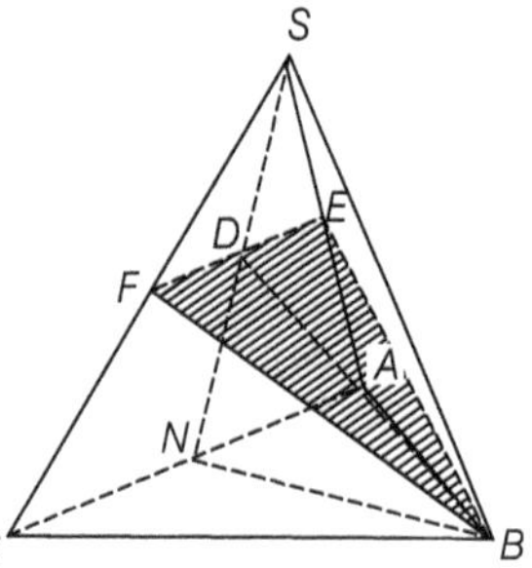

FIG. 148

Using formula (1) for the square of a median of a triangle derived in the solution of Problem 370, we find $CN = \frac{\sqrt{b^2 + 2q^2}}{2}$.

Furthermore, $KC = \frac{AC}{2} = q \sin \frac{3\pi}{10}$,

because $\angle ABK = \frac{3\pi}{10}$. If KL is the line segment joining the midpoints of the bases of the trapezoid $ACNM$, then

$$KL = \sqrt{\frac{b^2 + 2q^2}{4} - \left(q \sin \frac{3\pi}{10} - \frac{q}{4}\right)^2}$$

$$= \sqrt{\frac{b^2 + 2q^2}{4} - q^2\left(\frac{\sqrt{5} + 1}{4} - \frac{1}{4}\right)^2} = \frac{\sqrt{4b^2 + 3q^2}}{4}$$

(we have used here the equality $\sin \frac{3\pi}{10} = \frac{\sqrt{5} + 1}{4}$). Thus, the sought-for area is $S_{sec} = \frac{1}{2}(MN + AC)\,KL = \frac{q}{16}(2 + \sqrt{5})\sqrt{4b^2 + 3q^2}$.

435. Let E and F be the midpoints of the lateral edges of the regular triangular pyramid $SABC$ shown in Fig. 148, and D the midpoint of the line segment EF. Since the cutting plane is perpendicular to the face CSA, the angle SDB is right. Extend SD to intersect the straight line AC at a point M and consider the triangle MBS. It is obvious that the point D bisects the line segment SM. Besides, $BD \perp MS$ and therefore MBS is an isosceles triangle in which $SB = MB$. Let the side of the base of the pyramid be equal to a. Then

$$SB = MB = \frac{a\sqrt{3}}{2}.$$

The slant height is given by the expression

$$SM = \sqrt{SC^2 - CM^2} = \frac{a\sqrt{2}}{2}.$$

Therefore, $S_{lat} = \frac{3a^2\sqrt{2}}{4}$, and since the area of the base is

$$S_{base} = \frac{a^2\sqrt{3}}{4}, \text{ we have } \frac{S_{lat}}{S_{base}} = \sqrt{6}.$$

436. Let a be the length of the side of the square which is the base of the prism, l the length of the lateral edge of the prism and d the diagonal of the lateral face (Fig. 149). Let S_{sec} denote the area of the section. It is easily seen that the total surface area of the prism is equal to $4(S - S_{sec})$; therefore it is sufficient to determine S_{sec}. We have

$$S_{sec} = \frac{1}{2} d^2 \sin\alpha,$$

$$a = d\sqrt{2}\sin\frac{\alpha}{2} \text{ and } l = \sqrt{d^2 - a^2} = d\sqrt{1 - 2\sin^2\frac{\alpha}{2}} = d\sqrt{\cos\alpha}.$$

Furthermore,

$$S = S_{sec} + \frac{a^2}{2} + 2\frac{la}{2}$$

$$= d^2\left(\frac{\sin\alpha}{2} + \sin^2\frac{\alpha}{2} + \sqrt{2}\sin\frac{\alpha}{2}\sqrt{\cos\alpha}\right),$$

and consequently $d^2 = \dfrac{2S}{\sin\alpha + 2\sin^2\frac{\alpha}{2} + 2\sqrt{2}\sin\frac{\alpha}{2}\sqrt{\cos\alpha}}$.

Thus, we receive $S_{sec} = \dfrac{S\sin\alpha}{\sin\alpha + 2\sin^2\frac{\alpha}{2} + 2\sqrt{2}\sin\frac{\alpha}{2}\sqrt{\cos\alpha}}$.

Finally, after some simplification we find that the total surface area of the prism is

$$S_{total}=4(S-S_{sec})=4S\frac{\sin\frac{\alpha}{2}+\sqrt{2\cos\alpha}}{\cos\frac{\alpha}{2}+\sin\frac{\alpha}{2}+\sqrt{2\cos\alpha}}.$$

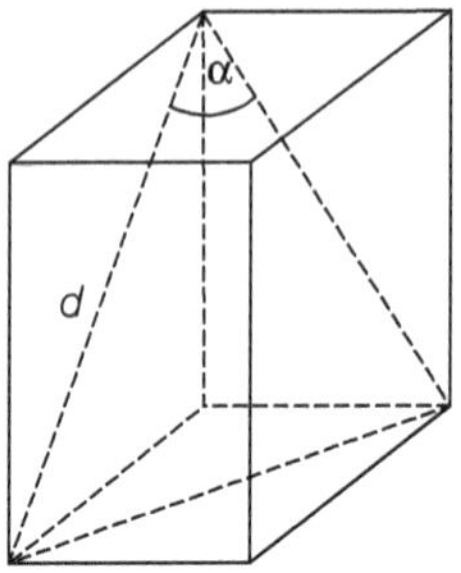

FIG. 149

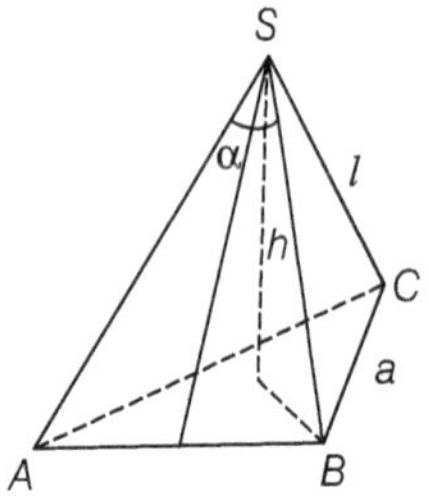

FIG. 150

437. By the well-known lemma by means of which the law of sines is deduced, the side of the base of the pyramid is equal to $a=2r\sin\alpha$. For the lateral edge (see Fig. 150) we have $l=\frac{a}{2}\frac{1}{\sin\frac{\alpha}{2}}=2r\cos\frac{\alpha}{2}$.

Therefore, the altitude of the pyramid is $h=\sqrt{l^2-\left(\frac{a\sqrt{3}}{3}\right)^2}$

$$=2r\sqrt{\cos^2\frac{\alpha}{2}-\frac{\sin^2\alpha}{3}},$$

and hence, the volume of the pyramid is $V=\frac{1}{3}h\frac{a^2\sqrt{3}}{4}$

$$=\frac{2}{3}r^3\sin^2\alpha\sqrt{3\cos^2\frac{\alpha}{2}-\sin^2\alpha}.$$

438. Let $ABC'D'$ be the indicated section of the given pyramid $OABCD$. Draw an auxiliary plane OPN through the vertex of the pyramid and the midpoints of its edges AB and CD (Fig. 151).

It is readily seen that the plane OPN is perpendicular to AB and CD, and the line segments OP and ON are equal.

Applying the law of sines to the triangle OPM we find $\frac{OM}{OP}=\frac{\sin\alpha}{\sin 3\alpha}$.

Since $D'C'\parallel DC$, we have $D'C'=DC\frac{OM}{ON}=a\frac{\sin\alpha}{\sin 3\alpha}$.

Now applying the law of sines to the triangle PMN we obtain $\frac{PM}{PN}=\frac{\sin 2\alpha}{\sin(\pi-3\alpha)}$, which yields $PM=a\frac{\sin 2\alpha}{\sin 3\alpha}$.

Thus we obtain the required area of the section $ABC'D'$:

$$S = \frac{1}{2}(AB + D'C')PM = \frac{1}{2}\left(a + a\frac{\sin\alpha}{\sin 3\alpha}\right)a\frac{\sin 2\alpha}{\sin 3\alpha} = a^2\frac{\sin^2 2\alpha \cos\alpha}{\sin^2 3\alpha}.$$

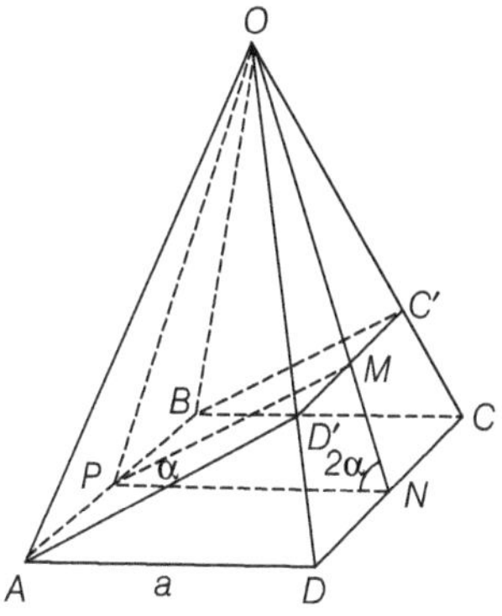

FIG. 151

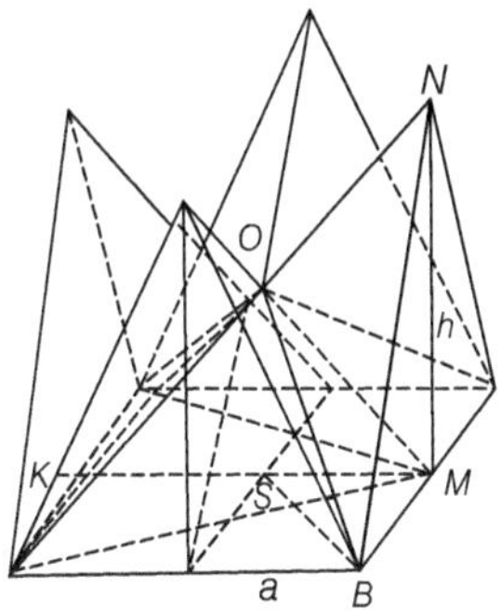

FIG. 152

439. We shall use the notation indicated in Fig. 152. Consider one eighth of the garret $OSBMN$ which consists of two pyramids. One of these pyramids with the base SBM and vertex O has the volume

$$V_1 = \frac{1}{3}SO \cdot S_{SBM} = \frac{a^2 h}{48}.$$

The volume of the other pyramid with the base BMN and vertex O is $V_2 = \frac{a^2 h}{24}$. Thus, the volume V of the garret is given by the formula

$$V = 8(V_1 + V_2) = \frac{a^2 h}{2}.$$

440. Let BM and CM be the perpendiculars dropped from the vertices B and C of the base (see Fig. 153) onto the lateral edge SA. The angle BMC formed by them is the required one. Designate it by β. Obviously, we have

$$\sin\frac{\beta}{2} = \frac{BK}{BM}. \qquad (1)$$

Let a be the side of the base of the pyramid. Then $SK = \frac{a\sqrt{3}}{6\cos\alpha}$ and

$$SB = \sqrt{\left(\frac{a\sqrt{3}}{6\cos\alpha}\right)^2 + \left(\frac{a}{2}\right)^2} = \frac{a}{6\cos\alpha}\sqrt{3(1 + 3\cos^2\alpha)}.$$

From the isosceles triangle ASB we easily find its altitude BM:

$$BM = \frac{a}{\sqrt{1 + 3\cos^2\alpha}}.$$

Thus, by virtue of (1), we obtain $\sin\frac{\beta}{2} = \frac{\sqrt{1+3\cos^2\alpha}}{2}$ and, hence,

$$\beta = 2\arcsin\frac{\sqrt{1+3\cos^2\alpha}}{2}.$$

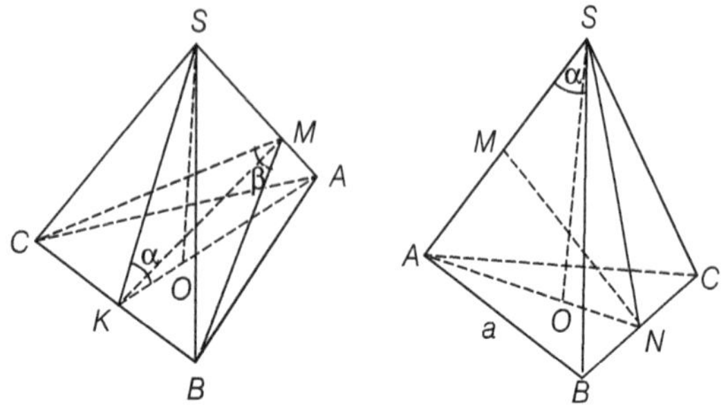

FIG. 153 FIG. 154

441. Draw a plane through the edge SA and the point N which is the foot of the perpendicular AN to the line segments BC (Fig. 154). Let NM be the altitude of the triangle ASN. The line segment NM is perpendicular to AS and BC and is obviously equal to d. Let a denote side of the base of the pyramid. Then $SA = \frac{a}{2\sin\frac{\alpha}{2}}$, and the altitude of the pyramid is $SO = \sqrt{SA^2 - AO^2} = \frac{a}{6\sin\frac{\alpha}{2}}\sqrt{9-12\sin^2\frac{\alpha}{2}}$.

Since $AN \cdot SO = AS \cdot d$, we have $a = \frac{6d}{\sqrt{3}\sqrt{9-12\sin^2\frac{\alpha}{2}}}$.

Finally, we obtain $V = \frac{1}{3}\frac{a^2\sqrt{3}}{4}SO = \frac{d^3}{3\left(3-4\sin^2\frac{\alpha}{2}\right)\sin\frac{\alpha}{2}}$.

442. Let $AD = a$ and $BC = b$ (Fig. 155). Draw the line segment EF joining the midpoints of the bases of the trapezoid. It is obvious that the dihedral angle with edge AD is less than the dihedral angle with edge BC. Let $\angle SEO = \alpha$; then $\angle SFO = 2\alpha$.

We have $$SO = OF \cdot \tan 2\alpha = OE \cdot \tan\alpha.$$

But $OF = \frac{b}{2}\tan\frac{\varphi}{2}$, $OE = \frac{a}{2}\tan\frac{\varphi}{2}$, and thus we obtain the equation

$a\tan\alpha = b\tan 2\alpha$ whose solution is $\tan\alpha = \sqrt{\frac{a-2b}{a}}^{*}$.

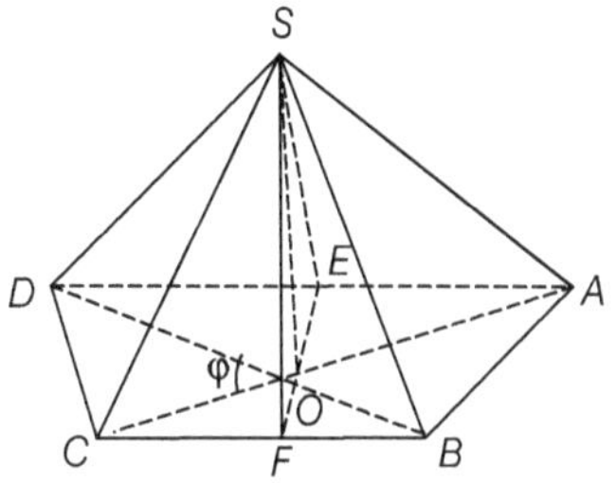

FIG. 155

Furthermore, we obtain $SO = OE \cdot \tan\alpha = \frac{a}{2}\tan\frac{\varphi}{2}\sqrt{\frac{a-2b}{a}}$

and $S_{base} = \frac{a+b}{2}(OE+OF) = \left(\frac{a+b}{2}\right)^2 \tan\frac{\varphi}{2}$.

Finally, the volume of the pyramid is $V = \frac{(a+b)^2}{24}\tan^2\frac{\varphi}{2}\sqrt{a(a-2b)}$.

443. Let $SL \perp AB$, $SK \perp AC$ and SM be the perpendicular to the plane P (Fig. 156). By the hypothesis, $SA = 25$ cm, $SL = 7$ cm and $SK = 20$ cm. Applying the Pythagorean theorem, we easily find that $AK = 15$ cm and $AL = 24$ cm. Extend the line segment KM to intersect the side AB at a point Q. It is readily seen that $\angle AQK = 30°$, and hence $AQ = 30$ cm. Therefore, $LQ = 6$cm, and $LM = 6\tan 30° = 2\sqrt{3}$cm.
From the right triangle SML we now find that $SM = \sqrt{7^2 - (2\sqrt{3})^2} = \sqrt{37}$ cm.

444. Let S be the vertex of the pyramid. SO the altitude and $BN = NC$ (Fig. 157). Designate the side of the base of the pyramid by a. Let us introduce

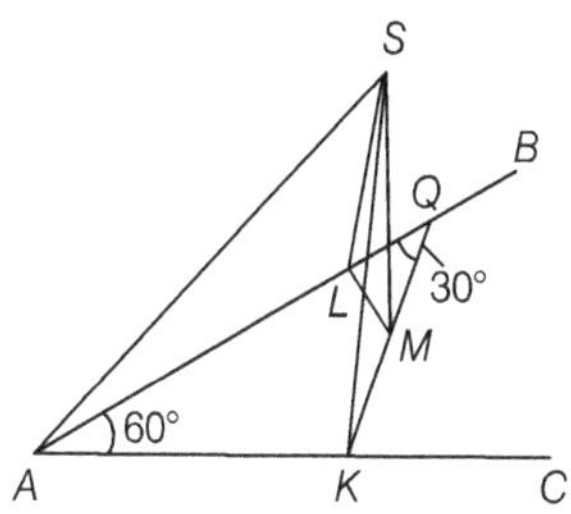

FIG. 156

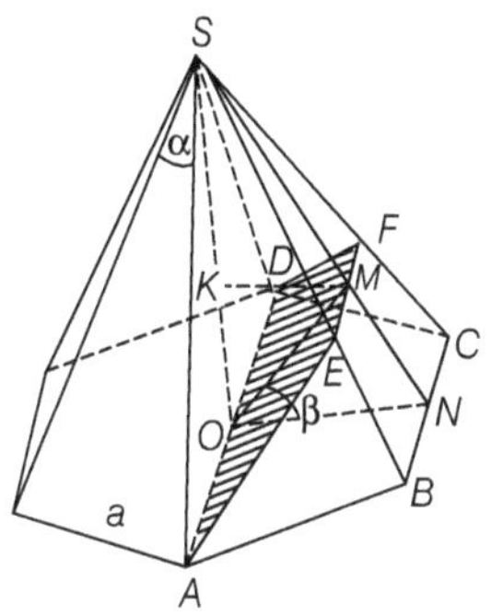

FIG. 157

* This result shows that for $a \leq 2b$ the problem has no solution.

the auxiliary parameter $\frac{SM}{SN}=\lambda$. The similarity of the triangle implies that $EF = a\lambda$, $KM = a\frac{\sqrt{3}}{2}\lambda$.

From ΔMKO we obtain $OM = \frac{KM}{\cos\beta} = \frac{a\lambda}{2\cos\beta}\sqrt{3}$.

The section area is equal to

$$\frac{1}{2}(AD+EF)OM = \frac{1}{2}(2a+\lambda a)\frac{\sqrt{3}}{2}\cdot\frac{\lambda}{\cos\beta}a = \frac{\sqrt{3}}{4\cos\beta}\lambda(\lambda+2)a^2.$$

The area of the base, as the area of a regular hexagon with side a, is equal to $6.\frac{a^2\sqrt{3}}{4}$. Thus, the sought-for ratio of the areas is equal to

$$\frac{1}{6\cos\beta}\lambda(\lambda+2). \qquad (2)$$

Consequently, the problem is now reduced to finding λ. For this purpose, put $\angle SNO=\varphi$. Then, by the law of sines, we obtain from ΔSOM the expression

$$SM = SO\frac{\sin\left(\frac{\pi}{2}-\beta\right)}{\sin(\beta+\varphi)} = SO\frac{\cos\beta}{\sin(\beta+\varphi)}.$$

Since $SO = SN\cdot\sin\varphi$. we can write

$$\lambda = \frac{SM}{SN} = \frac{\cos\beta\sin\varphi}{\sin(\beta+\varphi)} = \frac{1}{1+\tan\beta\cot\varphi}. \qquad (3)$$

Finally, we proceed to find $\cot\varphi$. To this end, not that

$$SN = \frac{a}{2}\cot\frac{\alpha}{2}, \qquad ON = \frac{a\sqrt{3}}{2},$$

$$SO = \sqrt{SN^2-ON^2} = \frac{a}{2}\sqrt{\cot^2\frac{\alpha}{2}-3}$$

and, hence,

$$\cot\varphi = \frac{ON}{SO} = \frac{\sqrt{3}}{\sqrt{\cot^2\frac{\alpha}{2}-3}}.$$

Substituting this value into formula (3), we obtain

$$\lambda = \frac{\sqrt{\cot^2\frac{\alpha}{2}-3}}{\sqrt{\cot^2\frac{\alpha}{2}-3}+\sqrt{3}\tan\beta}.$$

445. From a point S other than the vertex C of the trihedral angle (see Fig. 158) and lying on the edge of the trihedral angle which is not a side of the face angle α, drop the perpendiculars SB and SD onto the sides of this face angle. Also draw the perpendicular SA to the corresponding face. Denote the sought-for angles by β_1 and λ_1, that is $\angle SCB = \gamma_1$, $\angle SCD = \beta_1$.

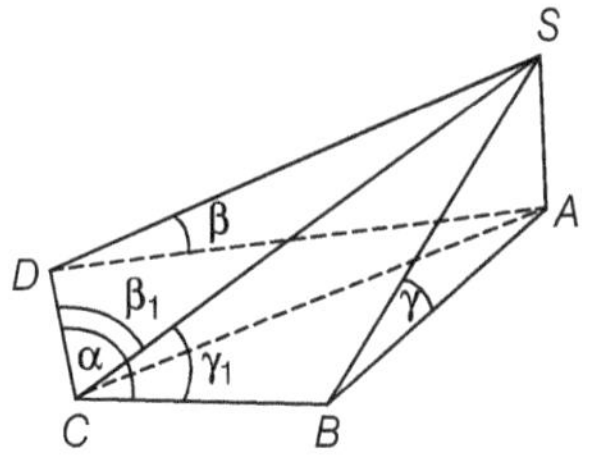

FIG. 158

Let then $\angle ABC = \alpha'$ and $\angle ACD = \alpha''$.

Putting $CA = a$, we find from the right triangles CBA, SBA and SBC the expression

$$\tan\gamma_1 = \frac{SB}{CB} = \frac{a\sin\alpha'}{a\cos\gamma\cos\alpha'} = \sec\gamma\tan\alpha'.$$

We similarly obtain $\tan\beta_1 = \sec\beta\tan\alpha''$.

The problem is thus reduced to finding $\tan\alpha'$ and $\tan\alpha''$. We have $\alpha' + \alpha'' = \alpha$. Computing the line segment SA by two different methods, we find $SA = a\sin\alpha'\tan\gamma$ and $SA = a\sin\alpha''\tan\beta$.

It follows that $\sin\alpha' = \sin\alpha''\tan\beta\cot\gamma$ and, hence,

$$\sin\alpha' = \sin(\alpha - \alpha')\frac{\tan\beta}{\tan\gamma} = (\sin\alpha\cos\alpha' - \cos\alpha\sin\alpha')\tan\beta\cot\gamma.$$

Dividing both members of the last equality by $\cos\alpha'$, we get

$$\tan\alpha' = \frac{\sin\alpha\tan\beta\cot\gamma}{1 + \cos\alpha\tan\beta\cot\gamma}.$$

Interchanging β and γ we find

$$\tan\alpha'' = \frac{\sin\alpha\tan\gamma\cot\beta}{1 + \cos\alpha\tan\gamma\cot\beta}.$$

We thus finally obtain

$$\tan\gamma_1 = \frac{\sin\alpha\tan\beta\csc\gamma}{1 + \cos\alpha\tan\beta\cot\gamma}$$

and

$$\tan\beta_1 = \frac{\sin\alpha\tan\gamma\csc\beta}{2 + \cos\alpha\tan\gamma\cot\beta}.$$

446. The sum of the interior angles in the regular polygon being equal to πn, the number of its sides is equal to $n+2$. Let PQ be the altitude of the pyramid

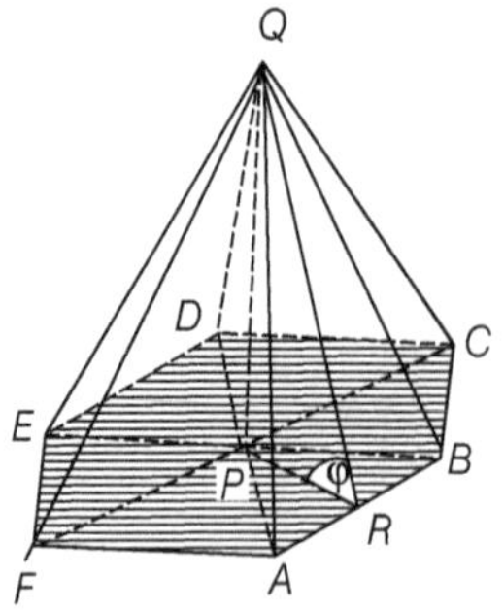

FIG. 159

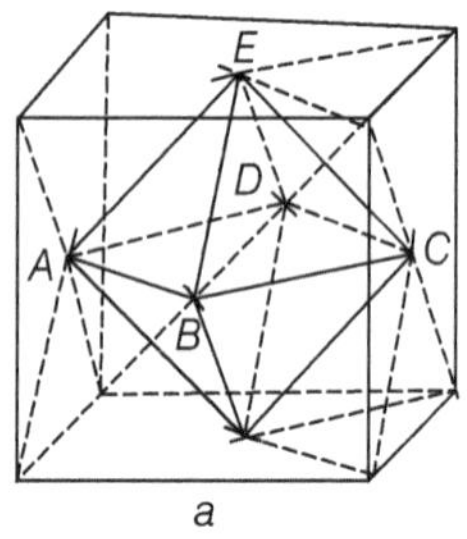

FIG. 160

(Fig. 159). Consider a lateral face of the pyramid, say ΔQAB, and its projection onto the base, i.e. ΔPAB. The conditions of the problem imply that
$$\frac{S_{\Delta PAB}}{S_{\Delta QAB}} = \frac{1}{k}.$$

The areas of the given triangles being in the ratio of their altitudes dropped onto the common base AB, for the cosine of the dihedral angle with edge AB we have $\cos\varphi = \frac{PR}{QR} = \frac{1}{k}$, whence it follows that the apothem of the base of the pyramid is equal to
$$d = h\cot\varphi = h\frac{1}{\sqrt{k^2-1}}.$$

We then find the side of the base: $a = \frac{2h}{\sqrt{k^2-1}}\tan\frac{\pi}{n+2}$.

Since the area of the base is determined by the formula $S = \frac{1}{2}(n+2)ad$, we see that the volume of the pyramid is
$$V = \frac{1}{3}Sh = \frac{1}{3}\frac{(n+2)h^3}{k^2-1}\tan\frac{\pi}{n+2}.$$

447. The solid in question is an octahedron whose vertices are the centres of symmetry of the faces of the cube (Fig. 160). The volume of the octahedron is twice the volume of the regular quadrangular pyramid $EABCD$ with altitude $\frac{a}{2}$, the area of the base $ABCD$ being equal to $\frac{1}{2}a^2$. Hence the required volume is equal to
$$2\times\frac{1}{3}\times\frac{a}{2}\times\frac{1}{2}a^2 = \frac{a^3}{6}.$$

448. It is obvious that the section is the isosceles trapezoid $ABCD$ (see Fig. 161). Let P be the midpoint of the side EF of the base of the pyramid. Consider ΔSPR containing the altitude SO of the pyramid. The line segment KO is apparently the altitude of the trapezoid $ABCD$. Since $KO \parallel SR$, we have $KO = \frac{1}{2}h$ where h is the slant height of the pyramid. It is also $AB = 2a$ where a is the length of the side of the base of the pyramid. We also have $DC = \frac{1}{2}EF = \frac{1}{2}a$ and therefore,

$$S_{tr} = \frac{1}{2}\left(2a + \frac{a}{2}\right).\frac{h}{2} = \frac{5ah}{8} = \frac{5}{4}\left(\frac{1}{2}ah\right)$$

and, hence, the sought-for ratio is equal to $\frac{5}{4}$.

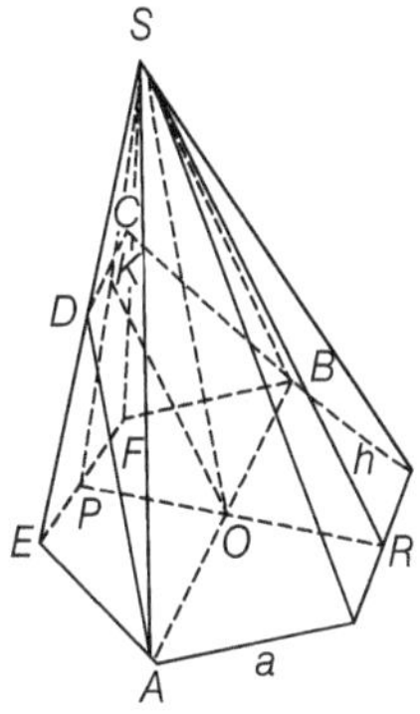

FIG. 161

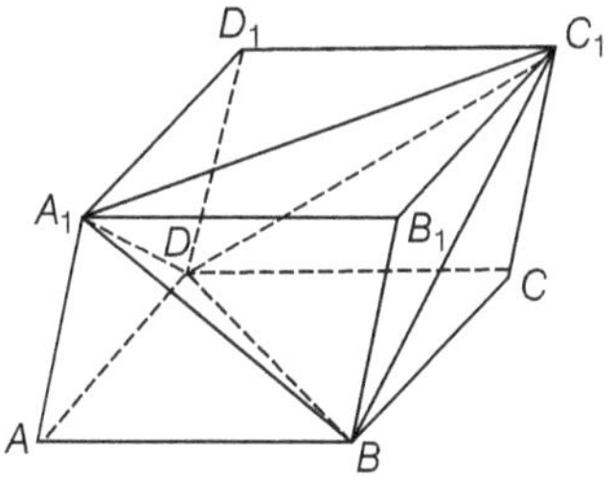

FIG. 162

449. Let A_1BC_1D be the given tetrahedron. $ABCD\,A_1B_1C_1D_1$ the parallelepiped obtained by the indicated construction (see Fig. 162). It is readily seen that the edges of the tetrahedron are the diagonals of the lateral faces of the parallelepiped. The tetrahedron can be obtained by cutting off and removing from the parallelepiped the four congruent pyramids $ABDA_1$ $BDCC_1$, $A_1B_1C_1B$ and $A_1D_1C_1D$. The volume of each pyramid being equal to $\frac{1}{6}$ of the volume of the parallelepiped, the ratio of the volume of the parallelepiped to that of the tetrahedron is equal to

$$\frac{V_p}{V_t} = \frac{V_p}{V_p - \frac{4}{6}V_p} = 3.$$

450. One can easily see that the vertices of the tetrahedrons not lying on the faces of the pyramid are the vertices of a square. To determine the length of the side of the square, draw, through the vertex S of the pyramid and one of those vertices, say A, of the tetrahedrons, a plane perpendicular to the base of the quadrangular pyramid (Fig. 163). This plane also passes through the foot O of the altitude of the pyramid, the foot Q of the altitude of the tetrahedron and the midpoint M of the edge KL. Drop the perpendicular AB onto the base of the pyramid and consider the quadrilateral $SOBA$. Its side OB is half the diagonal of the above square and is to be determined. However, it is easy to reveal that $SOBA$ is a rectangle. Indeed putting $\angle OMS = \alpha$ and $\angle ASM = \beta$, we find

and
$$\cos\alpha = \frac{OM}{MS} = \frac{\frac{1}{2}a}{\frac{\sqrt{3}}{2}a} = \frac{\sqrt{3}}{3}$$

$$\cos\beta = \frac{QS}{SA} = \frac{\frac{\sqrt{3}}{3}a}{a} = \frac{\sqrt{3}}{3}.$$

Therefore SA and OB are parallel and, hence,

$$OB = SA = a.$$

Thus, the sought-for distance is equal to $a\sqrt{2}$.

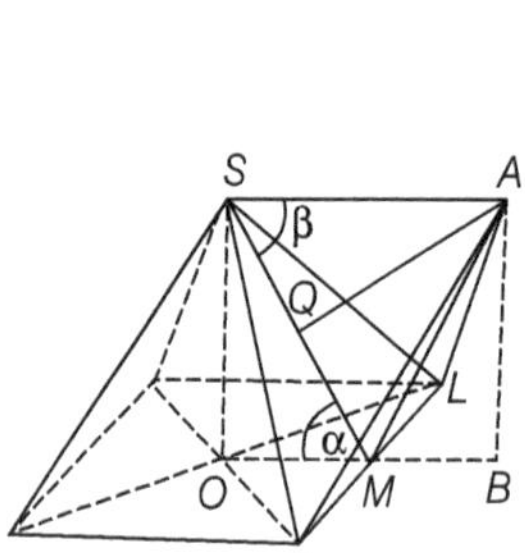

FIG. 163

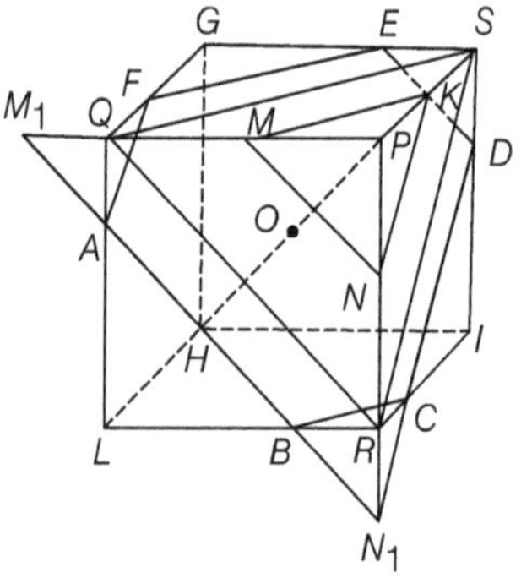

FIG. 164

451. Suppose that the cutting plane passes through a point of the diagonal HP of the given cube (Fig. 164). Let us first consider the sections which Intersect the diagonal at points belonging to the line segment OP. Take the plane section QRS passing through three vertices of the cube. It obviously is one of the indicated sections. This section is an equilateral triangle with side $a\sqrt{2}$. We can easily compute the distance from the centre of the cube to the chosen section which turns out to be equal to $\frac{a\sqrt{3}}{6}$. It is obvious, that for $x \geq \frac{a\sqrt{3}}{6}$ the sections are equilateral triangles. The sides of these triangles being in the ratio of their distances from the point P, we can write

$$\frac{MN}{QR} = \frac{OP - x}{OP - \frac{a\sqrt{3}}{6}}.$$

Now, taking into account that $QR = a\sqrt{2}$ and $OP = \frac{a\sqrt{3}}{2}$, we find

$$MN = \frac{3}{2}\sqrt{2}a - x\sqrt{6}. \qquad (1)$$

But if $\frac{a\sqrt{3}}{6} > x \geq 0$, then the section is a hexagon of the type $ABCDEF$.

The sides AB, FE and CD of the hexagon are, respectively, parallel to the sides QR, QS and RS of the equilateral triangle QRS. Therefore, when extended, these sides intersect and form angles of 60°. Furthermore, taking into account that $AF \parallel CD$ and so on, we arrive at a conclusion that each angle of the hexagon is equal to 120°. It is also readily seen that $AB = CD = EF$ and $BC = DE = AF$ (it should be noted that the sides of the hexagon cut off isosceles triangles from the faces of the cube).

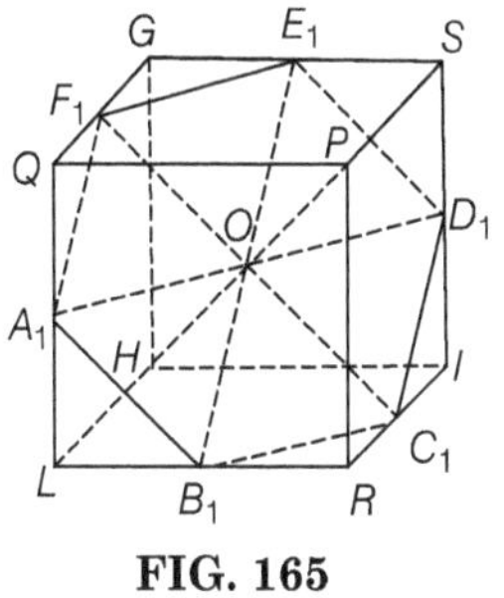

FIG. 165

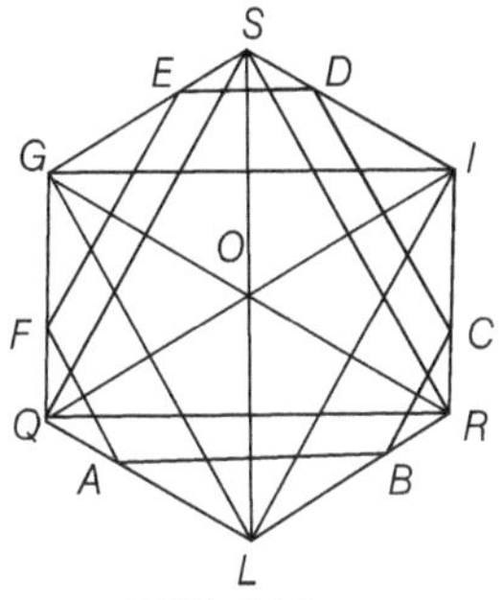

FIG. 166

To find the lengths of the sides of the hexagon let us extend its side AB to intersect the extension of the edges PQ and PR at points M_1 and N_1. It is apparent that the length of the line segment M_1N_1 can be computed by formula (1). Knowing M_1N_1, we find the line segment BN_1:

$$BN_1 = \left(\frac{\sqrt{2}}{2} M_1N_1 - a\right)\sqrt{2} = \frac{a}{2}\sqrt{2} - x\sqrt{6}.$$

It follows that

$$AB = M_1N_1 - 2BN_1 = \frac{a}{2}\sqrt{2} + x\sqrt{6}. \qquad (2)$$

The side BC can be determined in a similar way but it is clear that $BC = BN_1$, and, hence,

$$BC = \frac{a}{2}\sqrt{2} - x\sqrt{6}. \qquad (3)$$

Note, that the section obtained by the cutting plane π passing through the point O is a regular hexagon (consider formulas (2) and (3) for $x = 0$). The vertices of this hexagon are at the midpoints of the edges of the cube (Fig. 165).

It is obvious that if one of the two parts into which the cube is cut by the plane π is turned about the diagonal OP through an angle 60°, the hexagon goes into itself, and we thus obtain two polyhedrons symmetric with respect to the plane π. Consequently, the section intersecting the diagonal at a point of the line segment HO whose distance from the point O is x can be obtained from one of the sections that have been already considered by turning through 60°.

452. The projection is a regular hexagon with side $\frac{a\sqrt{6}}{3}$. To verify this assertion it is convenient to consider the projections of all the plane sections of the cube investigated in Problem 451 (see Fig. 164). All these sections, when projected, do not change their sizes, and thus we obtain the figure shown in Fig. 166.

Knowing that the side of the triangle RQS is equal to $a\sqrt{2}$, we find from the triangle GOS the relation $GS \cdot \frac{\sqrt{3}}{2} = \frac{a\sqrt{2}}{2}$,

which yields $GS = \frac{a\sqrt{6}}{3}$. The side of the regular hexagon $A_1B_1C_1D_1E_1F_1$ (see Fig. 165) being equal to $\frac{a\sqrt{2}}{2}$, the sought-for ratio of the areas is equal to

$$\left(\frac{a\sqrt{6}}{3}\right)^2 : \left(\frac{a\sqrt{2}}{2}\right)^2 = \frac{4}{3}.$$

453. Let $AEFD$ be the isosceles trapezoid obtained in the section, and G and H be the midpoints of its bases (see Fig. 167). Drop the perpendicular HK from the point H onto the base of pyramid. Since H is the midpoint of SN, we have

$$HK = \frac{h}{2}, KN = \frac{a}{4} \text{ and } GK = \frac{3a}{4}. \tag{1}$$

Now we shall determine the lengths of the line segments QO and QS. We have

$$\frac{QO}{HK} = \frac{GO}{GK},$$

and therefore, taking into account (1), we obtain

$$QO = \frac{h}{2} \cdot \frac{a}{2} \cdot \frac{4}{3a} = \frac{h}{3}.$$

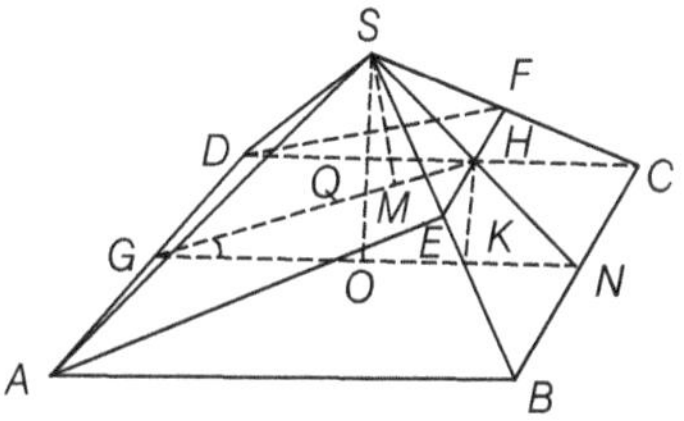

FIG.167

It follows that

$$QS = \frac{2}{3} h \text{ and } GQ = \sqrt{\left(\frac{a}{2}\right)^2 + \left(\frac{h}{3}\right)^2}. \tag{2}$$

Draw the perpendicular SM from the point S to GH. Then the similarity of the triangles SMQ and a GOQ implies that

$$\frac{SM}{QS} = \frac{GO}{GQ},$$

and, consequently, the sought-for distance is

$$SM = QS \cdot \frac{GO}{GQ} = \frac{2ah}{\sqrt{9a^2 + 4h^2}}.$$

454. The solid in question is made up of two pyramids with common base KMN (Fig. 168). We can easily find the altitude OR of the lower pyramid by dropping the perpendicular PD from the midpoint P of the side KN onto the base of the pyramid. The point D bisects the line segment QL. Taking advantage of this fact, we obtain, from ΔAPD, the relation

$$\frac{PD}{RQ} = \frac{DA}{QA} = \frac{5}{4},$$

whence we find $RQ = \frac{4}{5} PD$ and, hence,

$$OR = \frac{1}{5} PD = \frac{1}{5} \cdot \frac{1}{2} a \sqrt{\frac{2}{3}} = \frac{a\sqrt{6}}{30}.$$

Here we have taken advantage of the fact that in a regular tetrahedron with edge a the altitude is equal to $a\sqrt{\frac{2}{3}}$. The required volume is $V = \frac{a^3\sqrt{2}}{80}$.

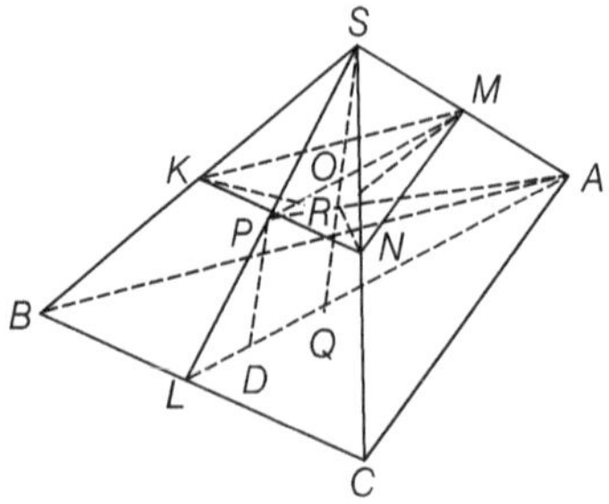

FIG. 168

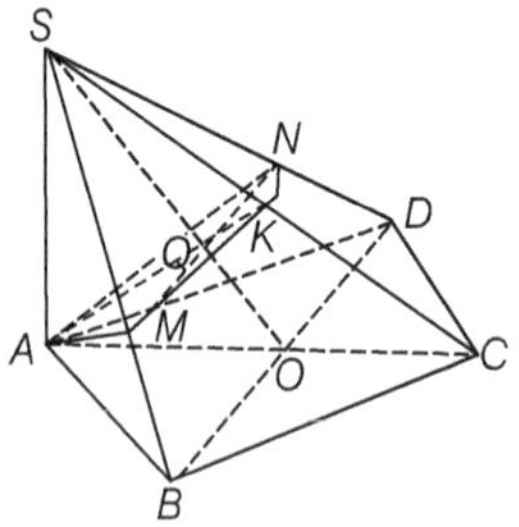

FIG. 169

455. Let $AMKN$ be the quadrilateral obtained in the section, and Q be the point of intersection of its diagonals (see Fig. 169). Considering ΔSAC, we readily note that Q lies in the point of intersection of the medians of this triangle. Therefore,

$$\frac{MN}{BD}=\frac{SQ}{SO}=\frac{2}{3},$$

and, hence, $MN=\frac{2}{3}b$. Furthermore, from the right triangle SAC we find

$$AK=\frac{1}{2}SC=\frac{1}{2}\sqrt{q^2+a^2}.$$

Since $AK\perp MN$, we have

$$S_{sec}=\frac{1}{2}AK.MN=\frac{b}{6}\sqrt{q^2+a^2}.$$

456. Let NQN_1Q_1 and LML_1M_1 be the parallel sections of the prism (Fig. 170), a the length of the diagonal AC of the base and H the length of the line segment KK_1. Then the area of the first section is

$$S=\frac{H}{2}\left(a+\frac{a}{2}\right)=\frac{3}{4}Ha.$$

The area of the other section is

$$S'=\frac{1}{2}PT(A_2C_2+LM)+\frac{1}{2}P_1T(A_2C_2+L_1M_1).$$

But we have

$$A_2C_2=a, LM=\frac{a}{4}, L_1M_1=\frac{3}{4}a, PT=\frac{3}{4}H \text{ and } P_1T=\frac{1}{4}H,$$

which is obviously implied by the similarity of the corresponding triangles. Therefore we obtain

$$S'=\frac{11}{16}aH$$

and, hence, $S'=\frac{11}{12}S.$

Note. This problem can also be solved in a simple way if we take into consideration the formula

$$S_{pr} = S\cos\varphi, \qquad (1)$$

where S is the area of a polygon in a plane P, S_{pr} is the area of the projection of this polygon on a plane Q and φ the angle between the planes P and Q. According to formula (1), the areas of the parallel sections in the problem are in the ratio of the areas of their projections. Therefore, the problem is reduced to finding the areas of two plane figures shown in Fig. 171, namely

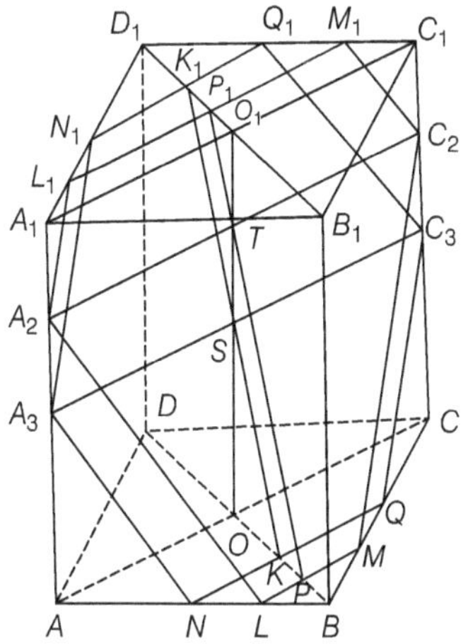

FIG. 170

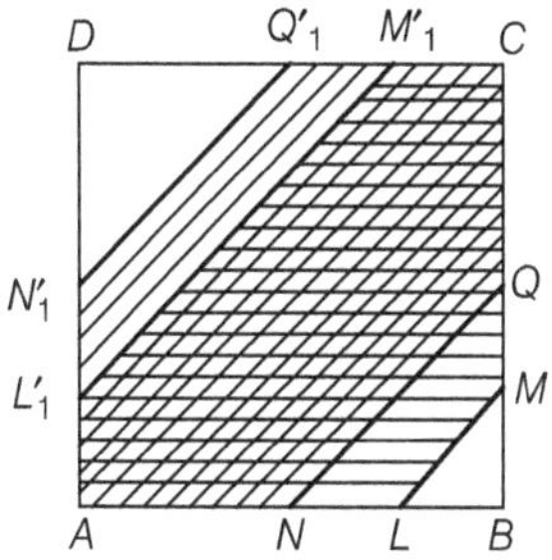

FIG. 171

$L'_1M'_1CMLA$ and $N'_1Q'_1CQNA$ (the primed letters denote the projections of the corresponding points onto the base of the prism).

457. Consider the pyramid $KAEF$ shown in Fig. 172 which is one polyhedrons. We suppose that

$$\frac{AE}{EB} = \frac{AF}{FC} = \frac{1}{2}.$$

Therefore,
$$\frac{AE}{AB} = \frac{AF}{AC} = \frac{1}{3}$$

and, hence, $S_{\Delta AEF} = \frac{1}{9} S_{\Delta ABC}$. (1)

Now, let KM and SN be the altitudes of the pyramids $KAEF$ and $SABC$. As is seen,

$$\frac{KM}{SN} = \frac{AK}{AS} = \frac{2}{3}.$$

Therefore, $KM = \frac{2}{3}$ SN and, taking into account (1), we obtain

$$V_{KAEF} = \frac{2}{27} V_{SABC}.$$

The sought-for ratio is equal to $\frac{2}{25}$.

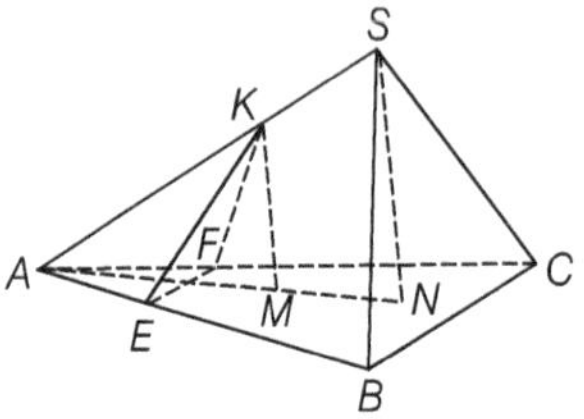

FIG. 172

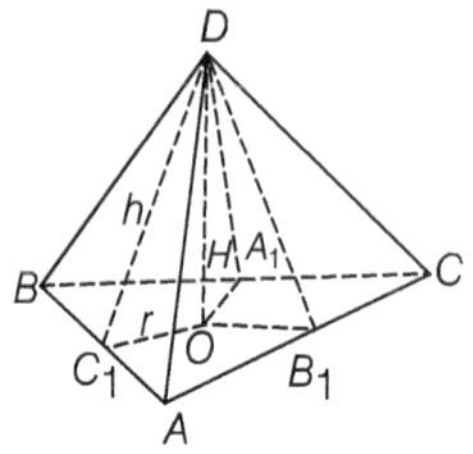

FIG. 173

458. Let the face of area S_0 be the base ABC of the given pyramid $ABCD$, DO the altitude of the pyramid, and DA_1, DB_1 and DC_1 the altitudes of the lateral faces (Fig. 173).

The line segments OC_1, OA_1 and OB_1 being, respectively, the projections of DC_1, DA_1 and DB_1 onto the base ABC, we have $OC_1 \perp AB$, $OA_1 \perp BC$ and $OB_1 \perp AC$, and therefore $\angle DC_1O, \angle DA_1O$ and $\angle DB_1O$ are the plane angles of the corresponding dihedral angles and, by the hypothesis, are equal. It follows that the triangles DOC_1, DOA_1 and DOB_1 are congruent. To facilitate the computation, let us introduce the following notation:

$$DO = H,\ DC_1 = DA_1 = DB_1 = h,$$
$$OC_1 = OA_1 = OB_1 = r,$$
$$S_1 + S_2 + S_3 = S.$$

It is obvious that r is the radius of the inscribed circle of $\Delta\, ABC$. The volume of the pyramid $ABCD$ is

$$V = \frac{1}{3} S_0 H. \tag{1}$$

From the right triangle DOC_1 we obtain

$$H = \sqrt{h^2 - r^2}. \tag{2}$$

Thus, the problem is reduced to finding h and r, From the formulas $S_1 = \frac{1}{2} BCh, S_2 = \frac{1}{2} ACh$ and $S_3 = \frac{1}{2} ABh$ we find the expressions for the sides of the triangle ABC:

$$AB = \frac{2S_3}{h}, BC = \frac{2S_1}{h}, AC = \frac{2S_2}{h}.$$

Hence, we have

$$p = \frac{1}{2}(AB + BC + AC) = \frac{S_3}{h} + \frac{S_1}{h} + \frac{S_2}{h} = \frac{S}{h}.$$

Furthermore,
$$p - AB = \frac{S}{h} - \frac{2S_3}{h} = \frac{S - 2S_3}{h},$$
$$p - BC = \frac{S - 2S_1}{h}, \quad p - AC = \frac{S - 2S_2}{h}.$$

and, hence by Heron's formula, we obtain

$$S_0^2 = p(p - AB)(p - BC)(p - AC)$$
$$= \frac{S}{h} \cdot \frac{S - 2S_1}{h} \cdot \frac{S - 2S_2}{h} \cdot \frac{S - 2S_3}{h}$$
$$= \frac{S(S - 2S_1)(S - 2S_2)(S - 2S_3)}{h^4}.$$

Consequently,

$$h = \frac{\sqrt[4]{S(S - 2S_1)(S - 2S_2)(S - 2S_3)}}{\sqrt{S_0}}. \qquad (3)$$

The radius r of the inscribed circle is found from the formula expressing the area S_0 of the triangle ABC in terms of this radius and $p = \frac{1}{2}(AB + BC + AC)$:

$$S_0 = pr = \frac{S}{h} r,$$

which yields $r = h \frac{S_0}{S}$.

Substituting this value into formula (2) we obtain

$$H = \sqrt{h^2 - h^2 \frac{S_0^2}{S^2}} = \frac{h}{S}\sqrt{S^2 - S_0^2}.$$

Now substituting the value of h determined by formula (3) into the expression of H and then the result thus obtained into formula (1), we finally receive

$$V = \frac{1}{3}\sqrt{S_0(S^2 - S_0^2)} \ \sqrt[4]{\frac{(S - 2S_1)(S - 2S_2)(S - 2S_3)}{S^3}}.$$

459. Cut the cube into two congruent parts by the plane perpendicular to the axis of revolution and turn the polyhedron thus obtained through 90°. The resulting geometric configuration is shown in Fig. 174.

The common portion is made up of the rectangular parallelepiped $ABCDD_1A_1B_1C_1$ and the regular pyramid $SABCD$. The altitude of the parallelepiped is found from ΔBB_1T.

$$h = B_1T = \frac{a\sqrt{2}}{2} - \frac{a}{2}.$$

The altitude of the pyramid is $H = \frac{a\sqrt{2}}{2} - h = \frac{a}{2}$.

The area of the common base is equal to a^2.

Thus, the sought for volume of the common portion is

$$V = 2\left[a^2\left(\frac{a\sqrt{2}}{2} - \frac{a}{2}\right) + a^2 \cdot \frac{a}{2} \cdot \frac{1}{3}\right]$$

that is $$V = a^3\left(\sqrt{2} - \frac{2}{3}\right).$$

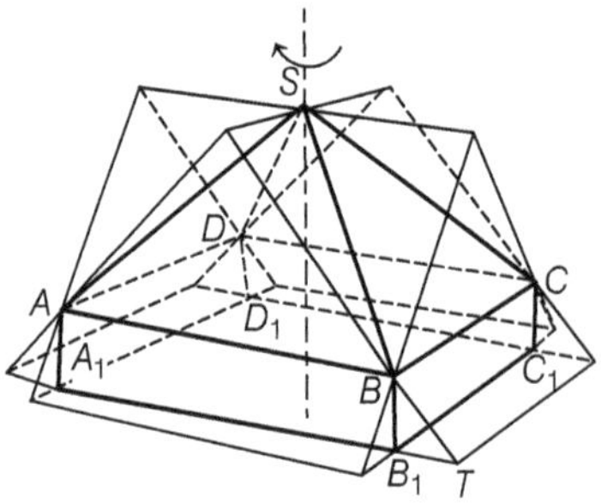

FIG. 174

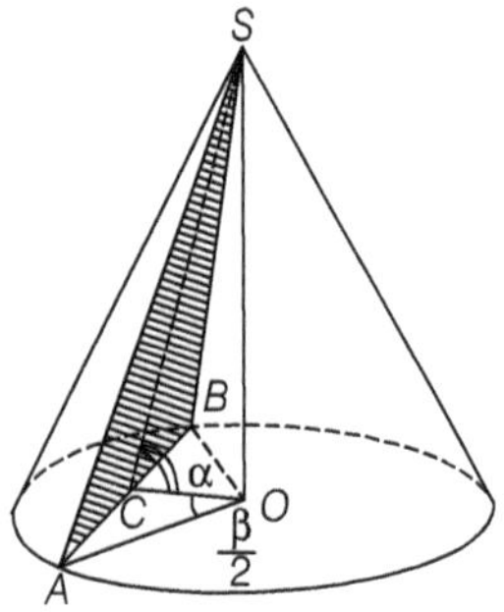

FIG. 175

460. Let S be the vertex of the cone, $SO = h$ the altitude of the cone, ASB the triangle obtained in the section, C the midpoint of the chord AB and $AO = r$ (Fig. 175). Noting that $\angle AOC = \frac{\beta}{2}$, we find

$$CO = \frac{a}{2}\cot\frac{\beta}{2},\ h = CO\tan\alpha = \frac{a}{2}\tan\alpha\cot\frac{\beta}{2},\ r = \frac{a}{2\sin\frac{\beta}{2}}.$$

Therefore the volume of the cone is

$$V = \frac{1}{3}\pi r^2 h = \frac{\pi a^3}{24}\,\frac{\tan\alpha\cos\frac{\beta}{2}}{\sin^3\frac{\beta}{2}}.$$

461. Let α be the required angle, l the length of the generator of the cylinder, l_1 the slant height of the cone, r the radius of the common base of the cone and cylinder (Fig. 176). By the hypothesis, we have

$$\frac{2\pi r(r+l)}{\pi r(r+l_1)} = \frac{7}{4}$$

and $$\frac{r+l}{r+l_1} = \frac{7}{8}.$$

Consequently, $$\frac{1+\frac{l}{r}}{1+\frac{l_1}{r}}=\frac{7}{8}, \text{ or } \frac{1+\cot\alpha}{1+\csc\alpha}=\frac{7}{8},$$

and, hence, $\sin\alpha+8\cos\alpha-7=0.$

Solving this equation we find

$$\sin\alpha=\frac{3}{5}, \alpha=\text{arc sin}\frac{3}{5}.$$

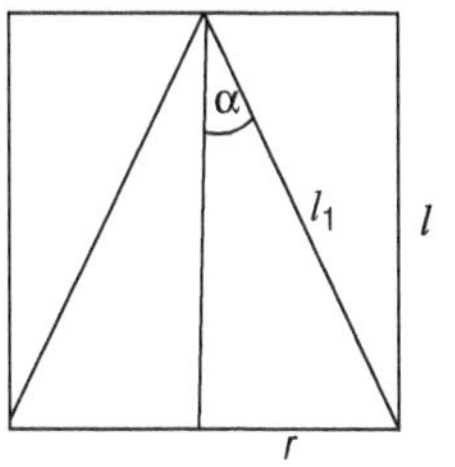

FIG. 176

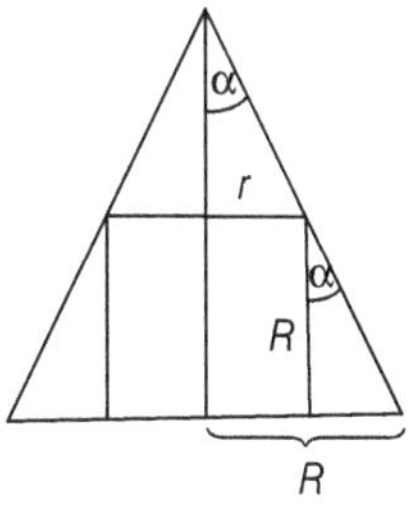

FIG. 177

462. Let α be the sought-for angle, R the radius of the base of the cone and r the radius of the base of the cylinder (Fig. 177). We have

$$\frac{2\pi r^2+2\pi rR}{\pi R^2}=2\left(1+\frac{r}{R}\right)\frac{r}{R}=\frac{3}{2}.$$

But $\frac{R-r}{R}=\tan\alpha$ and, hence, $\frac{r}{R}=1-\tan\alpha$. Thus we obtain the following equation with respect to $\tan\alpha$:

$$4\tan^2\alpha-12\tan\alpha+5=0.$$

Solving it, we find $\tan\alpha=\frac{5}{2}$ or $\tan\alpha=\frac{1}{2}$.

But it is easily seen that $\tan\alpha=\frac{R-r}{R}<1$, therefore $\tan\alpha=\frac{1}{2}$, and , hence, $\alpha=\text{arc}\tan\frac{1}{2}$.

463. Let l be the slant height of the cone and R the radius of its base, x the length of the edge of the prism, r the radius of the circle circumscribed about the base of the prism (Fig. 178). Consider the triangle formed by the altitude of the cone, an element of the cone passing through a vertex of the prism and the projection of that element onto the base of the cone. We have $\frac{l\sin\alpha}{l\sin\alpha-x}=\frac{R}{r}$.

Since $r=\frac{x}{2\sin\frac{\pi}{n}}$ and $R=l\cos\alpha$, we obtain

$$x = \frac{2l \sin\alpha \sin\dfrac{\pi}{n}}{2\sin\dfrac{\pi}{n} + \tan\alpha}.$$

Consequently, the total surface area of the prism is

$$S = \frac{1}{2} nx^2 \cot\frac{\pi}{n} + nx^2 = n\left(\frac{2l\sin\alpha\sin\dfrac{\pi}{n}}{2\sin\dfrac{\pi}{n} + \tan\alpha}\right)^2 \left(1 + \frac{1}{2}\cot\frac{\pi}{n}\right).$$

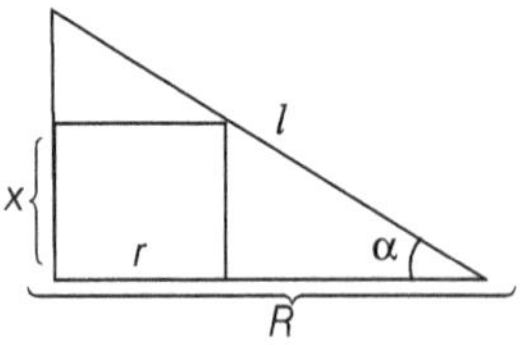

FIG. 178

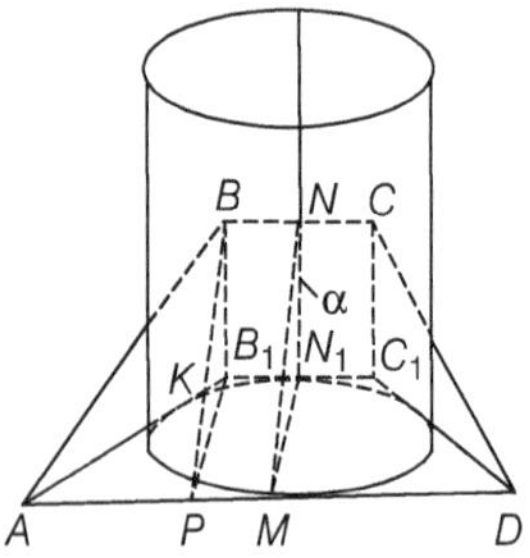

FIG.179

464. Consider the isosceles trapezoid AB_1C_1D which is the projection of the given trapezoid $ABCD$ onto the plane perpendicular to the axis of the cylinder shown in Fig. 179. The projected trapezoid is circumscribed about a circle, and , hence,

$$AB_1 = AK + KB_1 = AM + B_1N_1 = \frac{a+b}{2}.$$

From the right triangle APB_1 we obtain

$$\left(\frac{a+b}{2}\right)^2 = \left(\frac{a-b}{2}\right)^2 + h^2\sin^2\alpha.$$

It follows that

$$\sin\alpha = \frac{\sqrt{ab}}{h} \text{ and } \alpha = \arcsin\frac{\sqrt{ab}}{h}.$$

465. Let R be the radius of the sphere, and a, b and c be, respectively, the legs and the hypotenuse of the triangle ABC which is the base of the prism (Fig. 180). We have

$$a = \frac{h}{\cos\alpha}, b = \frac{h}{\sin\alpha},$$

$$c = \frac{a}{\sin\alpha} = \frac{h}{\cos\alpha\sin\alpha}.$$

The radius R is obviously equal to the radius of the inscribed circle of ΔABC.

Therefore,

$$R = \frac{2S_{\Delta ABC}}{a+b+c} = \frac{ab}{a+b+c} = \frac{h}{1+\sin\alpha+\cos\alpha},$$

and, hence, the volume of the prism is

$$V = S_{\Delta ABC} 2R = \frac{2h^3}{\sin 2\alpha(1+\sin\alpha+\cos\alpha)}.$$

466. The volume of the pyramid is equal to the sum of the volumes of the pyramids which are obtained by joining the centre of the inscribed sphere O to the vertices of the pyramid. The altitude of each constituent pyramid is equal to the radius r of the sphere inscribed in the given pyramid. If S is the

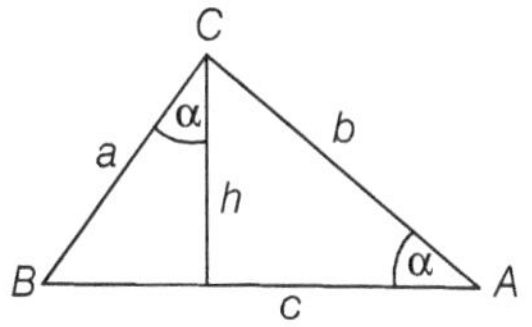

FIG. 180

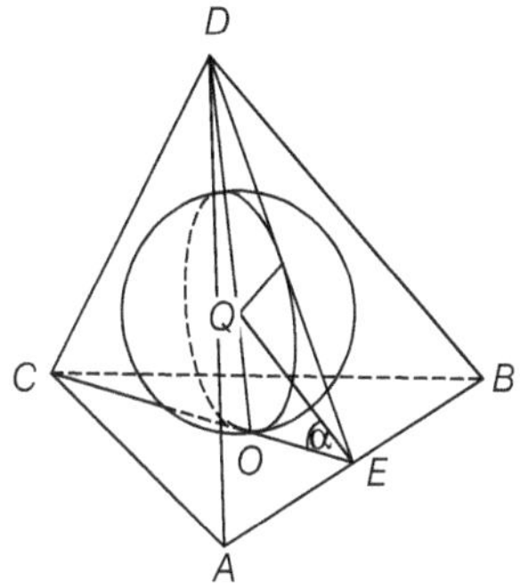

FIG. 181

area of the base of the pyramid and S_1 the lateral area, the volume of the pyramid will be

$$V = \frac{1}{3}(S_1 + S)r. \qquad (1)$$

on the other hand, we have

$$V = \frac{1}{3}hS,$$

and thus we obtain the following formula for r:

$$r = \frac{hS}{S_1 + S}. \qquad (2)$$

From the conditions of the problem it follows that

$$S = \frac{na^2}{4}\cot\frac{\pi}{n},$$

$$S_1 = \frac{na}{2}\sqrt{b^2 - \frac{a^2}{4}}$$

and
$$h = \sqrt{b^2 - \frac{a^2}{4\sin^2\frac{\pi}{n}}}.$$

Substituting these expressions into (2) we find

$$r = \frac{na^2 \cot\frac{\pi}{n}\sqrt{b^2 - \frac{a^2}{4\sin^2\frac{\pi}{n}}}}{4\left(\frac{na^2}{4}\cot\frac{\pi}{n} + \frac{na}{2}\sqrt{b^2 - \frac{a^2}{4}}\right)} = \frac{a\sqrt{4b^2 - a^2\csc^2\frac{\pi}{n}}}{2\left(a + \tan\frac{\pi}{n}\sqrt{4b^2 - a^2}\right)}$$

467. Let us denote by r the radius of the inscribed sphere and by a the length of the line segment OE (Fig. 181). Then

$$r = a\tan\alpha,$$

where α is half the sought-for angle (see Fig. 181). Hence, the volume of the sphere is

$$V_{sph} = \frac{4}{3}\pi a^3 \tan^3\alpha.$$

Since $DO = a\tan 2\alpha$ and $AB = 2\sqrt{3}a$, the volume of the pyramid is

$$V_{pyr} = \frac{1}{3}DO\frac{\sqrt{3}}{4}AB^2 = \sqrt{3}a^3\tan 2\alpha.$$

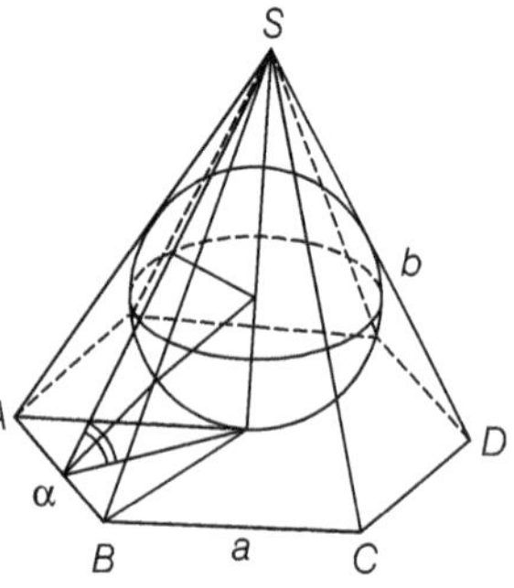

(a)

By the hypothesis,

we have
$$\frac{V_{pyr}}{V_{sph}} = \frac{27\sqrt{3}}{4\pi}.$$

Expressing $\tan 2\alpha$ in terms of $\tan\alpha$ we get the equation

$$\tan^2\alpha(1 - \tan^2\alpha) = \frac{2}{9}.$$

It follows that $(\tan\alpha)_1^2 = \frac{1}{3}$ and $(\tan\alpha)_2^2 = \frac{2}{3}$.

Taking into consideration that α is an acute angle, we find

$$\alpha_1 = \frac{\pi}{6}$$

and $\alpha_2 = \arctan\sqrt{\frac{2}{3}}$.

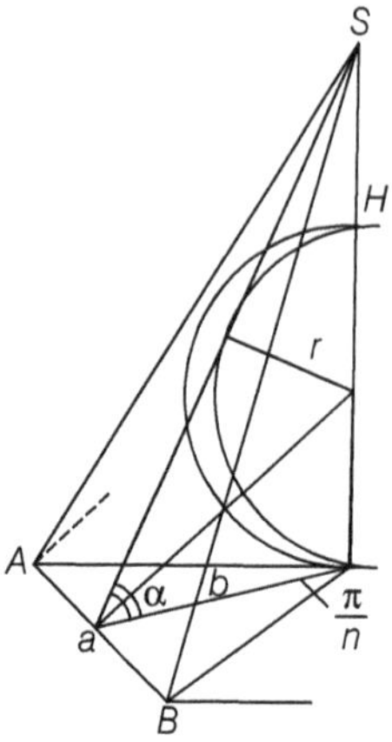

(b) **FIG. 182**

468. Let a be the side and b the apothem of the regular n-gon which is the base of the pyramid, and H be the altitude of the pyramid. Then (see Fig. 182, a and b) we have

$$b = r\cot\frac{\alpha}{2} \quad \text{and} \quad a = 2b\tan\frac{\pi}{n} = 2r\cot\frac{\alpha}{2}\tan\frac{\pi}{n}.$$

The area of the base is $\quad S_{base} = n\dfrac{ab}{2} = nr^2\tan\dfrac{\pi}{n}\cot^2\dfrac{\alpha}{2}.$

Furthermore, $H = b\tan\alpha = r\tan\alpha\cot\dfrac{\alpha}{2}$, and hence the volume of the pyramid is $\quad V_{pyr} = \dfrac{1}{3}nr^3\cot^3\dfrac{\alpha}{2}\tan\alpha\tan\dfrac{\pi}{n}.$

Since the volume of the sphere is $V_{sph} = \dfrac{4}{3}\pi r^3$, we can write

$$\frac{V_{sph}}{V_{pyr}} = \frac{4\pi}{n}\tan^3\frac{\alpha}{2}\cot\alpha\cot\frac{\pi}{n}.$$

469. Let a be the side of the base of the pyramid, b the apothem of the base, R the radius of the circumscribed circle of the base, h the altitude of the pyramid, r the radius of the sphere inscribed in the pyramid, y the slant height of the pyramid (see Fig. 183, a and b)

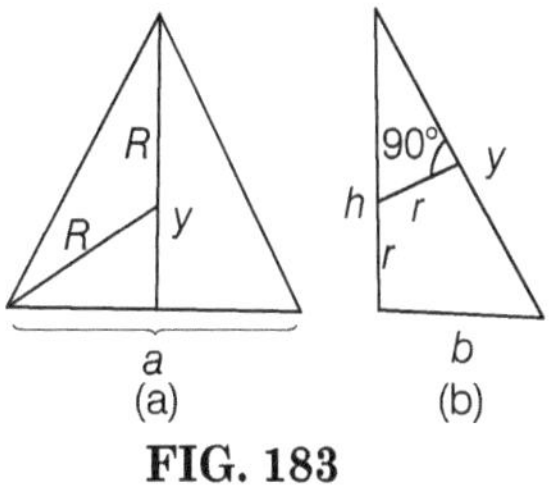

FIG. 183

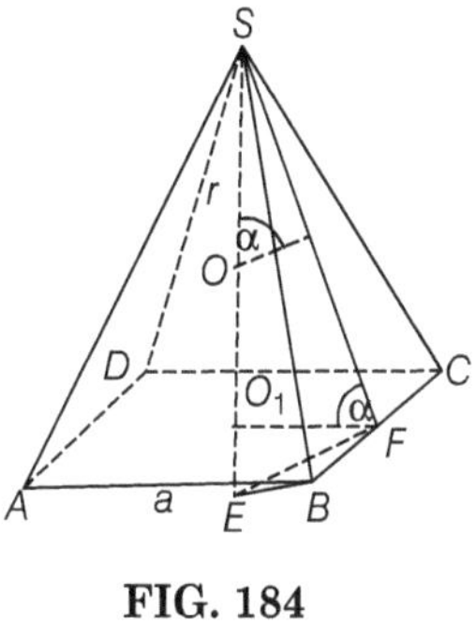

FIG. 184

Then $a = 2R\sin\dfrac{\pi}{n}$, $b = R\cos\dfrac{\pi}{n}$

and, besides,

$$y = R + \sqrt{R^2 - \frac{a^2}{4}} = R\left(1 + \cos\frac{\pi}{n}\right) \quad \text{and} \quad h = \sqrt{y^2 - b^2} = R\sqrt{1 + 2\cos\frac{\pi}{n}}.$$

From the equation $\dfrac{r}{h-r} = \dfrac{b}{y}$ (see Fig. 183, b), we find

$$r = \frac{hb}{y+b} = \frac{R\cos\frac{\pi}{n}\sqrt{1+2\cos\frac{\pi}{n}}}{1+2\cos\frac{\pi}{n}}.$$

Hence, the sought-for ratio is equal to

$$\frac{\frac{1}{3}h\cdot\frac{1}{2}nab}{\frac{4}{3}\pi r^3}=\frac{n\sin\frac{\pi}{n}\left(1+2\cos\frac{\pi}{n}\right)^2}{4\pi\cos^2\frac{\pi}{n}}.$$

470. Let a be the side of the base of the given pyramid $SABCD$, h the altitude of the pyramid, r the radius of the sphere circumscribed about the pyramid (Fig. 184).

Then $\quad V=\frac{4}{3}\pi r^3$ and $r=\left(\frac{3V}{4\pi}\right)^{\frac{1}{3}}.$

If SE is the diameter of the circumscribed sphere, then from the right triangle SBE we find $\left(\frac{a\sqrt{2}}{2}\right)^2=h(2r-h).$

However, from the triangle FO_1S we have $\frac{a}{2}=h\cot\alpha$, and therefore, eliminating a, we receive

$$h=\frac{2r}{2\cot^2\alpha+1}=\frac{1}{1+2\cot^2\alpha}\left(\frac{6V}{\pi}\right)^{\frac{1}{3}}.$$

471. Taking advantage of the equality of the dihedral angles we can readily show, as in Problem 458, that the perpendicular dropped from the vertex onto the base is projected in the centre of symmetry of the rhombus. It is also obvious that the centre of the inscribed sphere lies on that perpendicular.

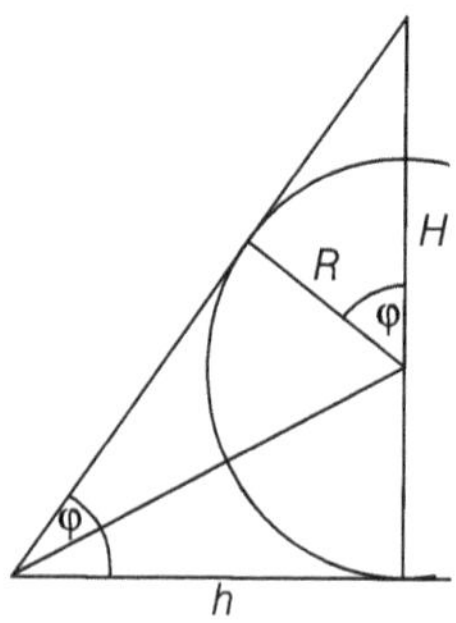

FIG. 185

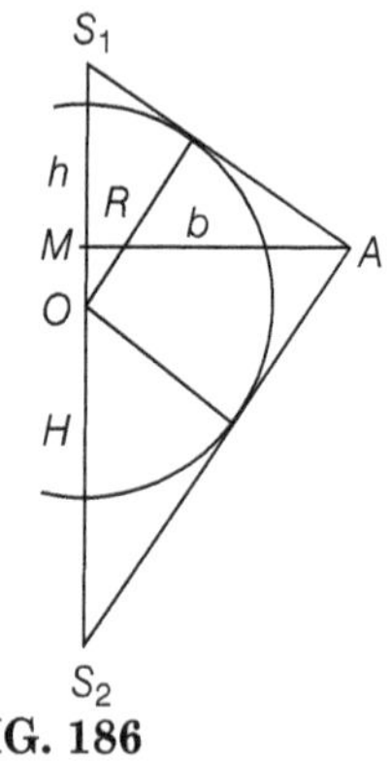

FIG. 186

Let a be the side of the rhombus, $2h$ the altitude of the rhombus, and H the altitude of the pyramid (Fig. 185). Then the area of the base is $S=a^2\sin\alpha$, and thus, since $a=\frac{2h}{\sin\alpha}$,

we obtain
$$S = \frac{4h^2}{\sin\alpha}.$$

But $h = R\cot\frac{\psi}{2}$ (see the section passing through the altitude of the pyramid and altitude of the rhombus shown in Fig. 185). It is also clear that

$$H = R + \frac{R}{\cos\psi} = R\frac{2\cos^2\frac{\psi}{2}}{\cos\psi}.$$

We finally obtain the volume of the prism:

$$V = \frac{8}{3}R^3\frac{\cos^4\frac{\psi}{2}}{\sin\alpha\cos\psi\sin^2\frac{\psi}{2}}.$$

472. Draw a plane through the vertices S_1 and S_2 of the pyramids and the midpoint A of a side of the base (Fig. 186). The radius of the semicircle inscribed in the triangle AS_1S_2 so that its diameter lies on S_1S_2 is obviously equal to the radius of the inscribed sphere. Let O be the centre of the semicircle. Denote by b the aititude in the triangle AS_1S_2 dropped onto the side S_1S_2. Since b is the apothem of the regular n-gon, we have

$$b = \frac{a}{2}\cot\frac{\pi}{n}.$$

Computing the area S of the triangle AS_1S_2 by means of the two methods indicated below we can find the radius of the sphere R. Indeed, on one hand, we have

$$S = \frac{b}{2}(H+h),$$

and, on the other hand,

$$S = \frac{R}{2}S_1A + \frac{R}{2}S_2A = \frac{R}{2}(\sqrt{h^2+b^2} + \sqrt{H^2+b^2}).$$

This result in the final formula

$$R = \frac{\frac{1}{2}a(H+h)\cot\frac{\pi}{n}}{\sqrt{h^2+\frac{a^2}{4}\cot\frac{\pi}{n}} + \sqrt{H^2+\frac{a^2}{4}\cot^2\frac{\pi}{n}}}$$

473. Let h_1 and h_2 be the altitudes of the pyramids, and r the radius of the circle circumscribed about the base (Fig. 187). Then we have

$$\frac{a}{2} = r\sin\frac{\pi}{n}.$$

From the right triangle S_1AS_2 whose vertices are the vertices of the given pyramids and one of the vertices of the base we find

$$h_1 h_2 = r^2 = \frac{a^2}{4\sin^2\frac{\pi}{n}}.$$

But $\qquad h_1 + h_2 = 2R,$

and, hence, $\qquad h_1 = R + \sqrt{R^2 - \frac{a^2}{4\sin^2\frac{\pi}{n}}},\quad h_1 = R - \sqrt{R^2 - \frac{a^2}{4\sin^2\frac{\pi}{n}}}.$

The problem is solvable if $R \geq \frac{a}{2\sin\frac{\pi}{n}}$.

474. It can easily be proved that the midpoint of the line segment joining the centres of the bases of the prism is the centre of the inscribed and circumscribed spheres. The radius of the circle inscribed in the base is equal to the radius of the inscribed sphere. Let r be the radius of the inscribed sphere and R the radius of the circumscribed sphere. Consider the right triangle whose vertices are one of the vertices of the base, the centre of the base and the centre of the spheres. We have $R^2 = r^2 + r_1^2$

where $r_1 = \frac{r}{\cos\frac{\pi}{n}}$. It follows that

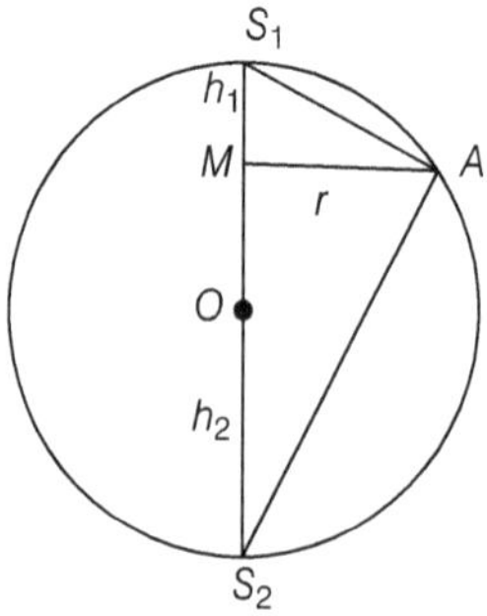

FIG. 187

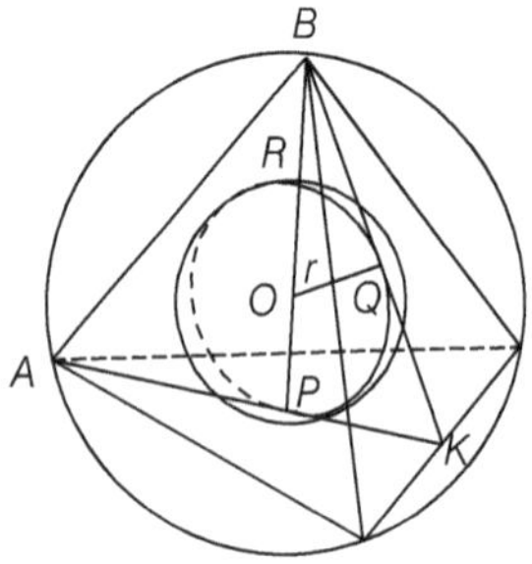

FIG. 188

$$R = r\sqrt{1 + \frac{1}{\cos^2\frac{\pi}{n}}}.$$

The ratio of the volume of the circumscribed sphere to that of the inscribed sphere is

$$\frac{R^3}{r^3} = \left(1 + \frac{1}{\cos^2\frac{\pi}{n}}\right)^{\frac{3}{2}}.$$

475. The radii of the circumscribed and inscribed spheres are equal to the segments of the altitude of the tetrahedron into which it is divided by the common centre of these spheres. It can easily be revealed that the ratio of these segments is 3:1. Indeed, from the similar triangles BQO and BPK (Fig. 188) we have $\frac{R}{r} = \frac{BK}{PK}$.

But $$\frac{BK}{PK} = \frac{BK}{QK} = 3,$$

and since the surface areas of the spheres are in the ratio of the squares of their radii the sought-for ratio is equal to 9.

476. The volumes of the regular tetrahedrons are in the ratio of the cubes of the radii of their inscribed spheres. The sphere inscribed in the larger tetrahedron being at the same time the circumscribed sphere of the smaller tetrahedron, the ratio of those radii of the inscribed sphere is equal to 3:1 (see the solution of Problem 475). Hence, the sought-for ratio of the volumes is equal to $3^3 = 27$.

477. Suppose that the problem is solvable. Draw a plane $A_1B_1C_1$ (see Fig. 189, a) tangent to the smaller sphere and parallel to the base ABC of the given tetrahedron. The tetrahedron $S\,A_1\,B_1\,C_1$ is circumscribed about the sphere of radius r, it is easy to show that its height is $SQ_1 = 4r$ (see Problem 475).

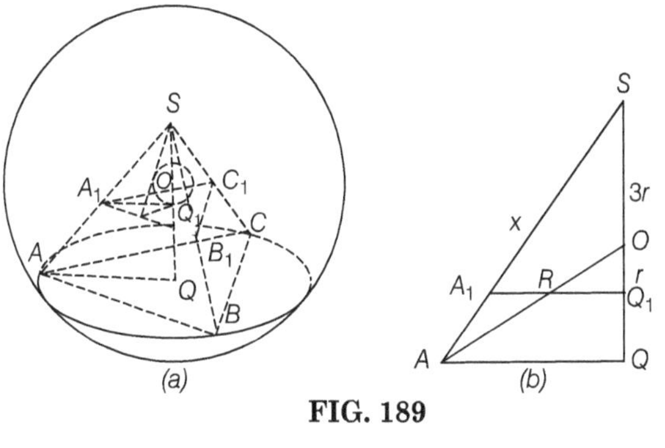

FIG. 189

Let the length of the edge of the tetrahedron $SABC$ be equal to x. Then the line segment AQ is equal to $\frac{x\sqrt{3}}{3}$, and the altitude SQ is equal to $x\frac{\sqrt{6}}{3}$. Furthermore (see Fig. 189, b), we have $QO = \frac{x\sqrt{6}}{3} - 3r$, and from the right triangle AQO it follows that

$$\left(\frac{x\sqrt{3}}{3}\right)^2 + \left(x\frac{\sqrt{6}}{3} - 3r\right)^2 = R^2.$$

Solving the quadiratic equation we find

$$x_{1,2} = r\sqrt{6} \pm \sqrt{R^2 - 3r^2}.$$

Here we must only take the root with the plus sign, because SA is in any case greater than $3r$, and $3r > r\sqrt{6}$. It is obvious that the problem is solvable if $R \geq \sqrt{3}\,r$.

478. Let $A_1B_1C_1D_1E_1F_1$ be the regular hexagon in the section of the cube by the cutting plane. The problem is reduced to determining the radius of the inscribed sphere of the regular hexagonal pyramid $SA_1B_1C_1D_1E_1F_1$ (Fig. 190). The side of the base of the pyramid is equal to $\frac{a\sqrt{2}}{2}$, and the altitude to $\frac{a\sqrt{3}}{2}$. Since the radius of the sphere inscribed in a regular pyramid is three times the volume of the pyramid divided by its total area (see formula (1) in the solution of Problem 466), we find

$$r = \frac{a(3-\sqrt{3})}{4}.$$

Hence, the required ratio is equal to $\frac{2(3+\sqrt{3})^3}{9\pi}$.

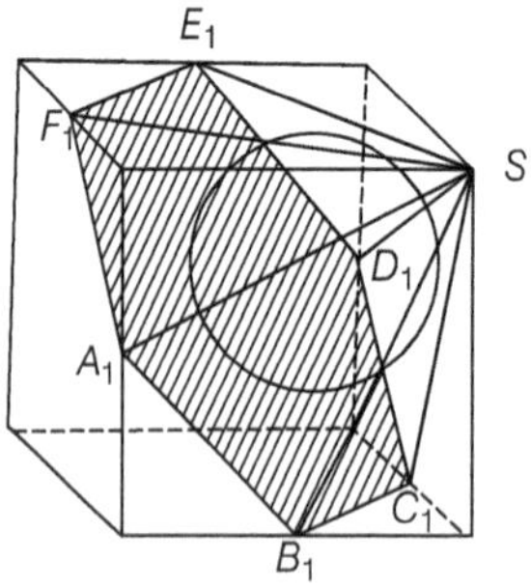

FIG. 190

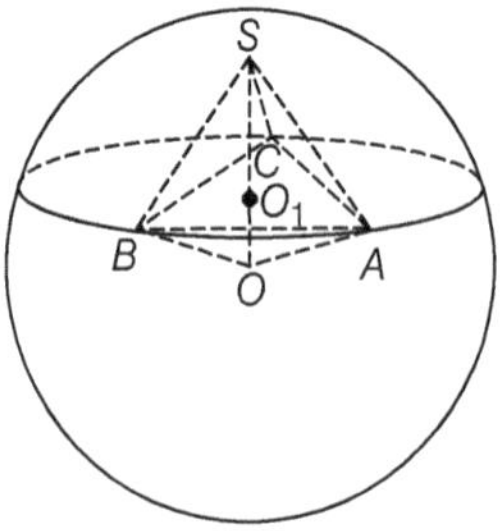

FIG.191

479. Let O be the centre of the sphere, and AS, BS and CS the given chords. As is obvious, the triangle ABC is equilateral (Fig. 191). It is also easily seen that the extension of the perpendicular SO_1 to the plane ABC passes through the centre O of the sphere because the point O_1 is the centre of the circle circumscribed about ΔABC.

Now let us denote the sought-for length of the chords by d. From the triangle SAB we find

$$AB = 2d\sin\frac{\alpha}{2},$$

and, hence, $O_1A = AB\cdot\frac{\sqrt{3}}{3} = \frac{2}{3}\sqrt{3}d\sin\frac{\alpha}{2}$.

Computing the area of the isosceles triangle SOA in two different ways, we get

$$\frac{1}{2}R\frac{2}{3}\sqrt{3}d\sin\frac{\alpha}{2}=\frac{1}{2}d\sqrt{R^2-\frac{d^2}{4}},$$

whence we find $d=2R\sqrt{1-\frac{4}{3}\sin^2\frac{\alpha}{2}}$.

480. The radius of the inscribed sphere r is found by the formula (cf. formula (1) in the solution of problem 466)

$$r=\frac{3V}{S},$$

where V is the volume of the pyramid and S its total area. We shall first find the volume of the pyramid. To this end, note that the right triangles BSC and BSA (Fig. 192) are congruent since they have equal hypotenuses and a common leg. Due to this, the right triangle ASC is isosceles. Since

$$AS=CS=\sqrt{a^2-b^2},$$

we have $V=\frac{1}{3}BS.S_{\Delta ASC}=\frac{1}{3}b\frac{(a^2-b^2)}{2}$.

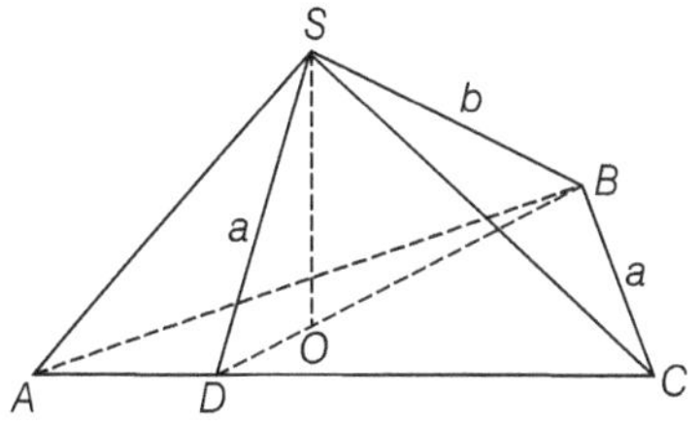

FIG. 192

It is also clear that

$$AD=\sqrt{a^2-b^2}\,\frac{\sqrt{2}}{2}$$

and $BD=\sqrt{AB^2-AD^2}=\frac{\sqrt{2}}{2}\sqrt{a^2+b^2}$,

and, hence,

$$S\ \Delta ABC=\frac{1}{2}\sqrt{a^4-b^4}.$$

Now substituting the necessary expressions into the above formula of r and simplifying the result we finally obtain

$$r=\frac{b\sqrt{a^2-b^2}}{\sqrt{a^2+b^2}+2b+\sqrt{a^2-b^2}}.$$

481. Let r and R be the radii of the inscribed and the circumscribed spheres.

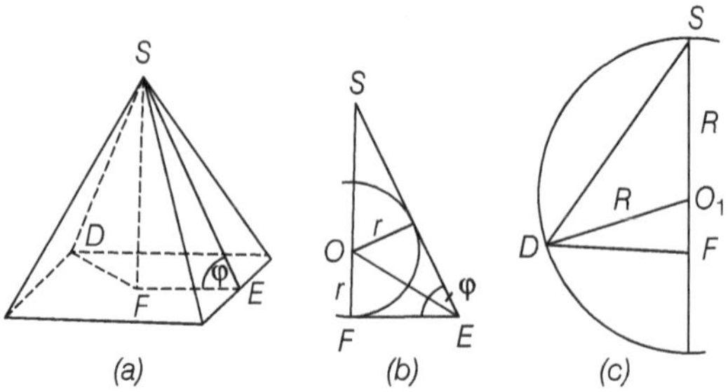

FIG. 193

We shall first consider the triangle SFE whose side SF is the altitude of the pyramid, the side SE being the slant height of the pyramid (Fig. 193, a). Let O be the centre of the inscribed sphere. In the triangles SFE and OFE (Fig. 193, b) we have

$$FE = r\cot\frac{\varphi}{2}$$

and

$$SF = r\cot\frac{\varphi}{2}\tan\varphi.$$

Furthermore, it is obvious, that

$$DF = EF\cdot\sqrt{2} = r\cot\frac{\varphi}{2}\sqrt{2}.$$

From Fig. 193, c showing the section passing through the axis of the pyramid and its lateral edge we easily find

$$DO_1^2 = O_1F^2 + DF^2,$$

that is $$R^2 = (SF - R)^2 + DF^2.$$

It follows that $$R = \frac{SF^2 + DF^2}{2SF}. \qquad ...(1)$$

We have $R = 3r$, and therefore, substituting the above expression for SF and DF, we obtain the following equation for φ:

$$3r = \frac{r^2\cot^2\frac{\varphi}{2}\tan^2\varphi + r^2\cot^2\frac{\varphi}{2}\cdot 2}{2r\cot\frac{\varphi}{2}\tan\varphi}.$$

Simplifying this equation we write

$$6\tan\frac{\varphi}{2}\tan\varphi = 2 + \tan^2\varphi.$$

Now put $\tan\frac{\varphi}{2} = z$. Noting that $\tan\varphi = \frac{2z}{1-z^2}$, we arrive at the equation $7z^4 - 6z^2 + 1 = 0$

from which we find $z_{1,2} = \pm\sqrt{\frac{3\pm\sqrt{2}}{7}}$.

But $z > 0$, and hence only the two following answers are possible:

$$\tan\frac{\varphi_1}{2} = \sqrt{\frac{3+\sqrt{2}}{7}}$$

and

$$\tan\frac{\varphi_2}{2} = \sqrt{\frac{3-\sqrt{2}}{7}}.$$

482. We have the total of six lunes (according to the number of the edges) and four triangles (Fig. 194). Let us denote the area of each triangle by S_1, and the area of each lune by S_2. We then have

$$4S_1 + 6S_2 = 4\pi R^2. \qquad \ldots(1)$$

Let S_o be the sum of the areas of a triangle and the three adjacent lunes. S_0 is the area of a spherical segment cut off by the plane of the corresponding face of the tetrahedron. This area is equal to $2\pi Rh$ where h is the altitude of the segment. Since the altitude of the tetrahedron is divided by the centre of the sphere in the ratio $3:1$ (see Problem 475), we have

$$H = R + \frac{1}{3}R = \frac{4}{3}R,$$

which yields $$h = 2R - \frac{4}{3}R = \frac{2}{3}R.$$

Furthermore, we have

$$S_1 + 3S_2 = 2\pi R\cdot\frac{2}{3}R = \frac{4}{3}\pi R^2. \qquad (2)$$

Solving the system consisting of equations (1) and (2) with respect to the unknowns S_1 and S_2, we abtain $S_1 = \frac{2}{3}\pi R^2$, $S_2 = \frac{2}{9}\pi R^2$.

483. Let R be the radius of the base of the cone, α the angle between the axis of the cone and its element, and r the radius of the inscribed sphere. The axial section of the cone shown in Fig. 195 is an isosceles triangle ABC. The radius of the inscribed circle of this triangle is equal to the radius r of the sphere inscribed in the cone. Let O be the centre of the sphere and $\angle OCA = \beta$. Then it is obvious that $\tan\beta = \frac{r}{R}$.

But, by the hypothesis,

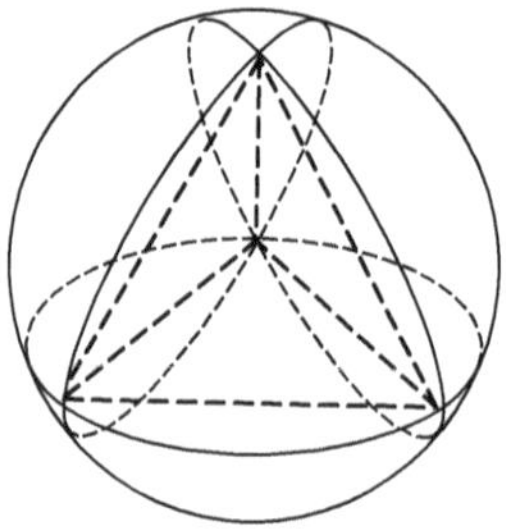

FIG. 194

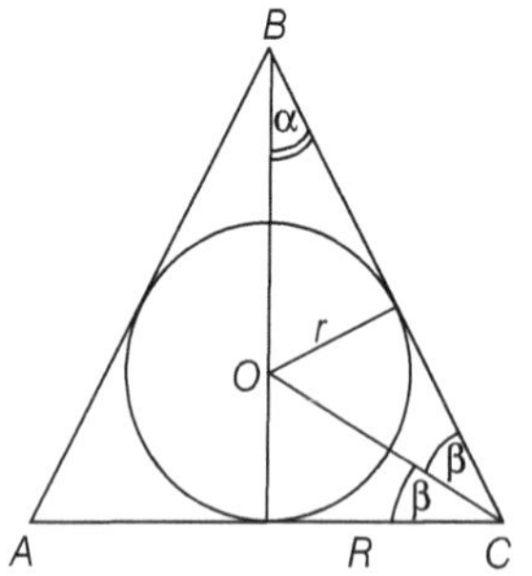

FIG. 195

$$\frac{4\pi r^2}{\pi R^2} = 4\left(\frac{r}{R}\right)^2 = \frac{4}{3}.$$

It follows that $\frac{r}{R} = \frac{1}{\sqrt{3}}$, and, hence, $\beta = \frac{\pi}{6}$. Besides, we have $\alpha + 2\beta = \frac{\pi}{2}$, and therefore $\alpha = \frac{\pi}{6}$. Consequently, the sought-for angle is equal to $2\alpha = \frac{\pi}{3}$.

484. Let r be the radius of the hemisphere, R the radius of the base of the cone, l the slant height of the cone, and α the angle between the axis of the cone and its element.

By the hypothesis, we have

$$\frac{\pi R(l+R)}{2\pi r^2} = \frac{18}{5}. \tag{1}$$

Let us introduce the angle α into this equality. For this purpose, consider the isosceles ΔABC (Fig. 196) obtained in the axial section of the cone. From ΔABC, we find

$R = l\sin\alpha, r = R\cos\alpha = l\sin\alpha\cos\alpha.$

Substituting these expressions in the left-hand side of (1), we get

$$\frac{1}{2}\frac{1+\sin\alpha}{\sin\alpha\cos^2\alpha} = \frac{18}{5}.$$

We have $\cos^2\alpha = 1-\sin^2\alpha$ and therefore, cancelling out $1+\sin\alpha$, we receive

$$36\sin^2\alpha - 36\sin\alpha + 5 = 0,$$

which gives us

$$\sin\alpha_1 = \frac{5}{6} \text{ and } \sin\alpha_2 = \frac{1}{6}.$$

Hence, the sought-for vertex angle of the axial section of the cone is equal to 2 arc sin $\frac{5}{6}$, that is to 2 arc sin $\frac{1}{6}$.

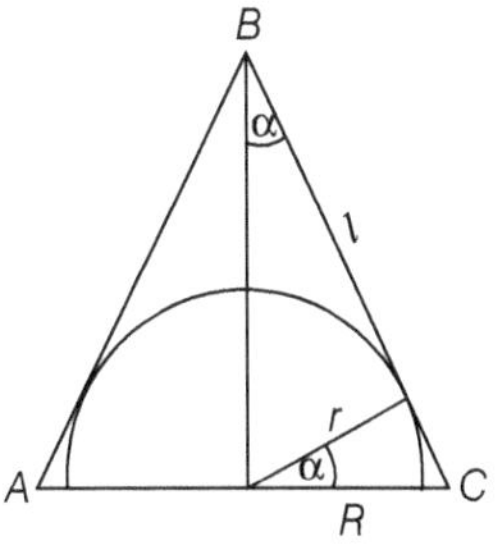

FIG. 196

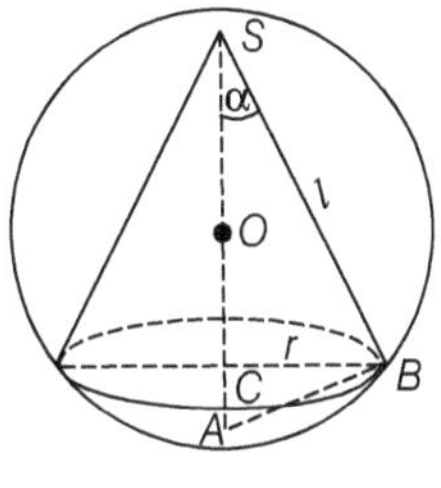

FIG. 197

485. Let h be the altitude of the cone, r the radius of the base, l the slant height of the cone, α the angle between the altitude of the cone and its element (Fig. 197). By the hypothesis, we have $\pi r l = k\pi r^2$ which yields $l = kr$ and, hence $\sin\alpha = \frac{1}{k}$. From the right triangle ABS we get

$$r = 2R\cos\alpha\sin\alpha = 2R\frac{\sqrt{k^2-1}}{k^2}$$

and

$$h = 2R\cos\alpha\cos\alpha = 2R\frac{k^2-1}{k^2}.$$

The sought-for volume of the cone is

$$V = \frac{1}{3}\pi r^2 h$$
$$= \frac{8}{3}\pi R^3\left(\frac{k^2-1}{k^3}\right)^2.$$

486. Let R be the radius of the sphere, h the altitude of the cone and r the radius of the base of the cone. The ratio of the volume of the cone to that of the sphere is equal to

$$x = \frac{r^2 h}{4R^3} = \frac{q}{4}\left(\frac{r}{R}\right)^2.$$

From the triangle SBA (Fig. 198) we have $r^2 = h(2R-h)$. It follows that

$$\frac{r^2}{R^2} = \frac{h}{R}\left(2-\frac{h}{R}\right) = q(2-q)$$

and, consequently, $x = \frac{q^2}{4}(2-q)$.

Obviously, the problem is only solvable if $0 < q < 2$.

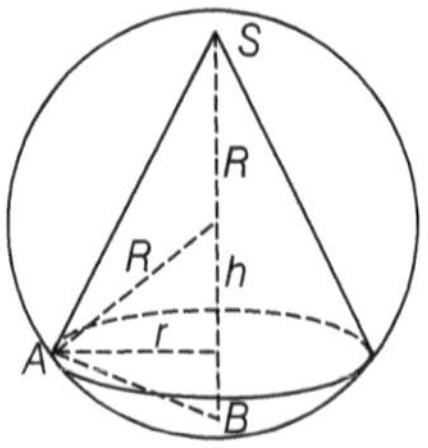

FIG. 198

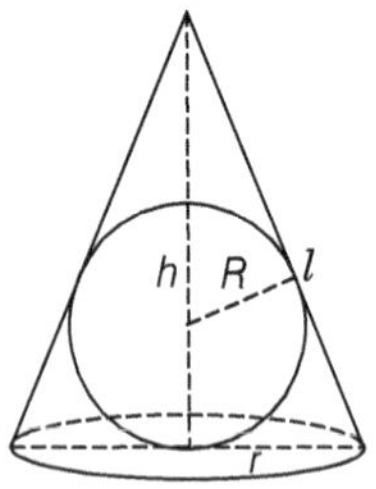

FIG. 199

487. Let R be the radius of the sphere, S_{sphere} and V_{sphere} the area and the volume of the sphere, S_{cone} and V_{cone} the total area and volume of the cone, h the altitude of the cone and r the radius of the base of the cone (Fig. 199). Then

$$\frac{V_{sphere}}{V_{cone}} = \frac{\frac{4}{3}\pi R^3}{\frac{1}{3}\pi r^2 h} = \frac{4R^3}{r^2 h} \quad \text{and} \quad \frac{S_{sphere}}{S_{cone}} = \frac{4\pi R^2}{\pi r(l+r)} = \frac{4R^2}{r(l+r)}.$$

However, let us note that $\frac{l}{r} = \frac{h-R}{R} = \frac{h}{R} - 1$

and, consequently, $\frac{l+r}{r} = \frac{h}{R}$.

Thus, we obtain $\frac{V_{sphere}}{V_{cone}} = \frac{S_{sphere}}{S_{cone}} = \frac{1}{n}$.

Note. The same result can be obtained in a simpler way by using the following

$$V_{cone} = \frac{1}{3} S_{cone} R. \tag{1}$$

where S_{cone} is the total area of a cone, and R the radius of its inscribed sphere. Formula (1) is readily obtained as the limiting case of the corresponding formula for a pyramid (see the solution of Problem 466). To obtain the result we take the obvious formula

$$V_{sphere} = \frac{1}{3} S_{sphere} \cdot R, \tag{2}$$

and then, dividing (2) by (1), obtain

$$\frac{V_{sphere}}{V_{cone}} = \frac{S_{sphere}}{S_{cone}} = \frac{1}{n}.$$

488. Let S be the total surface area of the frustum, S_1 the area of the sphere, r_1 and r the radii of the upper and lower base of the frustum, respectively, and l the slant height. Furthermore, let $CMDL$ be the trapezoid in the axial section of the frustum, O the centre of the inscribed sphere, $AB \perp LD$ and $OF \perp MD$ (Fig. 200). We have

$$\frac{S}{S_1} = \frac{\pi l(r + r_1) + \pi r_1^2 + \pi r^2}{4\pi R^2} = m. \quad (1)$$

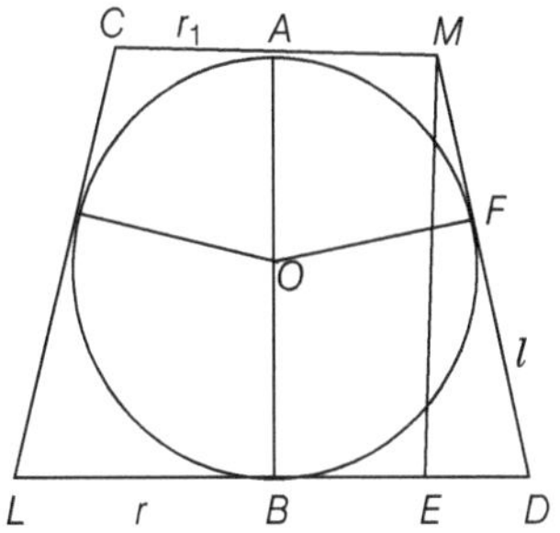

FIG. 200

It is obvious that $AM = MF$ and $BD = FD$ because O is the centre of the circle inscribed in the trapezoid and therefore

$$l = r + r_1. \quad (2)$$

Taking advantage of this equality, we obtain from equality (1) the relation

$$l^2 + r_1^2 + r^2 = 4mR^2. \quad (3)$$

It then follows from the triangle MED that

$$l^2 = (r - r_1)^2 + 4R^2. \quad (4)$$

Eliminating l from equalities (2) and (4) we find

$$rr_1 = R^2 \quad (5)$$

With the aid of this equality, eliminating l from (2) and (3), we obtain

$$r^2 + r_1^2 = R^2(2m - 1). \quad (6)$$

Solving the system of two equations (5) and (6), we find

$$r = \frac{R}{2}(\sqrt{2m+1} + \sqrt{2m-3}) \text{ and } r_1 = \frac{R}{2}(\sqrt{2m+1} - \sqrt{2m-3}).$$

Thus, for $m < \frac{3}{2}$ the problem has no solution; for $m = \frac{3}{2}$ the frustum of the cone turns into a cylinder.

489. There are two possible cases here, namely: (1) the vertex of the cone and the sphere lie on different sides of the tangent plane and (2) the vertex of the cone and the sphere lie on one side of the tangent plane.

Consider the first case. Draw a plane through the axis of the cone and its element BC mentioned in the statement of the problem (Fig. 201). The section of the cone by this plane is a triangle ABC and the section at the sphere is a circle with centre at O. Furthermore, this plane intersects the plane perpendicular to BC along a straight line ME, M being the point of tangency. Draw $BD \perp AC$ and $OF \perp BC$. Let $BD = h, OD = OF = r$ and $CD = R$. The figure $OMEF$ is obviously a square, and therefore

$$h = r + \sqrt{r^2 + (d+r)^2}.$$

Furthermore, $$\frac{R}{h} = \frac{r}{d+r}, R = \frac{hr}{d+r}.$$

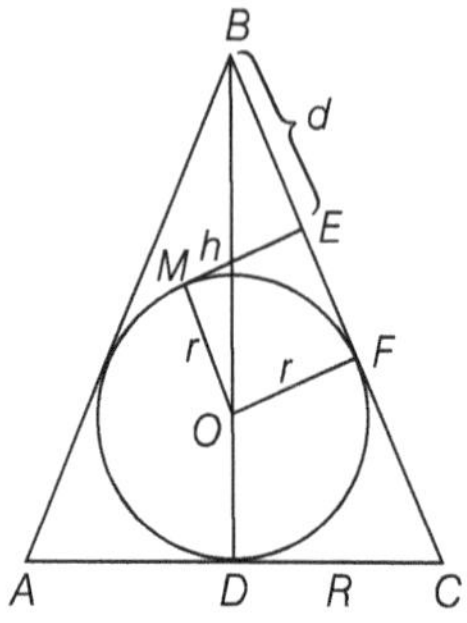

FIG. 201

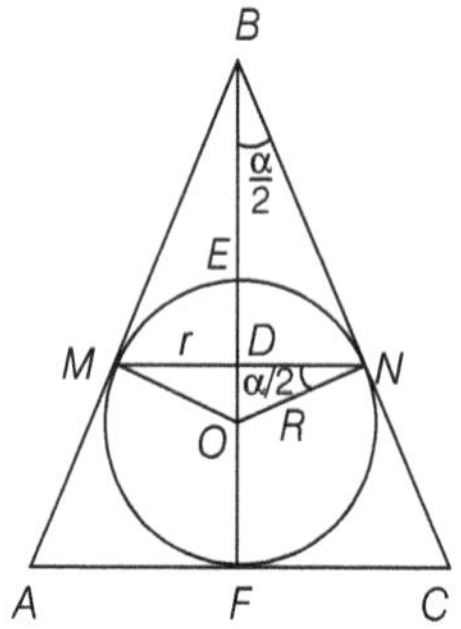

FIG. 202

Thus, in the first case the volume of the cone is
$V = \frac{1}{3}\pi R^2 h = \frac{1}{3}\pi \frac{h^3 r^2}{(d+r)^2} = \frac{\pi r^2 (r + \sqrt{r^2 + (d+r)^2})^3}{3(d+r)^2}$. In the second case the problem is solved analogously. The volume of the cone turns out to be equal to

$$\frac{\pi r^2 (r + \sqrt{r^2 + (d-r)^2})^3}{3(d-r)^2}.$$

490. Consider the axial section ABC of the cone shown in Fig. 202. Let BF be the altitude in the triangle ABC, N and M the points of tangency of the circle inscribed in the triangle ABC with the sides AB and BC, O the centre of the circle, E the point of intersection of the smaller arc MN and the line segment BF and D the point of intersection of the line segments MN and BF. Put $DM = r, DE = H$ and $BD = h$. The desired volume is

$$V = \frac{1}{3}\pi r^2 h - \frac{1}{3}\pi H^2 (3R - H).$$

But $$h = r\cot\frac{\alpha}{2} = R\cos\frac{\alpha}{2}\cot\frac{\alpha}{2} = R\frac{\cos^2\frac{\alpha}{2}}{\sin\frac{\alpha}{2}}$$

and $$H = R - R\sin\frac{\alpha}{2}.$$

Consequently, $$V = \frac{1}{3}\pi R^3 \left[\frac{\cos^4\frac{\alpha}{2}}{\sin\frac{\alpha}{2}} - \left(1 - \sin\frac{\alpha}{2}\right)^2 \left(2 + \sin\frac{\alpha}{2}\right)\right].$$

491. Denote the radii of the sphere by r and r_1 and consider the sections of the spheres by a plane passing through their centres O and O_1 (see Fig. 203). Let $AA_1 = 2a, KS = R$ and $AS = x$. Then $A_1S = 2a - x$. The total area of the lens is equal to

$$2xar_1 + (2a - x)2ar = S. \qquad (1)$$

From the triangle OKS we have

$$r^2 = R^2 + [r - (2a - x)]^2,$$

that is

$$R^2 - 2r(2a - x) + (2a - x)^2 = 0. \qquad (2)$$

Analogously, from the triangle O_1KS we have

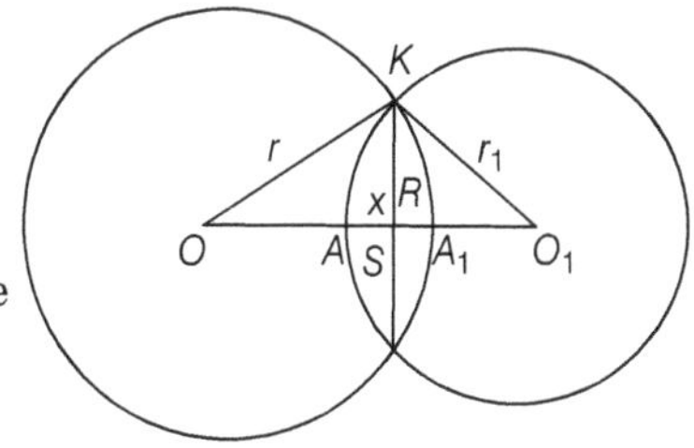

FIG. 203

$$r_1^2 = R^2 + (r_1 - x)^2,$$

that is

$$R^2 - 2r_1x + x^2 = 0. \qquad (3)$$

From (2) and (3) we find

$$r = \frac{R^2 + (2a - x)^2}{2\,(2a - x)}, r_1 = \frac{R^2 + x^2}{2x}. \qquad (4)$$

Substituting these expressions in equality (1), we get the equation

$$\pi(R^2 + x^2) + \pi\,[R^2 + (2a - x)^2] = S,$$

which can be rewritten in the form

$$x^2 - 2ax + R^2 + 2a^2 - \frac{S}{2\pi} = 0.$$

Solving the equation we receive

$$x = a + \sqrt{\frac{S}{2\pi} - R^2 - a^2}. \qquad (5)$$

Substituting this value of x in formulas (4) and simplifying the result we obtain

$$r = \frac{\frac{S}{4\pi} - a\sqrt{\frac{S}{2\pi} - R^2 - a^2}}{a - \sqrt{\frac{S}{2\pi} - R^2 - a^2}}.$$

$$r_1 = \frac{\frac{S}{4\pi} + a\sqrt{\frac{S}{2\pi} - R^2 - a^2}}{a + \sqrt{\frac{S}{2\pi} - R^2 - a^2}}.$$

The choice of the minus sign in front of the radical in (5) is equivalent to interchanging the letters r and r_1 which designate the radii.

492. Let V_1 and V_2 be, respectively, the volumes of the smaller and larger spherical segments into which the sphere is divided by the plane passing through the line of tangency of the sphere and the cone. Denote by R the radius of the sphere, by h the altitude of the smaller segment, by H the altitude of the cone, and by r the radius of its base (Fig. 204). Then $V_1 = \frac{1}{3}\pi h^2(3R-h)$, $V_2 = \frac{4}{3}\pi R^3 - \frac{\pi}{3}h^2(3R-h)$. The problem is reduced to finding the ratio $\frac{h}{R}$. Denoting the angle between the axis of the cone and its element by α, we find from ΔPKO that

$$\frac{R-h}{R} = \sin\alpha, \text{ whence } \frac{h}{R} = 1-\sin\alpha,$$

Furthermore, by the hypothesis, we have

$$k = \frac{\frac{1}{3}\pi r^2 H}{\frac{4}{3}\pi R^3} = \frac{1}{4}\frac{r^2 H}{R^3}.$$

Let us now express r and H in terms of R and α. We have

$$H = \frac{R}{\sin\alpha} + R = R\cdot\frac{1+\sin\alpha}{\sin\alpha},$$

$$r = H\cdot\tan\alpha + R\cdot\frac{1+\sin\alpha}{\cos\alpha}.$$

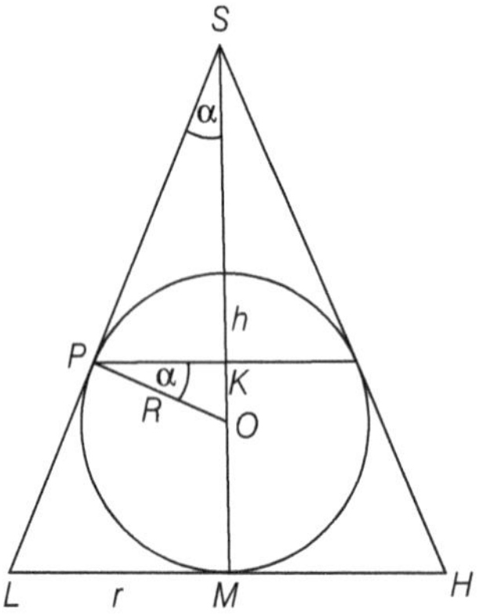

FIG. 204

Hence,

$$k = \frac{1}{4}\cdot\frac{(1+\sin\alpha)^3}{\sin\alpha(1-\sin^2\alpha)} = \frac{1}{4}\cdot\frac{(1+\sin\alpha)^2}{\sin\alpha(1-\sin\alpha)}.$$

Substituting $\sin\alpha = 1-\frac{h}{R}$ into this relation we get the following equation for $\frac{h}{r} = z$: $k = \frac{1}{4}\frac{(2-z)^2}{(1-z)z}$.

Simplifying the equation we receive

$$z^2(4k+1) - 4(k+1)z + 4 = 0$$

and then, solving it, we obtain

$$z_{1,2} = \frac{2\,(k+1) \pm 2\sqrt{k(k-2)}}{4k+1}. \quad (1)$$

Finally we find $\frac{V_1}{V_2} = \frac{z_{1,2}^2(3-z_{1,2})}{4-z_{1,2}^2(3-z_{1,2})}$.

The problem has two solutions for $k > 2$ because both roots of the quadratic equation can then be taken.

493. We shall find radius r of each of the eight inscribed spheres by considering the triangle AOC (shown in Fig. 205, a) in the plane passing through the centres of these spheres and the centre O of the sphere S. We have

$$\frac{AB}{AO} = \frac{r}{R-r} = \sin\frac{\pi}{8}.$$

It follows that
$$r = R\frac{\sin\frac{\pi}{8}}{\sin\frac{\pi}{8}+1}.$$

Draw the plane section through the centre O of the sphere S, the centre O_1 of the sphere S_1 and the centres of the two spheres of radius r shown in Fig. 205. b, which lie on a diameter of the sphere O. From the right triangle AOO_1 we find

$$AO_1^2 = AO^2 + OO_1^2,$$

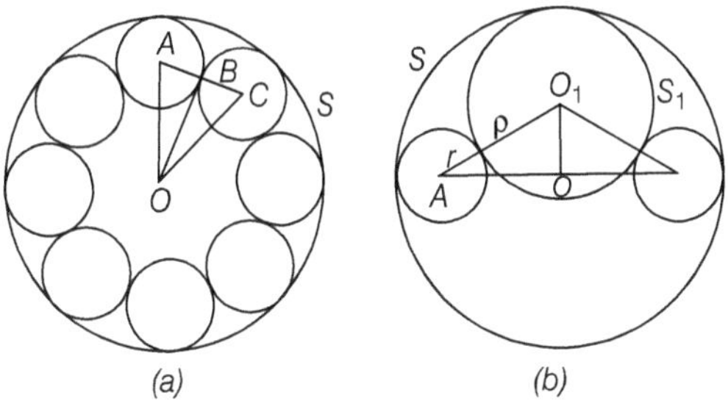

FIG. 205

that is $(r+\rho)^2 = (R-r)^2 + (R-\rho)^2.$

It follows that $\rho = R\cdot\dfrac{R-r}{R+r}$,

which results in $\rho = R\cdot\dfrac{1}{2\sin\frac{\pi}{8}+1} = \dfrac{R}{\sqrt{2-\sqrt{2}}+1}$

494. The inscribed spheres being congruent, their centres are equidistant from the centre O of the sphere S. Consequently, the centre of symmetry of the cube indicated in the problem coincides with the centre O of the sphere S (Fig. 206). Let x be the sought-for radius of the spheres. It is readily seen that the edge of the cube is then $AB = 2x$, and half the diagonal of the cube is $AO = CO - CA = R - x$. On the other hand, we have $AO = \frac{1}{2}\cdot 2x\sqrt{3}$,

and therefore we get the equation $R - x = x\sqrt{3}$, whence we find

$$x = \frac{R}{\sqrt{3}+1}.$$

495. Let r be the radius of the base of each of the two inscribed cones whose common portion consists of two congruent frustums of the cones. Let r_1 and r_2 be, respectively, the radii of the upper and lower bases of each frustum, and H its altitude. The sought-for ratio of the volumes is equal to

$$q = \frac{H(r_1^2 + r_1 r_2 + r_2^2)}{2R^3}$$

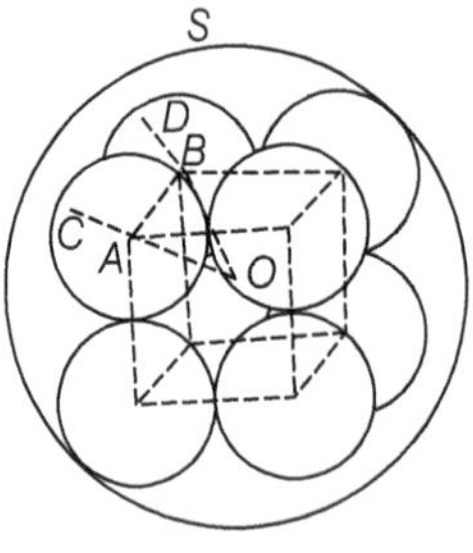

FIG. 206

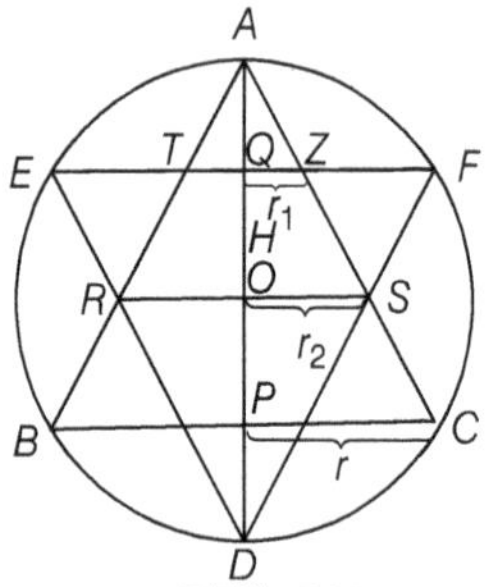

FIG. 207

The similarity of the triangles AQZ, AOS and APC (Fig. 207) implies

$$\frac{r_1}{r_2} = \frac{R-H}{R} \text{ and } \frac{r_2}{r} = \frac{R}{h}.$$

Besides, $H = h - R$ and

$$r = \sqrt{R^2 - H^2} = \sqrt{2Rh - h^2}.$$

Therefore, the two foregoing equalities enable us to express r_1 and r_2 in terms of R and h;

$$r_2 = \frac{R\sqrt{2Rh-h^2}}{h}, r_1 = r_2 \frac{2R-h}{R}.$$

By the hypothesis, we have $\frac{h}{R} = k$ and consequently

$$q = \frac{(h-R)\left\{r_2^2 \frac{(2R-h)^2}{R^2} + r_2^2 \frac{2R-h}{R} + r_2^2\right\}}{2R^3}$$

$$= \frac{1}{2}(k-1)\left(\frac{2}{k}-1\right)(k^2-5k+7).$$

496. Let the radii of the circular sections with areas S_1 and S_2 be equal to R_1 and R_2, respectively, and the distances from the centre of the sphere to these sections be equal to l_1 and $l_2 (l_1 < l_2)$. Let R be the radius of the sphere, r the radius of the section in question and l the distance between this section and the centre of the sphere. Then we have (see Fig. 208).

$$l_2 - l_1 = d \qquad (1)$$

and $l_1^2 + R_1^2 = l_2^2 + R^2 = R_2^2$. From these two equations we find

$$l_2 + l_1 = \frac{R_1^2 - R_2^2}{d}$$

and, hence,

$$l_2 + l_1 = \frac{S_1 - S_2}{\pi d}. \qquad (2)$$

From equations (1) and (2) we obtain

$$l_2 = \frac{S_1 - S_2}{2\pi d} + \frac{d}{2}, \quad t = \frac{S_1 - S_2}{2\pi d}$$

Therefore, the sought-for area is

$$S = \pi r^2 = \pi (R^2 - l^2) = \pi (R_2^2 + l_2^2 - l^2) = \frac{1}{2}\left(S_1 + S_2 + \frac{1}{2}\pi d^2\right).$$

497. Let us denote the sought-for radius of the base of the cone by r. Consider the section passing through the centre of one of the spheres and the axis of the cone (Fig. 209). Note that the distance between the centres of two congruent spheres tangent to one another is equal to $2R$. It can readily be proved that the centre A of the base of the cone is equidistant from all the three points of tangency of the spheres with the plane P. Based on this fast, we find

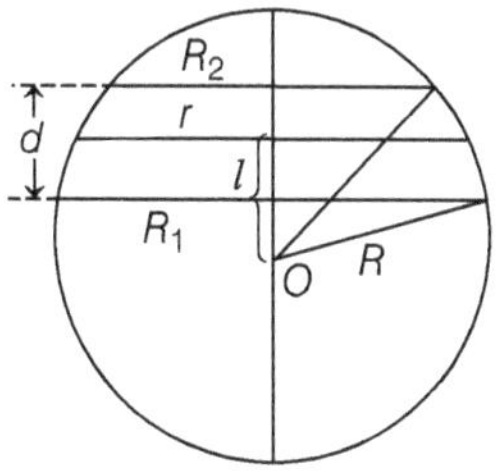

FIG. 208

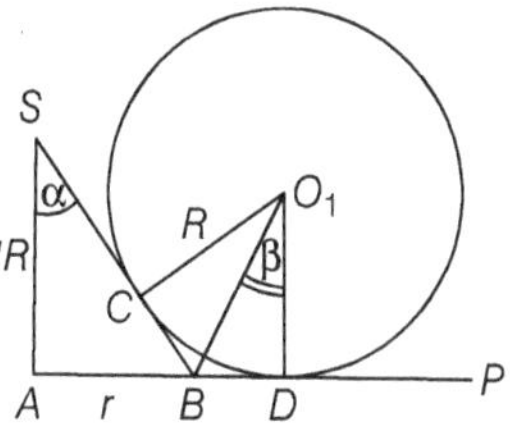

FIG. 209

$AD = \frac{2\sqrt{3}}{3}R$. It is evident that $\angle SBA = \angle CO_1D = 2\beta$ and

consequently, $\qquad 2\beta = \frac{\pi}{2} - \alpha.$

Taking the tangents of the angles on both sides of this equality, we obtain

$$\frac{2\tan\beta}{1 - \tan^2\beta} = \frac{1}{\tan\alpha}. \qquad (1)$$

From Fig. 209 we see that $\tan\beta = \left(\frac{2\sqrt{3}}{3}R - r\right) : R$ and $\tan\alpha = r : qR$. If now we put $\frac{r}{R} = x$, equality (1) leads to the following equation for x:

$3(q-2)x^2 - 4\sqrt{3}(q-1)\,x + q = 0.$

For $q = 2$ we obtain from this equation $x = \frac{\sqrt{3}}{6}$, and, hence, $r = \frac{\sqrt{3}}{6}R$. If $q \neq 2$, then
$$x_{1,2} = \frac{2\sqrt{3}(q-1) \mp \sqrt{9q^2 - 18q + 12}}{3(q-2)}.$$

Since $0 < x < \frac{2\sqrt{3}}{3}$, the above formula should be taken with the minus sign. It can easily be shown that for $q > 2$ the root with the plus sign is greater than $\frac{2\sqrt{3}}{3}$ and corresponds to a cone externally tangent to the spheres; for $q < 2$ this root is negative.

498. The centres of the first four spheres lie at the vertices of a regular tetrahedron, since the distance between the centres of any two congruent spheres tangent to one another is equal to $2R$. It is easy to show that the centres of the fifth and sixth spheres coincide with the centre of gravity of the tetrahedron (Fig. 210). Let r be the radius of the fifth (larger) sphere, and ρ the radius of the sixth sphere. As is obvious,

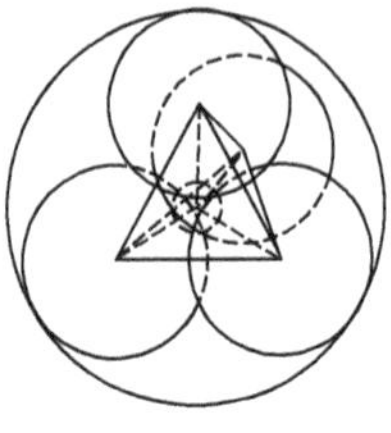

FIG. 210

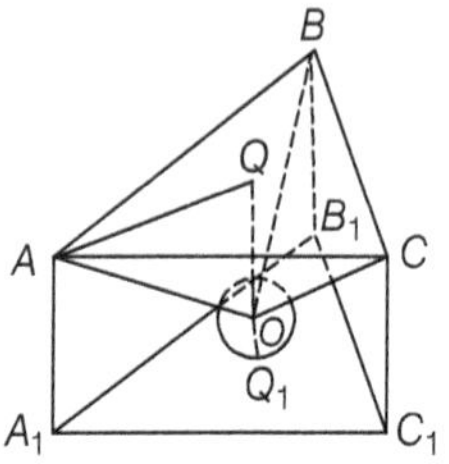

FIG. 211

$$r = \rho + 2R. \tag{1}$$

Since the distance from the centre of gravity and the vertex of the tetrahedron in question is equal to $\frac{\sqrt{6}}{2}R$, we obtain
$$\rho + R = \frac{\sqrt{6}}{2}R. \tag{2}$$

Hence, $\rho = R\left(\frac{\sqrt{6}}{2} - 1\right)$, and from formula (1) we find $r = R\left(\frac{\sqrt{6}}{2} + 1\right)$.

Thus the sought-for ratio of the volumes is
$$\frac{V_6}{V_5} = \left(\frac{\rho}{r}\right)^3 = \left(\frac{\sqrt{6}-2}{\sqrt{6}+2}\right)^3 = (5 - 2\sqrt{6})^3 = 485 - 198\sqrt{6}.$$

499. Let A, B and C be the centres of the spheres of radius R and A_1, B_1 and C_1 the projections of these centres onto the plane. Denote by O the centre of the fourth sphere whose radius r is to be found (Fig. 211). Joining the centres of all the spheres we obviously obtain a regular triangular pyramid $OABC$ in which $AB = BC = AC = 2R$, $AO = BO = CO = R + r$ and $OQ = R - r$. The line segment AQ is the radius of the circumscribed circle of ΔABC and therefore

$$AQ = \frac{AB}{\sqrt{3}} = \frac{2R}{\sqrt{3}}.$$

Applying the Pythagorean theorem to the triangle AQO we find that

$$\left(\frac{2R}{\sqrt{3}}\right)^2 + (R - r)^2 = (R + r)^2.$$

Solving this equation, we obtain $r = \frac{R}{3}$.

500. Let A, B, C and D be the centres of the larger spheres. Consider the projections of all the spheres onto the plane containing A, B, C and D (Fig. 212). The centres of the smaller spheres are equidistant from the centres of the corresponding larger spheres and therefore they are projected into the centres of gravity O_1 and O_2 of the equilateral triangles ABC and BCD. Besides, the radii of the smaller spheres are equal, by the hypothesis, and therefore the line segment joining their centres is parallel to the plane under consideration and is bisected by the point of tangency of the spheres. Therefore, the projection of that point is on the line segment BC. It follows that the smaller spheres are projected into circles inscribed in the triangles ABC and BCD. Therefore, the radius of the smaller spheres is equal

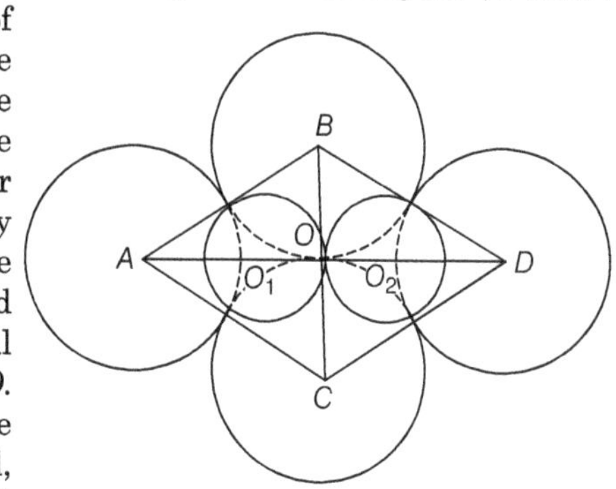

FIG. 212

to
$$r = \frac{AB\sqrt{3}}{6} = \frac{2R\sqrt{3}}{6}$$

which yields
$$\frac{R}{r} = \sqrt{3}.$$

2. Proof Problems

501. Let E and F be the midpoints of the bases of the trapezoid $ABCD$ in the axial section of the cone shown in Fig. 213. Through the midpoint O of the line segment EF draw the straight lines $OM \perp CD$, $ON \perp EF$ and $CP \perp AD$. For brevity, let us introduce the following notation: $CD = l$, $EF = h$, $OM = x$, $EC = r, DF = R$ and $\angle MON = \angle PCD = \alpha$.

For the assertion to be proved it is sufficient to show that $x = \frac{h}{2}$. By the hypothesis, we have $\pi l(R + r) = \pi l^2$,

and, consequently, $R + r = l$. However, from the triangles OMN and CPD we obtain

$$x = \frac{R + r}{2} \cos\alpha \text{ and } h = l\cos\alpha.$$

and, hence, $x = \frac{h}{2}$ which is what we set out to prove.

502. Consider the trapezoid $ABCD$ in the axial section of the cone shown in Fig. 213. Let E and F be the midpoints of its bases, and O the midpoint of EF. We also construct $OM \perp CD, ON \perp EF$ and $CP \perp AD$, and $\angle MON = \angle PCD = \alpha$.

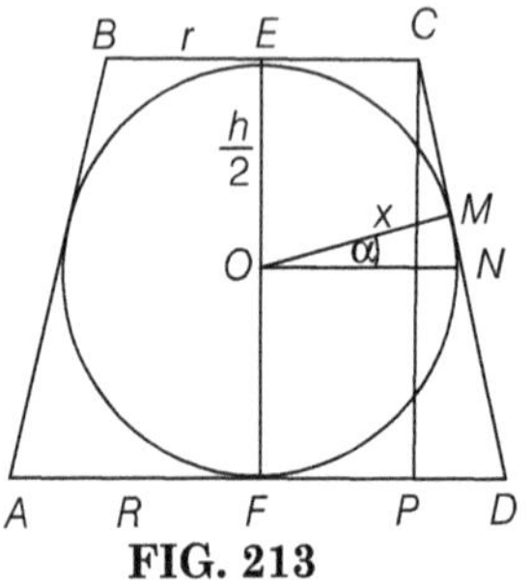

FIG. 213

To solve the problem it is sufficient to prove that $OM = OE$. Let us introduce the notation $EC = r$, $DF = R$, $OM = x$ and $OE = \frac{h}{2}$.

Then we have $x = ON \cos\alpha = \frac{R + r}{2} \cos\alpha$.

For the triangle CPD we can write

$$h = CD\cos\alpha = \sqrt{(R - r)^2 + CP^2}\, \cos\alpha$$

But, by the hypothesis, $CP^2 = 4Rr$ and therefore

$$h = \sqrt{(R - r^2) + 4Rr}\, \cos\alpha = (R + r)\cos\alpha.$$

Thus, $x = \frac{h}{2}$ which is what we set out to prove.

503. Let SD be the altitude of a regular tetrahedron $SABC$, O the midpoint of the altitude and E the midpoint of the line segment BC whose length is designated by a (Fig. 214).

We have
$$DE = \frac{a\sqrt{3}}{6};$$
$$SD = \sqrt{SE^2 - DE^2} = \frac{a\sqrt{6}}{3};$$
$$OD = \frac{a\sqrt{6}}{6},$$
whence
$$OE = \sqrt{OD^2 + DE^2} = \frac{a}{2}.$$
Consequently, $OE = BE = EC$ and, hence, $\angle BOC = 90°$.

504. Let a be the side of the base of the given pyramid $SABCD$, α the plane angle of the dihedral angle with edge BC and H the length of the altitude. SO of the pyramid (Fig. 215). Then we have
$$r = \frac{a}{2}\tan\frac{\alpha}{2}.$$
Besides, according to formula (1) in the solution of Problem 481 we can write
$$R = \frac{H^2 + \left(\frac{a\sqrt{2}}{2}\right)^2}{2H}.$$
Consequently,
$$R = \frac{a}{4}\,\frac{\tan^2\alpha + 2}{\tan\alpha},$$
and, hence,
$$\frac{R}{r} = \frac{\tan^2\alpha + 2}{2\tan\alpha\tan\frac{\alpha}{2}}.$$
Putting $\tan\frac{\alpha}{2} = x$ we obtain
$$\frac{R}{r} = \frac{1 + x^4}{2x^2(1 - x^2)}.$$

---

* From the above equality $R + r = l$ it follows that $2R + 2r = l + l$. This means that the sums of the opposite sides of the considered quadrillateral are equal. This is sufficient for the possibility of inscribing a circle in the quadrilateral. But here we do not take advantage of this fact.

Introducting the notation $x^2 = t$ we reduce the problem to proving the inequality

$$\frac{l+t^2}{2t(1-t)} \geq 1+\sqrt{2}$$

or $$0 < t < 1.$$

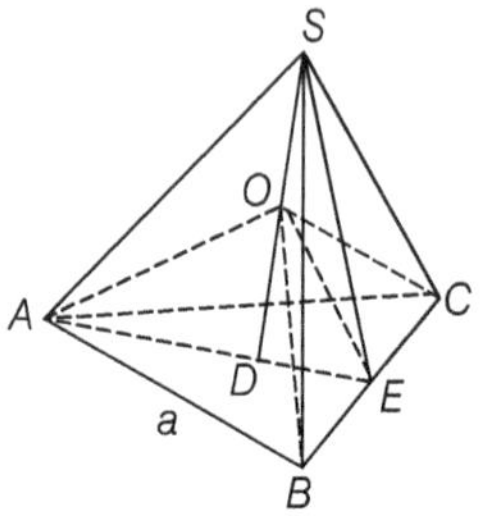

FIG. 214

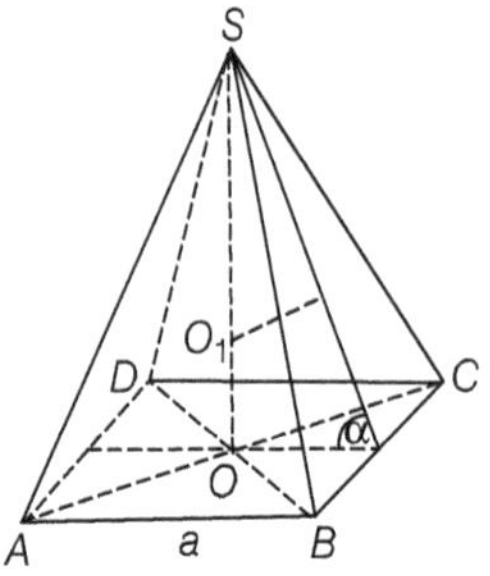

FIG. 215

Multiplying both members of the inequality by the denominator and opening the brackets we obtain the quadratic inequality

$$(2\sqrt{2}+3)\,t^2 - 2\,(\sqrt{2}+1)\,t + 1 \geq 0.$$

Computing the discriminant of the trinomial, we find out that it is equal to zero. Consequently, the trinomial retains its sign for all values of t. The value of the trinomial for $t = 0$ being positive, the inequality has thus been proved.

505. The pyramids $ASBC$ and $OSBC$ have a common base SBC (Fig. 216), and therefore their volumes are in the ratio of their altitudes dropped onto that common base. Since $OA' \parallel AS$, the ratio of the altitudes of the pyramids $ASBC$ and $OSBC$ drawn to the base SBC is equal to the ratio of SA to OA'. Hence, the ratio of the volumes is

$$\frac{V_{OSBC}}{V_{ASBC}} = \frac{OA'}{SA}.$$

Analogously, $$\frac{V_{OSCA}}{V_{ASBC}} = \frac{OC'}{SC}, \frac{V_{OSAB}}{V_{ASBC}} = \frac{OB'}{SB}.$$

Adding together these equalities, we obtain

$$\frac{OA'}{SA} + \frac{OB'}{SB} + \frac{OC'}{SC} = 1.$$

506. Let P be the plane of the triangle ABC, P_1 the plane of the triangle $A_1B_1C_1$ and l the line of intersection of P and P_1 (Fig. 217). Denote by Q_{AB} the plane passing through A, B and O. The straight line A_1B_1 is in the plane Q_{AB}. The straight lines A_1B_1 and AB are nonparallel and, hence, they intersect at a point T_{AB}. This point lies in the planes P and P_1 and thus on the line l. We similarly prove that the straight lines BC and B_1C_1 intersect at a point T_{BC} lying on l, and the straight lines AC and A_1C_1 at a point T_{AC} also belonging to l.

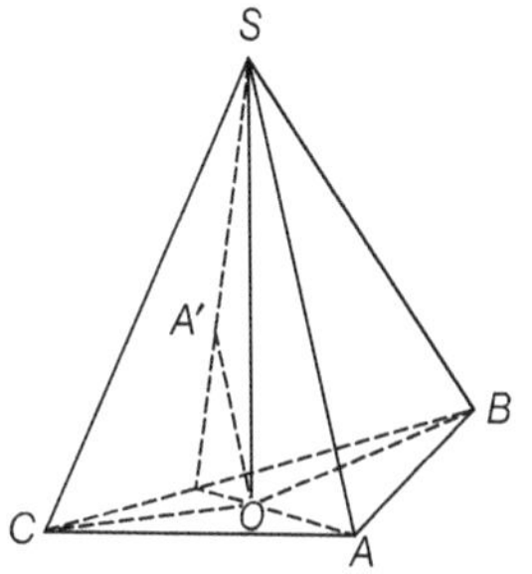

FIG. 216

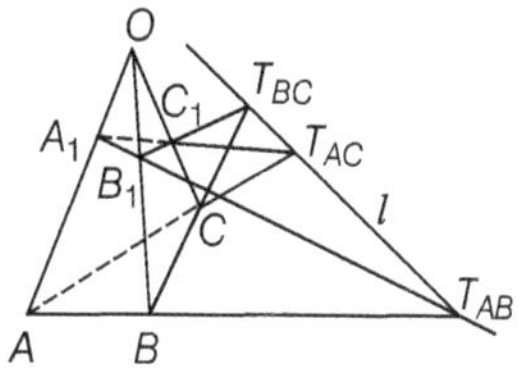

FIG. 217

507. Let O_1 be the centre of gravity of the face ASC of a triangular pyramid $SABC$ (see Fig. 218) and BO_1 one of the line segments considered in the problem. Take another face, for instance BSC. We shall designate its centre of gravity by O_2 and prove that the line segment AO_2 intersects the line segments BO_1, point of intersection O of these segments dividing the line segment BO_1 into the parts OO_1 and O_1B which are in the ratio 1:3. Indeed, if M_1 and M_2 are the mid points of the line segments AC and BC, then it is obvious that $AB \| M_1M_2$; it is also clear that $O_1O_2 \| M_1M_2$, since the points O_1 and O_2 divide, respectively, the line segments M_1S and M_2S in one and the same ratio. Therefore, $AB \| O_1O_2$ and the figure ABO_2O_1 is a trapezoid. Consequently, its diagonals BO_1 and AO_2 intersect. Let us denote the point of intersection of the diagonals by O. We have

$$\frac{M_1M_2}{AB}=\frac{1}{2},\ \frac{O_1O_2}{M_1M_2}=\frac{2}{3}.$$

Multiplying these equalities termwise, we get $\frac{O_1O_2}{AB}=\frac{1}{3}$. But the similarity of the triangles AOB and O_1OO_2 implies $\frac{O_1O}{OB}=\frac{O_1O_2}{AB}$. Thus, we have in fact

$$\frac{O_1O}{OB}=\frac{1}{3}.$$

If now we take the centre of gravity of another face and construct the corresponding line segment, then, by virtue of the above, it also intersects the line segment BO_1, the point of intersection dividing this segment in the ratio 1:3. Hence, this point coincides with the point O. Consequently, all the line segments in question intersect at the point O. It is also evident that the point O divides each of them in the ratio 1:3 which is what we set out to prove.

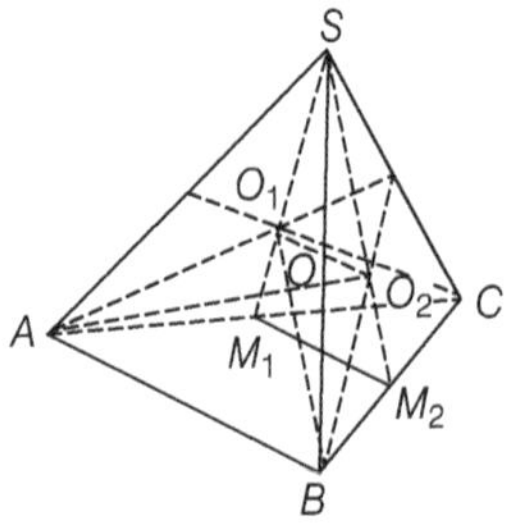

FIG. 218

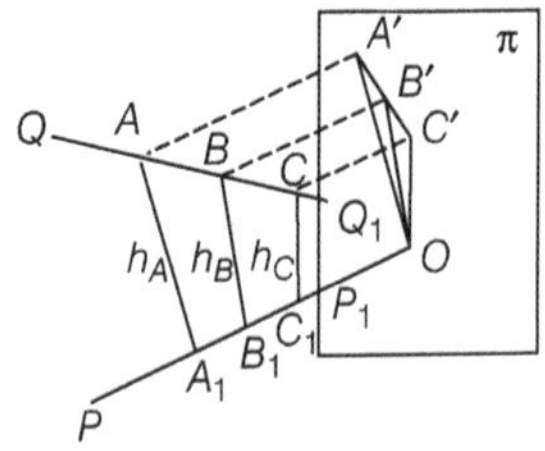

FIG. 219

508. We shall precede the proof with an auxiliary argument. Let PP_1 and QQ_1 be two skew lines and points A, B and C lie on QQ_1, the point B being between the points A and C. Also let A_1, B_1 and C_1 be the feet of the perpendiculars dropped from the points A, B, C onto PP_1. Denote, respectively, by h_A, h_B and h_C the distances from the points A, B and C to the straight line PP_1. We shall prove that h_B is less than atleast one of the distances h_A or h_C.

To this end, project the configuration shown in Fig. 219 onto a plane π perpendicular to the straight line PP_1. Then the straight line PP_1 is projected into a point O, and the line segments AA_1, BB_1 and CC_1, when projected, do not change their size because they all are parallel to the plane π. The point B' is then between the points A' and C'. Now taking the triangle $A'OC'$, we can assert that the inclined line OB' is shorter than one of the inclined lines OA' or OC'. Indeed, dropping from the point O the perpendicular to $A'C'$. (which is not shown in Fig. 219), we see that the point B' is closer to the foot of that perpendicular than one of the other two points A' and C'. It follows that h_B is shorter than h_A or h_C.

Let now $ABCD$ be an arbitrary triangular pyramid, and EFG a triangular section such that atleast one of its vertices, say F, is not a vertex of the pyramid. Let us prove that the area of the triangle EFG is then less than the area of one of the triangles AEG or DEG (Fig. 220).

In fact, all the three triangles have a common side EG, and, as has been proved, the distance from F to the straight line EG is less than the distance from A or D and this line. If $S_{\Delta EFG} < S_{\Delta AFG}$, then the assertion has been proved. If $S_{\Delta EFG} < S_{\Delta DEG}$ and, for instance, the point E is not a vertex of the pyramid, then we apply the above argument to ΔDEG and compare its area with the areas of the triangles DGA and BDG. If necessary, again applying the same argument to the triangle BDG we prove the assertion of the problem. It is clear from this solution that if a section of the pyramid does not coincide with its face, then the area of the section is strictly less than the area of one of the faces.

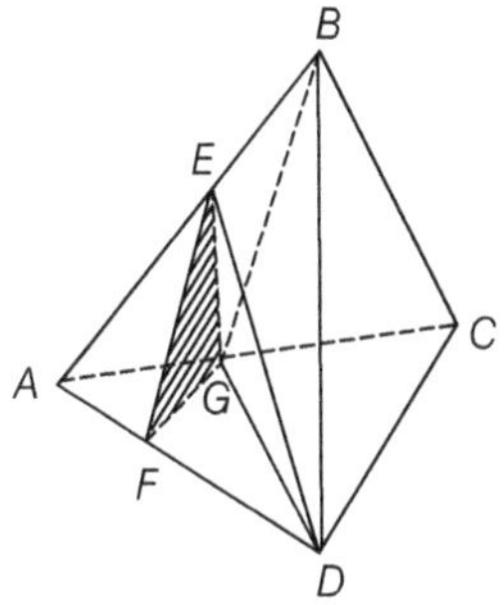

FIG. 220

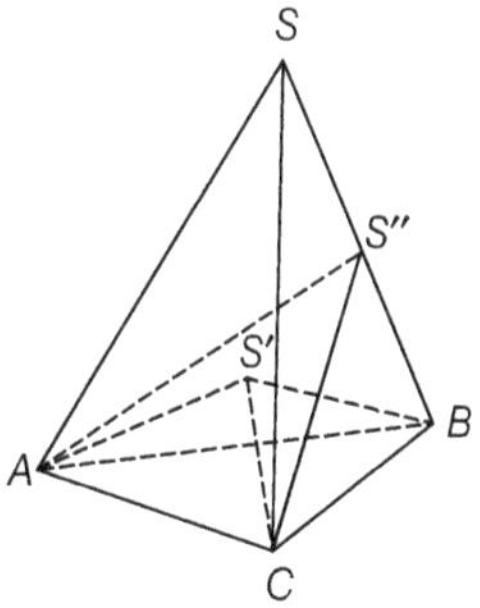

FIG. 221

509. Instead of comparing the sums of the face angles at the vertices S and S' we shall compare the sums of the base angles of the lateral faces of both pyramids adjacent to each of the three vertices of their common base. We shall prove that for the outer pyramid every sum of this kind is greater than the corresponding sum for the inner pyramid.

For instance, we shall prove below that

$$\angle ACS + \angle SCB > \angle ACS' + \angle S'CB \tag{1}$$

(see Fig. 221).

From (1) and analogous inequalities for the vertices A and B we obtain the solution of the problem. Indeed, adding together these three inequalities we find out that the sum Σ' of all the six base angles of the lateral faces of the outer pyramid is greater than the corresponding sum Σ' for the inner pyramid, that is we have the inequality

$$\Sigma > \Sigma'. \tag{2}$$

But the quantities we are interested in are, respectively, equal to the differences $180°\cdot 3 - \Sigma = 540° - \Sigma$ and $180°\cdot 3 - \Sigma' = 540° - \Sigma'$, and, consequently, they satisfy the opposite inequality. Thus, to solve the problem, we must only prove inequality (1).

Extend the plane ACS' to intersect the outer pyramid. Considering the trihedral angle $CS'S''B$, we conclude that

$$\angle S'CS'' + \angle S''CB > \angle S'CB. \tag{3}$$

Adding $\angle ACS'$ to both members of this inequality we obtain

$$\angle ACS'' + \angle S''CB > \angle ACS' + \angle S'CB. \tag{4}$$

But for the trihedral angle $CASS''$ we have

$$\angle ACS + \angle SCS'' > \angle ACS''. \tag{5}$$

Based on (5), we substitute the larger quantity $\angle ACS + \angle SCS''$ for $\angle ACS''$ in inequality (4) and thus obtain

$$\angle ACS + (\angle SCS'' + \angle S''CB) > \angle ACS' + \angle S'CB.$$

i.e. inequality (1).

510. Let O_1, O_2, O_3 and O_4 be the centres of the given spheres and P_{ik} the plane tangent to the spheres with centres O_i and O_k $(i < k)$. Thus, we consider the six planes $P_{12}, P_{13}, P_{23}, P_{14}, P_{24}$ and P_{34}. Let us first prove that the planes P_{12}, P_{13} and P_{23} have a common straight line. Indeed, each of the planes is perpendicular to the plane $O_1O_2O_3$ because it is perpendicular to the centre line of the corresponding spheres, this centre line lying in that plane.

Besides, it is evident that the planes under consideration (Fig. 222) pass through the point Q_4 of intersection of the bisectors of $\Delta O_1O_2O_3$. Thus, the planes P_{12}, P_{13} and P_{23} in fact intersect along a straight line which, as we have incidentally proved, is perpendicular to the plane of the centres $O_1O_2O_3$ and passes through the centre of the inrcribed circle of the triangle $O_1\,O_2\,O_3$. Let us designate this line by L_4.

We similarly prove that the planes P_{23}, P_{24} and P_{34} have a common straight line L_1 which is perpendicular to the plane of the triangle $O_2O_3O_4$ and passes through the centre of its inscribed circle and so on. Therefore we arrive at the following auxiliary problem (Fig. 223): a circle is inscribed in each face of the triangular pyramid $O_1O_2O_3O_4$, and the perpendicular is drawn through its centre to this face. It is necessary to prove that all four perpendiculars L_1, L_2, L_3 and L_4 have a point in common provided that the points of tangency of every two circles with the corresponding edge of the pyramid coincide.

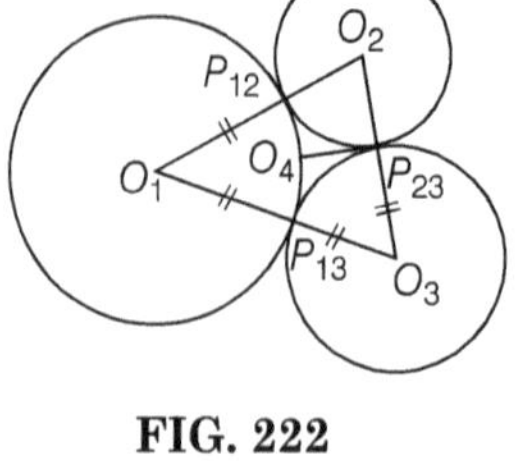

FIG. 222

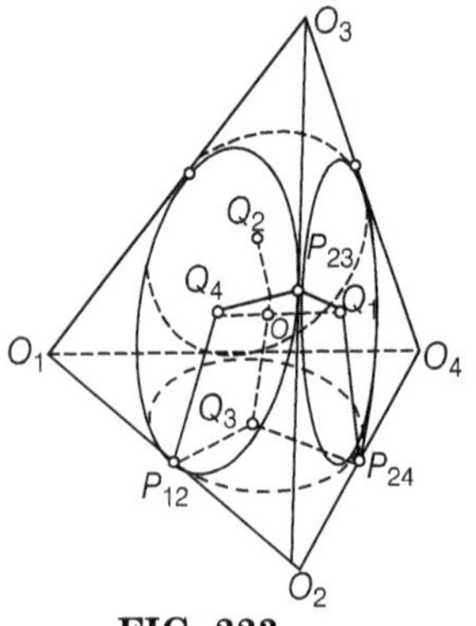

FIG. 223

This fact is almost apparent. Let O be the point of intersection of the straight lines L_1 and L_4; the latter intersect because they are in the plane P_{23} and are not parallel. Let us now prove that the straight lines L_3 and L_2 also pass through the point O. Indeed, the point O lies on the line of intersection of the planes P_{12} and P_{24} because the straight line L_4 belongs to the plane P_{12}, and the line L_1 to the plane P_{24}. But the line of intersection of P_{12} and P_{24} is the straight line L_3, and hence the latter passes through the point O. We analogously prove that the straight line L_2 passes through the point O.

511. If we are given three point A, B and C not lying in a straight line, then these points are the centres of three pairwise tangent spheres. Indeed, if P is the point of intersection of the bisectors of the interior angles in ΔABC, and P_1, P_2 and P_3 are the feet of the perpendiculars dropped from P to the corresponding sides AB, BC and CA, then

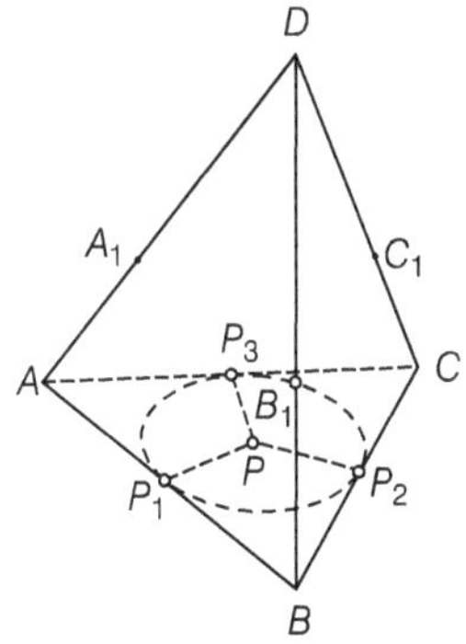

FIG. 224

$$AP_1 = AP_3, \quad BP_1 = BP_2, \quad CP_2 = CP_3,$$

and the spheres with the centres A, B and C whose radii are respectively equal to

$$r_A = AP_1, \quad r_B = BP_2, \quad r_C = CP_3$$

are pairwise tangent to one another.

Let $ABCD$ be the given pyramid (Fig. 224). Consider the three spheres of radii r_A, r_B and r_C with. centres at A, B and C which are pairwise tangent to one another. Let us denote the points at which the spheres intersect the edges AD, BD and CD by A_1, B_1 and C_1. We shall prove that $A_1D = B_1D = C_1D$.

By the hypothesis indicated in the problem, we have

$$AD + BC = BD + AC.$$

By the above construction, we can write

$$AD = r_A + A_1D_1, \quad BC = r_B + r_C,$$

$$BD = r_B + B_1D, \quad AC = r_A + r_C,$$

Substituting the last four expressions in the foregoing equality we obtain

$$A_1D = B_1D.$$

Similarly, using the equality $BD + AC = CD + AB$,

we deduce

$$B_1D = C_1D.$$

Consequently, the sphere with centre D and radius $r_D = A_1D = B_1D = C_1D$ is tangent to each of the first three spheres and, hence, the four constructed spheres are pairwise tangent to one another.

512. Let us denote by r_1, r_2 and r_3 the radii of the spheres. We shall suppose that $r_1 \geq r_2 \geq r_3$. Draw a tangent plane to the first two spheres. In addition, through the centres of these spheres draw a plane perpendicular to this tangent plane, and consider the circle of radius r tangent to the two great circles in the section and to their common tangent line (Fig. 225). It is obvious that the third sphere can be tangent to the first two spheres and to their common tangent plane if it is "not too small", namely, if $r_3 \geq r$. We have (see Fig. 225)

$$\sqrt{O_1O_2^2 - O_1C^2} = AO + OB,$$

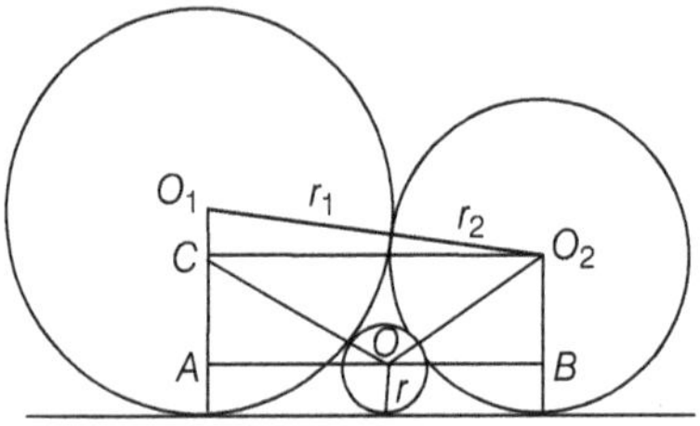

FIG. 225

that is

$$\sqrt{(r_1 + r_2)^2 - (r_1 - r_2)^2} = \sqrt{(r_1 + r)^2 - (r_1 - r)^2} + \sqrt{(r_2 + r)^2 - (r_2 - r)^2}.$$

From this equation we find

$$r = \frac{r_1 r_2}{(\sqrt{r_1} + \sqrt{r_2})^2}.$$

Consequently, the radii of the spheres must satisfy the relation $r_3 \geq \dfrac{r_1\, r_2}{(\sqrt{r_1} + \sqrt{r_2})^2}$.

513. Let n be the number of lateral faces of the pyramid in question. Join an arbitrary point O lying in the plane of the base to all the vertices. We thus obtain n triangular pyramids with common vertex at the point O. It is obvious. That the volume V of the given pyramid is equal to the sum of the volumes of the smaller triangular pyramids. We have

$$V = \frac{1}{3} S(r_1 + r_2 + \ldots + r_n).$$

where $r_1, r_2, \ldots, r_n$ are the distances from the point O to the lateral faces of the pyramid, and S is the area of its lateral face.

Hence, the quantity $r_1 + r_2 + \ldots + r_n = \dfrac{3V}{S}$ is a constant independent of the position of the point O in the plane of the base which is what we set out to prove.

514. Consider the configuration shown in Fig. 226 where we see two shaded planes and the triangle ADE in the plane P passing through the vertices A, D, H and E of the given parallelepiped. The plane P intersects the plane of ΔBCD along the straight line KD which passes through the point K of intersection of the diagonals of the parallelogram $ABEC$. Consequently, the line

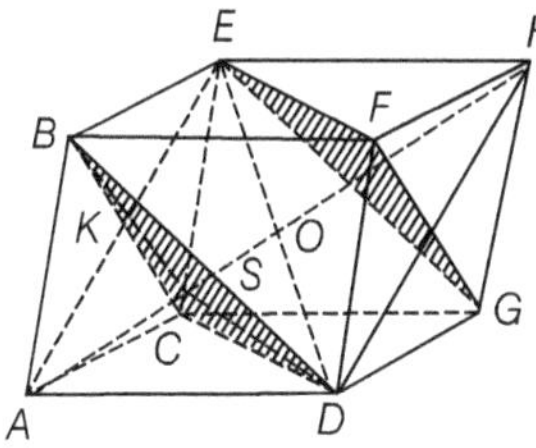

FIG. 226

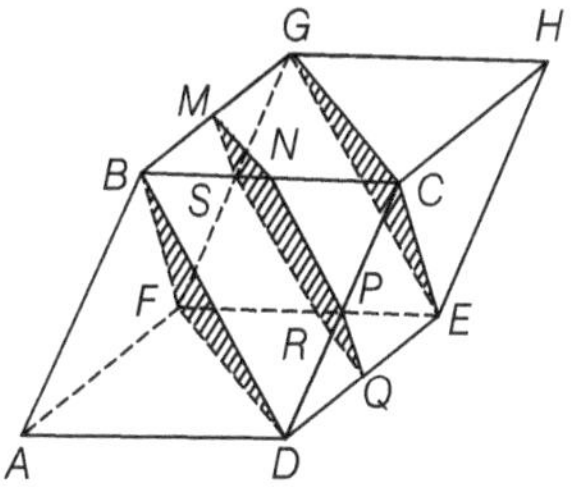

FIG. 227

segment KD is a median of ΔAED. As is obvious, AO is also a median of ΔAED. Therefore, S is the point of intersection of the medians of ΔAED, and, hence, we arrive at the required result:

$$AS = \frac{2}{3} AO = \frac{1}{3} AH.$$

515. Let us draw the plane indicated in the problem through the vertices B, D and F (Fig. 227) and a plane parallel to it through the vertices C, E and G. These planes give in the sections two congruent equilateral triangles. Let a be the length of the sides of these triangles. If now we draw a plane parallel to the above planes through the midpoint of one of the six edges joining the vertices of the two triangles, for example, through the midpoint N of the edge BC, then the section of the parallelepiped by this plane is a hexagon $MNPQRS$ whose all sides are obviously equal to $\frac{a}{2}$. Furthermore, note that $MN \parallel DF$ and $NP \parallel BD$. Therefore, MNP and BDF are supplementary angles and, consequently, $\angle MNP = 120°$. We similarly prove that the other angles of the hexagon are also equal to 120°.

516. Let $SABC$ be the given tetrahedron, P and Q the midpoints of the opposite edges AC and SB. Consider a section $MPNQ$ of the tetrahedron containing the line segment PQ (Fig. 228). Let us take the plane section SPB which obviously divides the tetrahedron into two parts of the same volume. The solution of the problem reduces to proving that the volumes of the pyramids $SPQN$ and $MPQB$ are equal.

Drop the perpendiculars from the points M and N onto the plane SPB, and designate their feet by K and L, respectively.

The triangles PQB and SPQ are of the same area, and therefore to solve the problem it is sufficient to show that $LN = MK$. We shall prove this equality establishing the relation

$$MO = NO. \tag{1}$$

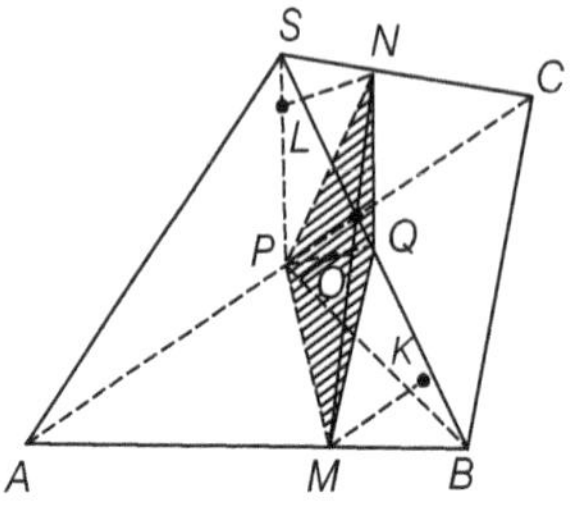

FIG. 228

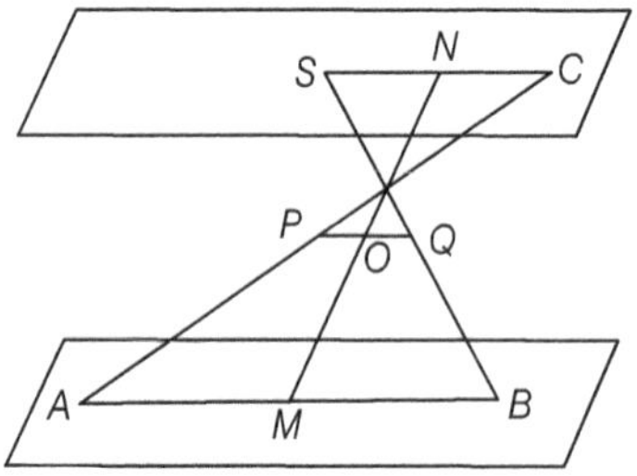

FIG. 229

For this purpose, let us consider a pair of parallel planes containing the skew lines SC and AB (Fig. 229). The line segment PQ joining the midpoints of the segments AC and SB, we see that PQ is in the plane parallel to the given planes and equidistant from them. Therefore, the line segments PQ and MN intersect, the point of intersection bisecting MN.

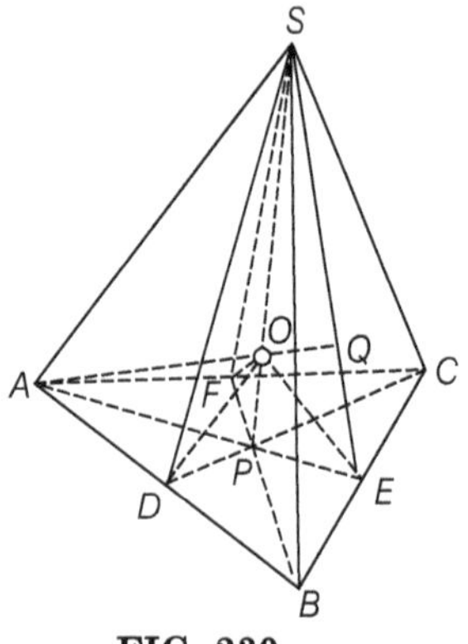

FIG. 230

517. Let $SABC$ be the given pyramid (Fig. 230). Draw the altitude SP from the vertex S to the face ABC and also the altitudes SD, SE and SF from the same vertex to the bases AC, AB and BC of the other three faces. It is readily seen that the triangles SPD, SPE and SPF are equal because $\angle SDP = \angle SEP = \angle SFP$ (cf. Problem 458).

Then we draw through the edges AB, BC and AC the planes bisecting the corresponding dihedral angles. These planes intersect at a point O equidistant from all four faces of the pyramid. Therefore,

O is the centre of the inscribed sphere of the pyramid. It is evident that in the case under consideration the point O is on the altitude SP of the pyramid because, as has been shown, the above triangles are congruent. Repeating this argument we establish that all the altitudes of the pyramid intersect at the point O. Based on this fact, we can assert that, for instance, the triangles APS and SPE lie in one plane, and, consequently, the line segments AP and PE are in a straight line. Therefore, in ΔABC, the straight line AE is a bisector of the angle A and, simultaneously, the altitude drawn to BC. Analogously, the other bisectors of $\Delta\ ABC$ are its altitudes. Hence, ABC is an equilateral triangle. Repeating this argument we establish that all the faces of the pyramid are equilateral triangles which is what we set out to prove.

518. Let the line segment AB be in a plane Q (see Fig. 231) and the line segment CD in a plane P, these planes being parallel. Through the point A draw a straight line parallel to CD, and lay off the line segment $AA_1 = CD$. Construct a parallelogram ABB_1A_1 on the sides AB and AA_1. Make an analogous construction in the plane P.

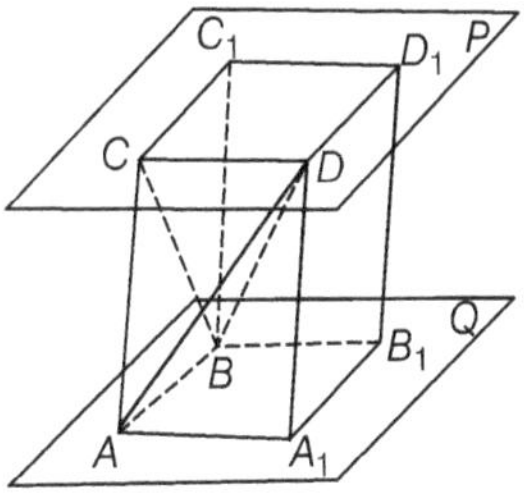

FIG. 231

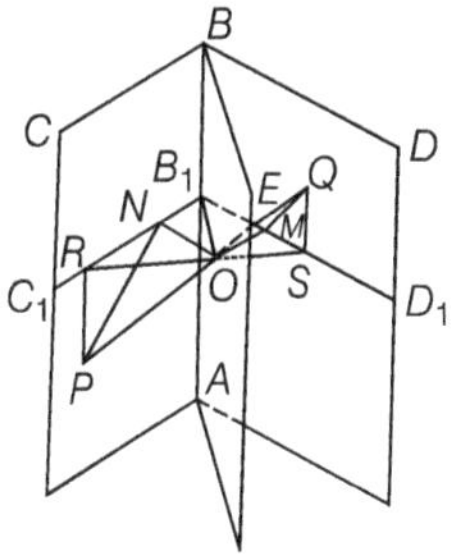

FIG. 232

Joining A with C, B with C_1, A_1 with D and B_1 with D_1 we obtain a parallelepiped $ABB_1A_1DCC_1D_1$. Considering the face ACB as the base of the pyramid $DACB$, we see that the volume of the pyramid is equal to $\frac{1}{6}$ of the volume of the parallelepiped. However, the volume of the parallelepiped is retained when AB and CD are translated in their planes P and Q because the area of the base ABB_1A_1 and the altitude (which is the distance between the planes P and Q) remain unchanged. Therefore the volume of the pyramid is also retained.

519. Let P and Q be the points of intersection of a given line with the faces CBA and DBA of a given dihedral angle (Fig. 232). Draw through the edge AB the plane ABE bisecting the dihedral angle and then through the point O at which the straight line PQ intersects ABE draw the plane $C_1B_1D_1$ perpendicular to the edge AB. Furthermore, let $OM \perp B_1D_1, ON \perp B_1C_1$ and SR be the projection of PQ onto the plane $D_1B_1C_1$ so that $QS \perp B_1D_1$ and $PR \perp B_1C_1$. If the points P and Q are equidistant from the edge, i.e.

$$B_1R = B_1S, \quad (1)$$

then B_1RS is an isosceles triangle, $SO = RO$ and, hence, $QO = PO$, i.e. the line segments QO and PO are congruent as inclined lines with equal projections. Also taking into account that, by the construction, we have

$$MO = NO, \quad (2)$$

we conclude that $\Delta\, OMQ$ and $\Delta\, ONP$ are right and congruent. It follows that

$$\angle\, MQO = \angle\, NPO. \quad (3)$$

Thus, we have proved that condition (1) implies (3).

Conversely, let it be given that condition (3) expressing the equality of the angles is fulfilled. Then, by virtue of (2) the triangles QMO and PNO are congruent. It follows that $QO = PO$, and,hence, $SO = OR$ which implies (1).

520. Join the point B with C and A with D (Fig. 233). Through the point A draw a straight line parallel to MN to intersect the line passing through B and N at a point K. Note that $AK = 2MN$, since MN is a midline in the triangle ABK. Furthermore, we have $\Delta\, BNC = \Delta\, KND$ because $BN = NK$, $CN = ND$ and $\angle\, BNC = \angle\, KND$. Therefore, $DK = BC$. From the triangle ADK it follows that

$$DK + AD > AK = 2MN.$$

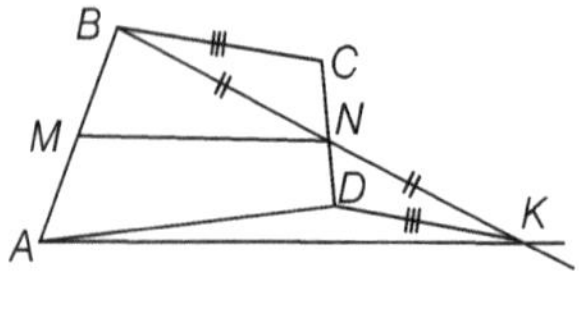

FIG. 233

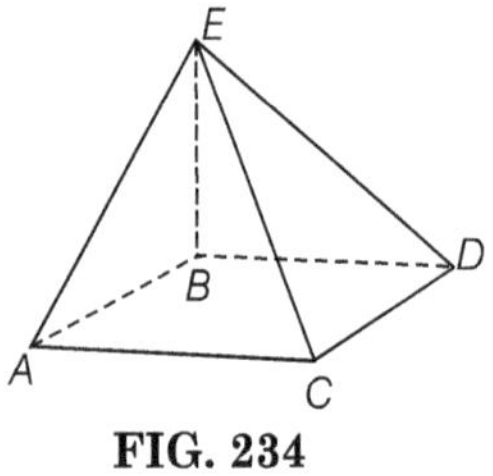

FIG. 234

(It is essential here that the point D is not in the straight line AK because, if otherwise, we must put the sign $\geq$). Thus, we obtain the required result:

$$BC + AD > 2MN.$$

521. Let A, B, C and D be arbitrary points lying on the edges of a tetrahedral angle with vertex E (Fig. 234). We shall prove, for instance, that

$$\angle\, CED < \angle\, CEA + \angle\, AEB + \angle\, BED. \quad (1)$$

Draw the plane CEB. By the property of the face angles of a trihedral angle, we have

$$\angle CED < \angle CEB + \angle BED, \tag{2}$$

and, by the same reason,

$$\angle CEB < \angle CEA + \angle AEB. \tag{3}$$

Inequalities (2) and (3) imply (1) and the desired inequality has thus been proved.

It is evident that the above argument also holds true when the tetrahedral angle is not convex, i. e. when the edge ED is on the other side of the plane CEB.

522. Suppose that we are given a convex tetrahedral angle with vertex S(Fig. 235.). The extensions of the planes BSC and ASD intersect along a straight line l_1 and the extensions of the planes ASB and DSC intersect along a straight line l_2. Obviously, the straight lines l_1 and l_2 do not coincide because, if otherwise, the extended faces pass through one straight line. Let P be the plane containing the straight lines l_1 and l_2. Taking advantage of the convexity of the tetrahedral angle, we can easily show that the plane P and the given angle have only one point in common, namely the point of intersection S.

Therefore, the whole angle lies on one side of the plane P (this fact is, however, almost obvious). Now let us show that every plane parallel to the plane P and intersecting the angle yields a parallelogram in the section.

Indeed, by the above, a plane of this type intersects all the edges of the tetrahedral angle. Denoting the points corresponding to intersection by A', B', C' and D' we see that $A'D' \parallel B'C'$ because each of these line segments is parallel to l_1. Analogously, we have $A'B' \parallel D'C'$.

Hence, we obtain the required result: the quadrilateral $A'B'C'D'$ is a parallelogram.

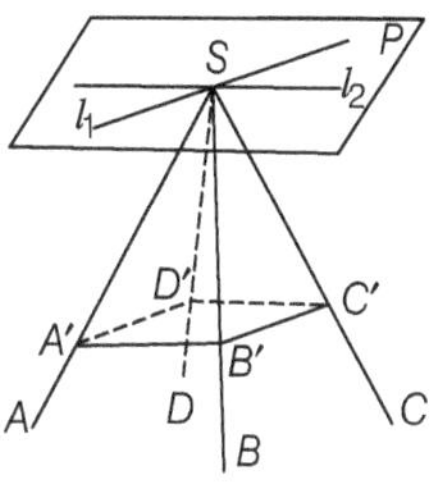

FIG. 235

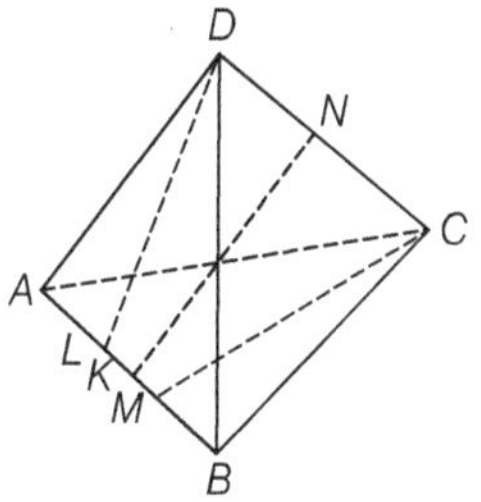

FIG. 236

523. Consider the configuration in Fig. 236. Let DL and CM be the altitudes of two triangles ADB and ACB drawn to their common base AB. The triangles are of the same area, and therefore $DL = CM$. Furthermore, let KN be the common perpendicular to the skew lines AB and DC.

Draw through the line segment KN a plane P perpendicular to the edge AB, and project the quadrilateral $LMCD$ onto the plane P (Fig. 237). The segments DL and CM being projected without changing their lengths (because they are parallel to the plane P), and the projection of the line segment LM being the point K, we obtain in the plane P the isosceles triangle KD_1C_1. By the construction, we have $KN \perp DC$ and, consequently, $KN \perp D_1C_1$. Therefore KN is an altitude in $\Delta\, KD_1C_1$. Consequently, N is the midpoint of the segment D_1C_1 and thus of the segment DC as well.

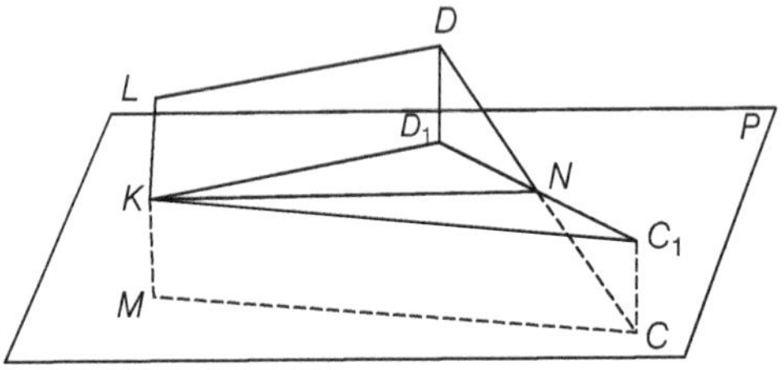

FIG. 237

We see that, under the assumptions of the problem, the common perpendicular KN to the two skew lines AB and DC bisects the edges AB and DC.

As is readily seen from Fig. 237, $LK = KM$ because $DD_1 = CC_1$. Therefore (see Fig. 236), $AL = BM$, and the congruence of the right triangles ALD and BMC implies that

$$AD = BC.$$

We analogously prove that $AC = BD$ and $AB = DC$. Consequently, all the faces are congruent as triangles with three equal corresponding sides.

3. Loci of Points

524. Let P be one of the planes passing through a given point A, and M the projection of another given point B on the plane P. Let C be the midpoint of the line segment AB (Fig. 238).

The triangle ABM being right, we have $CM = \frac{1}{2} AB$. Thus, all the points M which can be thus constructed are at the same distance $\frac{1}{2} AB$ from the point C and, consequently, are on the sphere of radius $\frac{1}{2} AB$ with centre at the point C. Besides, it is apparent that every point of this sphere coincides with one of the projections of the point B. The required locus is thus a sphere of diameter AB.

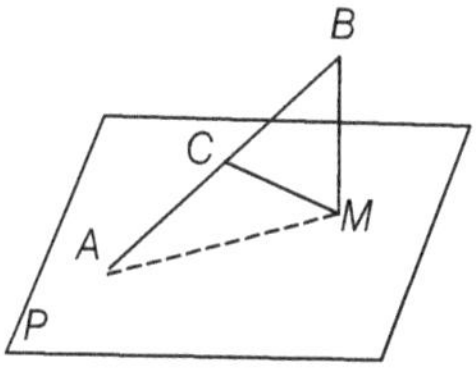

FIG. 238

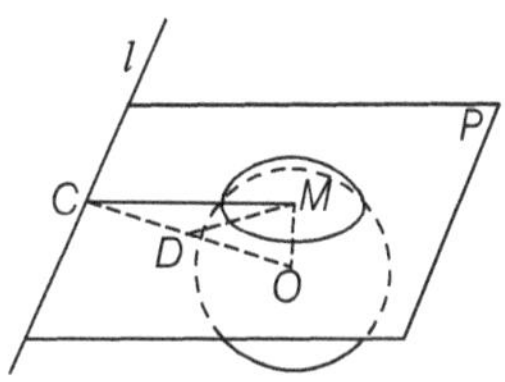

FIG. 239

525. Let O be the centre of the given sphere. Draw through the given straight line l a plane P intersecting the sphere in a circle with centre at a point M (Fig. 239). As is known, $OM \perp P$. Then draw through the point O a plane P_1 perpendicular to the straight line l. Denote the point of intersection of the plane P_1 and line l by C. The planes P_1 and P being mutually perpendicular, the line segment OM is in the plane P_1. Now consider the right triangle OMC. The point C is independent of the choice of the cutting plane P, and the hypotenuse OC of the right triangle OMC is invariable. If D is the midpoint of OC, then $MD = \frac{OC}{2}$. Consequently, if l and the sphere have no points in common, the sought-for locus is a portion of the circumference of a circle of radius $\frac{OC}{2}$ contained inside the sphere (this arc lies in the plane P_1 and passes through the centre of the sphere). If l is tangent to the sphere, then the sought-for locus is a circle of radius $\frac{R}{2}$ where R is the radius of the sphere. Finally, if l intersects the sphere at two points, the locus of points M is a circle of radius $\frac{OC}{2}$.

526. The required locus is a surface of revolution obtained by rotating an arc of a circle or an entire circle about its diameter OC (see the solution of the preceding problem).

527. We shall prove that the required locus is a sphere of radius $R\frac{\sqrt{6}}{2}$ and that the centre of this sphere coincides with the centre of the given sphere.

Let M be an arbitrary point of the required locus; the line segments MA, MB and MC (see Fig. 240) being the segments of the tangent lines drawn to the given sphere from a common point, their lengths are equal. Therefore, the right triangles AMC, CMB and AMB are congruent. Hence, ABC is an equilateral triangle. As is obviously seen , the line segment OM intersects $\Delta\ ABC$ at its centre of gravity O_1. Let $AM = a$, then $AC = a\sqrt{2}$ and $AO_1 = \frac{a\sqrt{6}}{3}$.

Substituting these values in the equality $OM \cdot AO_1 = OA \cdot AM$

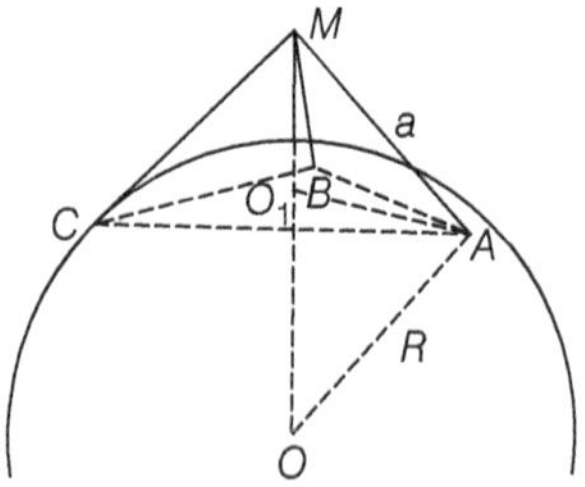

FIG. 240

(here we take advantage of the fact that OAM is a right triangle, and express its area in two different ways) we obtion

$$OM.a\frac{\sqrt{6}}{3} = Ra.$$

It follows that

$$OM = \frac{\sqrt{6}}{2}R.$$

Thus, the point M lies on the above-mentioned sphere. Rotating the given sphere, together with the tangents AM, CM and BM, about the centre O, we see that every point of the sphere belongs to the locus of points in question.

528. Let A be a given point in space, B the point of intersection of straight lines lying in a fixed plane, and C the foot of the perpendicular dropped from A on the plane.

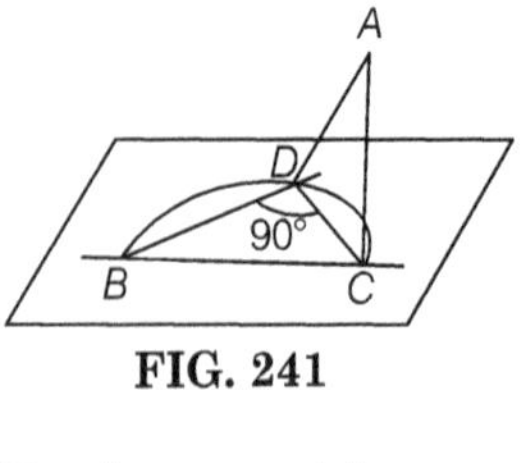

FIG. 241

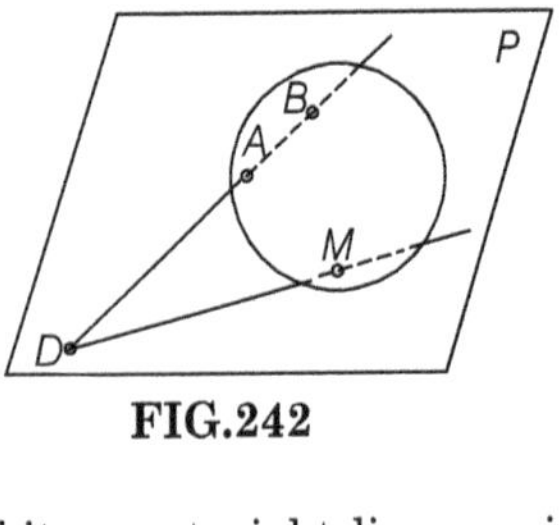

FIG.242

Furthermore, take an arbitrary straight line passing through the point B and draw the perpendicular AD to it (Fig. 241). Then, according to the well-known theorem, $CD \perp BD$.

Consequently, the point D lies in the sphere whose diameter is the line segment BC. It can easily be proved that, conversely, any point of the indicated circle is the foot of the perpendicular drawn from the point A to a straight line belonging to the family in question. Therefore, the sought-for locus is the circle in the given plane constructed on the line segment BC as diameter.

529. There are two possible cases here which are considered below.

(1) The straight line AB is not parallel to the plane P. Designate the corresponding point of intersection of AB and P by D (Fig. 242). Let M be the point of tangency of the plane with one of spheres belonging to the family in question. Draw the plane through the straight lines AB and DM. It intersects the sphere along a circle tangent to the straight line DM at the point M. By the well-known property of a tangent and secant drawn from one point to a circle, we have $DB \cdot DA = DM^2$. Consequently, the line segment DM has the constant length $\sqrt{DB \cdot DA}$ independent of the choice of the sphere, and, hence, all the points M lie in the circle of radius $r = \sqrt{DB \cdot DA}$ with centre at the point D. Let us denote this circle by C. Let now, conversely, M be a point of the circle C. We shall prove that it belongs to the locus of points under consideration.

Draw an auxiliary circle through the point A, B and M and denote its centre by O_1 (Fig. 243). According to the construction we have $DB \cdot DA = DM^2$, and therefore the straight line DM is tangent to this circle. Hence, $O_1M \perp DM$. Now erect at the point M the perpendicular to the plane P, and at the point O_1 the perpendicular to the plane of the auxiliary circle. The two perpendiculars lie in a plane perpendicular to DM at the point M and are not parallel to each other because, if otherwise, the point O_1, and the points A and B as well, are in the plane P. Therefore, these perpendiculars intersect at a point O. It is obvious that $OA = OB = OM$ because the projections O_1A, O_1B, and O_1M of these line segments are equal as radii of one circle. Therefore, the sphere with centre at the point O and radius OM is tangent to the plane P and passes through the points A and B. Thus, conversely, any point of the circle C belongs to the locus. Hence, the sought-for locus of points is the circle C.

(2) If the straight line AB is parallel to the plane, the required locus is a straight line which lies in the plane P, is perpendicular to the projection of the line segment AB on the plane P and bisects this projection.

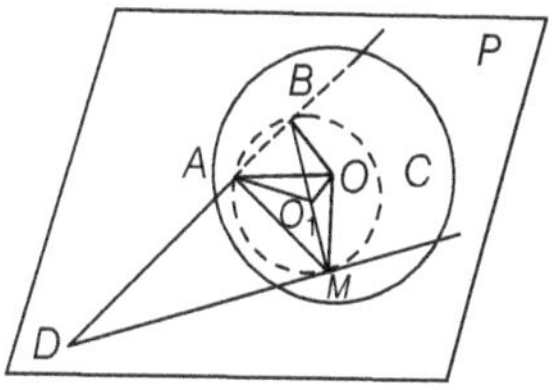

FIG. 243

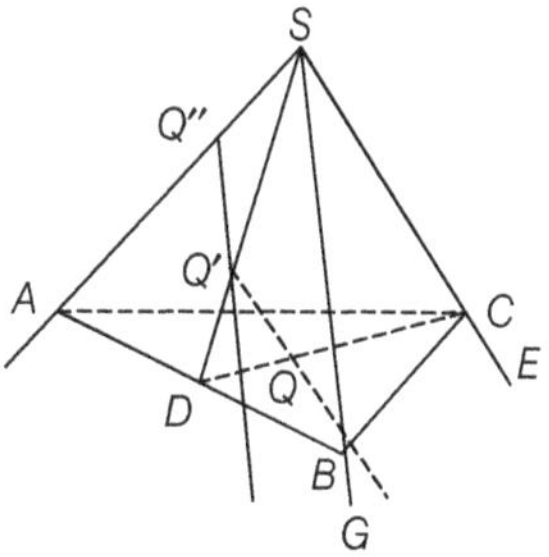

FIG. 244

530. *Case (a).* Let D be the midpoint of the line segment AB (Fig. 244), C the movable vertex, Q the centre of gravity of $\Delta\ ABC$ and Q' the centre of gravity of $\Delta\ ASB$. Since the point Q divides the line segment DC in the ratio 1:2, the locus of these points is obviously a ray parallel to the edge SE and passing through the point Q' which is the centre of gravity of $\Delta\ ASB$.

Case (b). If the point B is also moved along the edge SG, then the centres of gravity Q' of the triangles ASB are in the ray parallel to the edge SG and passing through the point Q'' which divides the line segment AS into the parts AQ'' and $Q''S$ which are in the ratio 2:1. The rays considered in the case (a), which correspond to every fixed position of the point B, cover the whole section of the trihedral angle by the plane passing through the point Q'' and parallel to the edges SG and SE.

4. The Greatest and Least Values

531. Without loss of generality, we may assume that the cutting plane intersects the edge CE of the cube shown in Fig. 245. It is evident that in the section we always obtain a parallelogram $AMBN$. The area S of the parallelogram can be found by the formula

$$S = AB \cdot MK,$$

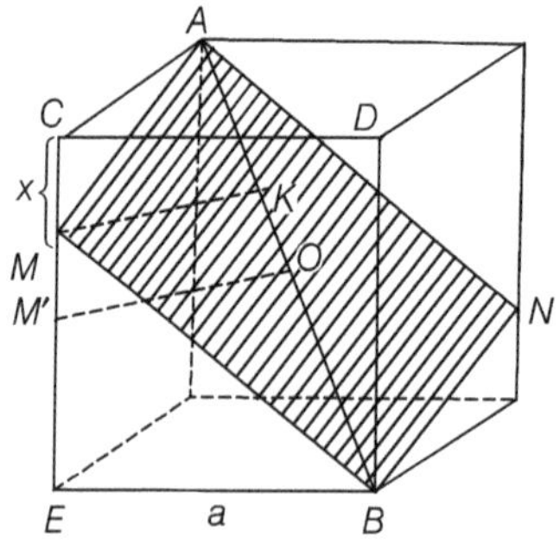

FIG. 245

Where MK is the perpendicular drawn from the point M of the edge CE to the diagonal AB. Thus, the area S is the least when the length of the line segment MK attains its minimal value. But, among the line segments joining the points of two skew lines CE and AB, the perpendicular common to these lines has the least length. It is readily seen that the common perpendicular to the indicated straight lines is the line segment $M'O$ joining the midpoints of the edge CE and of the diagonal AB. Indeed $AM'B$ is an isosceles triangle, and therefore $M'O \perp AB$. Since COE is also an isosceles triangle, we have $M'O \perp CE$. Thus, the section bisecting the edge CE has the least area $S = a\sqrt{3} \cdot \frac{a\sqrt{2}}{2} = \frac{a^2\sqrt{6}}{2}$.

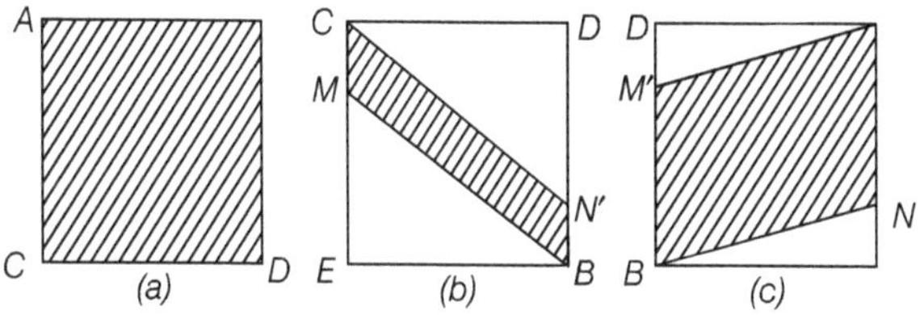

FIG. 246

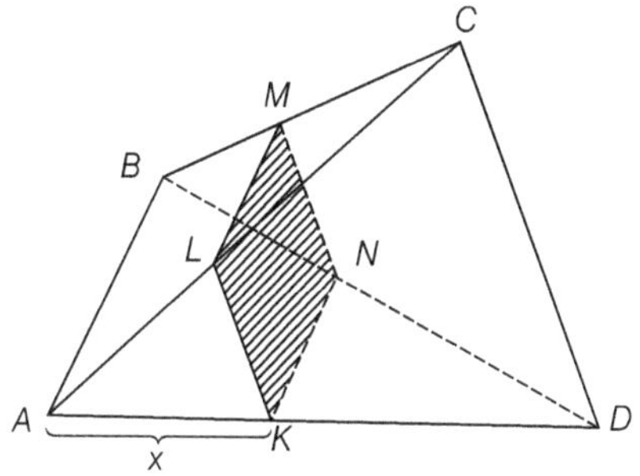

FIG. 247

This problem can also be solved by applying the following theorem: the square of the area of a plane polygon is equal to the sum of the squares of the areas of its projections on three mutually perpendicular planes. The theorem is easily proved on the basis of the formula by which the area of the projection of a plane polygon on a plane is equal to the area of the polygon multiplied by the cosine of the angle between the planes (see formula (1) in the solution of Problem 456).

Considering this theorem proved, let us denote the length of the line segment CM by x (see Fig. 245). The projections of the parallelogram we are interested in on the planes ACD, $ECDB$ and BDN are shown in Fig. 246, a, b, c. The areas of the projections are respectively equal to a^2, ax and $a^2 - ax$, and, by virtue of the above theorem, $S^2 = (a^2)^2 + (ax)^2 + (a^2 - ax)^2 = 2a^2(x^2 - ax + a^2)$. Rewriting the quadratic trinomial $x^2 - ax + a^2$ in the form $\left(x - \frac{a}{2}\right)^2 + \frac{3}{4}a^2$, we find (cf. (1), page 35) that S^2 takes on its least value for $x = \frac{a}{2}$, and the minimum area is $S_{min} = \sqrt{2a^2 \frac{3}{4} a^2} = \frac{a^2\sqrt{6}}{2}$.

532. Consider the configuration shown in Fig. 247. The quadrilateral $MNKL$ in the section of the pyramid $ABCD$ is a parallelogram because $LK \parallel CD$ and $MN \parallel CD$. Hence, $LK \parallel MN$ and, analogously, $LM \parallel KN$. If $\angle LKN = \alpha$, then the area of the parallelogram is equal to

$$S = KN \cdot KL \sin\alpha.$$

The angle LKN being equal to that between the skew lines AB and CD, its sine is a constant quantity for all the parallel sections under consideration. Thus, the section area only depends on the product magnitude of the $KN \cdot KL$. Let us denote the length of the line segment AK by x. Then, by the similarity of the triangles, we have

$$\frac{KN}{AB} = \frac{AD - x}{AD}, \quad \frac{KL}{CD} = \frac{x}{AD}.$$

Let us multiply these equalities. $KN \cdot KL = \dfrac{AB \cdot CD}{AD^2}(AD - x)x.$

Since $\dfrac{AB \cdot CD}{AD^2}$ is constant, it follows from the preceding formula that the product $KN \cdot KL$ attains its greatest value when the product $(AD - x)x$ is maximal.

Regarding this product as the quadratic trinomial $-x^2 + ADx$ and representing it in the form $-\left(x - \dfrac{AD}{2}\right)^2 + \left(\dfrac{AD}{2}\right)^2$, we see that its greatest value is attained for $x = \dfrac{AD}{2}$ (cf. (1), page 35).

TRIGONOMETRY

1. Transforming Expressions Containing Trigonometric Functions

533. Applying the formula

$a^3 + b^3 = (a + b)(a^2 - ab + b^2) = (a + b)[(a + b)^2 - 3ab]$,

we obtain

$$\sin^6 x + \cos^6 x = (\sin^2 x + \cos^2 x)[(\sin^2 x + \cos^2 x)^2 - 3\sin^2 x\cos^2 x]$$
$$= 1 - 3\sin^2 x\cos^2 x = 1 - \frac{3}{4}\sin^2 2x.$$

534. Denote the left member of the identity by S and, according to formula (14), page 62, substitute the sum $\cos(\alpha + \beta) + \cos(\alpha - \beta)$ for the product $2\cos\alpha\cos\beta$. Then S can be written in the form

$$S = \cos^2\alpha - \cos(\alpha + \beta)\cos(\alpha - \beta).$$

Again applying formula (14), we find

$$\cos(\alpha + \beta)\cos(\alpha - \beta) = \frac{1}{2}(\cos 2\alpha + \cos 2\beta).$$

If now we substitute $\frac{1 + \cos 2\alpha}{2}$ for $\cos^2\alpha$, then we obtain the required result

$$S = \frac{1 - \cos 2\beta}{2} = \sin^2\beta.$$

535. From the formula

$$\tan(\alpha + \beta) = \frac{\tan\alpha + \tan\beta}{1 - \tan\alpha\cdot\tan\beta}$$

it follows that

$$\tan\alpha + \tan\beta = \tan(\alpha + \beta)[1 - \tan\alpha\tan\beta],$$

whence

$$\tan\alpha + \tan\beta - \tan(\alpha + \beta) = -\tan\alpha\tan\beta\tan(\alpha + \beta).$$

Putting $\alpha = x$ and $\beta = 2x$ in the last relation we obtain the required formula.

536. We have

$$\tan x\tan\left(\frac{\pi}{3} - x\right)\tan\left(\frac{\pi}{3} + x\right)$$
$$= \tan x\frac{\sqrt{3} - \tan x}{1 + \sqrt{3}\tan x}\cdot\frac{\sqrt{3} + \tan x}{1 - \sqrt{3}\tan x} = \frac{\tan x(3 - \tan^2 x)}{1 - 3\tan^2 x}. \qquad (1)$$

On the other hand, applying the formula for the tangent of a sum of two angles repeatedly, we easily find that

$$\tan 3x = \frac{\tan x\,(3 - \tan^2 x)}{1 - 3\tan^2 x}. \qquad (2)$$

Comparing (1) with (2) we get the required result.

Note. Formula (2) can also be deduced from formulas (7) and (8) on page 62.

537. Applying the formulas for the sum and difference of sines, we represent the left member of the identity in the following form :

$$2\sin\frac{\alpha+\beta}{2}\cos\frac{\alpha-\beta}{2} - 2\cos\left(\gamma + \frac{\alpha+\beta}{2}\right)\sin\frac{\alpha+\beta}{2}$$

$$= 2\sin\frac{\alpha+\beta}{2}\cdot\left[\cos\frac{\alpha-\beta}{2} - \cos\left(\gamma+\frac{\alpha+\beta}{2}\right)\right].$$

Then, using the formula for the difference of cosines, we see that the left member of the identity coincides with the right one.

538. Using the identity of Problem 537, we obtain

$$\sin\alpha + \sin\beta + \sin\gamma = 4\sin\frac{\alpha+\beta}{2}\sin\frac{\beta+\gamma}{2}\sin\frac{\gamma+\alpha}{2}$$

$$= 4\cos\frac{\gamma}{2}\cos\frac{\alpha}{2}\cos\frac{\beta}{2},$$

because

$$\frac{\alpha+\beta}{2} = \frac{\pi}{2} - \frac{\gamma}{2}, \frac{\beta+\gamma}{2} = \frac{\pi}{2} - \frac{\alpha}{2}, \frac{\alpha+\gamma}{2} = \frac{\pi}{2} - \frac{\beta}{2}.$$

539. Using the identity indicated in Problem 537, we obtain

$$\sin 2n\alpha + \sin 2n\beta + \sin 2n\gamma$$

$$= 4\sin n(\alpha+\beta)\cdot\sin n(\beta+\gamma)\cdot\sin n(\gamma+\alpha). \qquad (1)$$

Furthermore, we have

$$\sin n(\alpha+\beta) = \sin n(\pi-\gamma) = (-1)^{n+1}\sin n\gamma.$$

Transforming analogously two other factors on the right-hand side of (1), we get the required result.

540. To prove the assertion, we multiply both sides of the equality $\cos(\alpha+\beta) = 0$ by $2\sin\beta$ and apply formula (15) on page 62.

541. The permissible values of the arguments are determined by the condition $\cos\alpha\cos(\alpha+\beta) \neq 0$. Note that the equality

$$\tan(\alpha+\beta) = 2\tan\alpha \qquad (1)$$

to be proved involves the arguments $\alpha+\beta$ and α. Therefore, it is natural to introduce the same arguments into the original equality. We have

$$\beta = (\alpha+\beta) - \alpha, 2\alpha+\beta = (\alpha+\beta)+\alpha.$$

Substituting these expressions of β and $2\alpha+\beta$ into the original equality

$$3\sin\beta = \sin(2\alpha+\beta) \qquad (2)$$

and using the formulas for the sines of a sum and difference of angles, we transform (2) to the following form :

$$\sin(\alpha+\beta)\cos\alpha = 2\cos(\alpha+\beta)\sin\alpha. \qquad (3)$$

Dividing both members of (3) by $\cos\alpha\cdot\cos(\alpha+\beta)$ we obtain (1).

542. All values of α and β are permissible here except those for which $\cos(\alpha+\beta)=0$ and $\cos\beta = A$. Noting that $\sin\alpha = \sin(\alpha+\beta-\beta)$, let us rewrite the original equality in the form

$$\sin(\alpha+\beta)\cos\beta - \cos(\alpha+\beta)\sin\beta = A\sin(\alpha+\beta). \qquad (1)$$

Dividing both members of (1) by $\cos(\alpha+\beta)\neq 0$, we obtain $\tan(\alpha+\beta)\times\cos\beta - \sin\beta = A\tan(\alpha+\beta)$. Expressing $\tan(\alpha+\beta)$ from the latter relation we arrive at the required equality.

543. It is readily seen that, by virtue of the conditions of the problem, we have $\sin\alpha\cos\alpha\cos\beta\neq 0$ because, if otherwise, we have $|m|\leq|n|$. Therefore, the equality to be proved makes sense. We represent this equality in the form

$$\frac{\tan\alpha+\tan\beta}{1-\tan\alpha\tan\beta} = \frac{m+n}{m-n}\tan\alpha, \qquad (1)$$

whence

$$\tan(\alpha+\beta) = \frac{m+n}{m-n}\tan\alpha. \qquad (2)$$

Replace in (2) the tangents of the angles α and $\alpha+\beta$ by the ratios of the corresponding sines and cosines, reduce the fractions to a common denominator and discard it. We then obtain

$$m[\cos\alpha\sin(\alpha+\beta) - \sin\alpha\cos(\alpha+\beta)]$$
$$-n[\sin\alpha\cos(\alpha+\beta)] + \cos\alpha\sin(\alpha+\beta) = 0, \qquad (3)$$

that is

$$m\sin\beta - n\sin(2\alpha+\beta) = 0. \qquad (4)$$

Thus, the proof is reduced to establishing relation (4). Since relation (4) is fulfilled by the hypothesis of the problem, we conclude that (3) holds true which implies the validity of (2).

But (2) implies (1), and (1), in its turn, implies the required relation

$$\frac{1+\dfrac{\tan\beta}{\tan\alpha}}{m+n} = \frac{1-\tan\alpha\tan\beta}{m-n}.$$

544. Consider the identity

$$\cos(x+y+z) = \cos(x+y)\cos z - \sin(x+y)\sin z$$

$$= \cos x\cos y\cos z - \cos z\sin x\sin y - \cos y\sin x\sin z - \cos x\sin y\sin z.$$

By the hypothesis of the problem, we have $\cos x\cos y\cos z \neq 0$, and therefore this identity implies

$$\cos(x+y+z) = \cos x\cos y\cos z\,(1 - \tan x\tan y - \tan y\tan z - \tan z\tan x).$$

545. *First solution.* By the hypothesis, we have

$$0 < \alpha < \pi, 0 < \beta < \pi, 0 < \gamma < \pi \text{ and } \alpha + \beta + \gamma = \pi. \quad (1)$$

Therefore, from (1) we conclude that

$$\tan\left(\frac{\beta+\gamma}{2}\right) = \tan\left(\frac{\pi}{2} - \frac{\alpha}{2}\right) = \frac{1}{\tan\frac{\alpha}{2}}. \quad (2)$$

On the other hand, by the formula for the tangent of a sum of two angles, we can write

$$\tan\left(\frac{\beta+\gamma}{2}\right) = \frac{\tan\frac{\beta}{2} + \tan\frac{\gamma}{2}}{1 - \tan\frac{\beta}{2}\tan\frac{\gamma}{2}}. \quad (3)$$

Equating the right-hand members of equalities (2) and (3), reducing the fractions to a common denominator and discarding the latter we obtain the required equality.

Second solution. From the formula

$$\cos\left(\frac{\alpha+\beta+\gamma}{2}\right) = \cos\frac{\alpha}{2}\cos\frac{\beta}{2}\cos\frac{\gamma}{2}\left(1 - \tan\frac{\alpha}{2}\tan\frac{\beta}{2} - \tan\frac{\beta}{2}\tan\frac{\gamma}{2} - \tan\frac{\gamma}{2}\tan\frac{\alpha}{2}\right)$$

proved in the preceding problem we immediately find that

$$1 - \tan\frac{\alpha}{2}\tan\frac{\beta}{2} - \tan\frac{\beta}{2}\tan\frac{\gamma}{2} - \tan\frac{\gamma}{2}\tan\frac{\alpha}{2} = 0,$$

because $\dfrac{\alpha+\beta+\gamma}{2} = \dfrac{\pi}{2}$.

546. The meaning of the expression considered in the problem indicates that $\cos x\cos y\cos z \neq 0$. Therefore, from the formula obtained in Problem 544 we find

$$\tan x\tan y + \tan y\tan z + \tan z\tan x$$

$$= 1 - \frac{\cos(x+y+z)}{\cos x\cos y\cos z} = 1 - \frac{\cos\frac{\pi}{2}k}{\cos x\cos y\cos z}.$$

If k is odd, then the investigated expression is equal to unity and is independent of x, y and z. For even values of k it depends on x, y and z.

547. *First solution.* Note first that $\tan\beta\tan\gamma \neq 1$ because, if otherwise, we have $\tan\beta + \tan\gamma = 0$ which contradicts the equality $\tan\beta\tan\gamma = 1$. Therefore, from the conditions of the problem it follows that

$$\tan\alpha = -\frac{\tan\beta + \tan\gamma}{1 - \tan\beta\tan\gamma} = -\tan(\beta+\gamma) = \tan(-\beta-\gamma),$$

whence we find $\alpha = k\pi - \beta - \gamma$, i.e. $\alpha + \beta + \gamma = k\pi$.

Second solution. In Problem 544 we obtained a formula for the cosine of a sum of three angles. We can analogously derive the formula

$$\sin(\alpha+\beta+\gamma) = \cos\alpha\cos\beta\cos\gamma\,(\tan\alpha + \tan\beta + \tan\gamma - \tan\alpha\tan\beta\tan\gamma)$$

assuming that $\cos\alpha\cos\beta\cos\gamma \neq 0$. From this formula we find that under the conditions of the problem we have

$$\sin(\alpha+\beta+\gamma) = 0, \text{ i.e. } \alpha+\beta+\gamma = k\pi.$$

548. Denote the given sum by S. Transform the first two terms in the following way :

$$\cot^2 2x - \tan^2 2x = \frac{\cos^2 2x}{\sin^2 2x} - \frac{\sin^2 2x}{\cos^2 2x} = \frac{\cos^4 2x - \sin^4 2x}{\sin^2 2x\cos^2 2x}$$

$$= \frac{\cos^2 2x - \sin^2 2x}{\frac{1}{4}\sin^2 4x} = \frac{4\cos 4x}{\sin^2 4x}.$$

Hence,

$$S = \frac{4\cos 4x}{\sin^2 4x}(1 - 2\sin 4x\cos 4x) = \frac{4\cos 4x}{\sin^2 4x}(1 - \sin 8x).$$

Since $1 - \sin 8x = 2\sin^2\left(\frac{\pi}{4} - 4x\right)$, we finally obtain

$$S = \frac{8\cos 4x\sin^2\left(\frac{\pi}{4} - 4x\right)}{\sin^2 4x}.$$

549. Denote the expression under consideration by S. Let us transform the first two summands according to formula (16), page 62, replace the product $\cos\alpha\cos\beta$ by a sum using formula (14), page 62, and, finally substitute $1 - \cos^2\gamma$ for $\sin^2\gamma$. We then obtain

$$S = -\frac{1}{2}(\cos 2\alpha + \cos 2\beta) - \cos^2\gamma + [\cos(\alpha+\beta) + \cos(\alpha-\beta)]\cos\gamma.$$

Transforming the sum $\cos 2\alpha + \cos 2\beta$ into a product and opening the square brackets we receive

$$S = -\cos(\alpha+\beta)\cos(\alpha-\beta) - \cos^2\gamma + \cos(\alpha+\beta)\cos\gamma + \cos(\alpha-\beta)\cos\gamma.$$

Grouping the terms, we find that

$$S = -[\cos(\alpha-\beta) - \cos\gamma][\cos(\alpha+\beta) - \cos\gamma].$$

Hence, $$S = 4\sin\frac{\alpha-\beta+\gamma}{2}\sin\frac{\gamma-\alpha+\beta}{2}\sin\frac{\alpha+\beta+\gamma}{2}\sin\frac{\alpha+\beta-\gamma}{2}.$$

550. The expression in question can be transformed in the following way (see (13), page 62):
$$\frac{1-4\sin 10°\sin 70°}{2\sin 10°} = \frac{1-2(\cos 60° - \cos 80°)}{2\sin 10°} = \frac{2\cos 80°}{2\cos 80°}.$$
Thus, $$\frac{1}{2\sin 10°} - 2\sin 70° = 1.$$

551. By virtue of formula (12) given on page 62, the left-hand member of the identity is equal to
$$2\sin\frac{\pi}{10}\sin\frac{3\pi}{10}. \qquad (1)$$
Multiplying and dividing (1) by $2\cos\frac{\pi}{10}\cos\frac{3\pi}{10}$, and applying the formula for $\sin 2\alpha$, we obtain
$$2\sin\frac{\pi}{10}\sin\frac{3\pi}{10} = \frac{\sin\frac{\pi}{5}\sin\frac{3\pi}{5}}{2\cos\frac{\pi}{10}\cos\frac{3\pi}{10}}.$$
Put $$\cos\frac{\pi}{10} = \sin\left(\frac{\pi}{2}+\frac{\pi}{10}\right) = \sin\frac{3\pi}{5}$$
and $$\cos\frac{3\pi}{10} = \sin\left(\frac{\pi}{2}-\frac{3\pi}{10}\right) = \sin\frac{\pi}{5}.$$
Hence, the left-hand side of the identity is equal to $\frac{1}{2}$.

552. Multiplying and dividing the left member of the identity by $2\sin\frac{\pi}{7}$ and making use of the formulas expressing products of trigonometric functions in terms of sums, we find
$$\cos\frac{2\pi}{7} + \cos\frac{4\pi}{7} + \cos\frac{6\pi}{7}$$
$$= \frac{2\cos\frac{2\pi}{7}\sin\frac{\pi}{7} + 2\cos\frac{4\pi}{7}\sin\frac{\pi}{7} + 2\cos\frac{6\pi}{7}\sin\frac{\pi}{7}}{2\sin\frac{\pi}{7}}$$
$$= \frac{\sin\frac{3\pi}{7} - \sin\frac{\pi}{7} + \sin\frac{5\pi}{7} - \sin\frac{3\pi}{7} + \sin\pi - \sin\frac{5\pi}{7}}{2\sin\frac{\pi}{7}}.$$
It follows that the sum under consideration is equal to $-\frac{1}{2}$.

553. Applying formula (16) to all the terms of the sum S, and then (17), page 62, we find that

$$S=\frac{3}{2}-\frac{1}{2}\left(\cos\frac{\pi}{8}+\cos\frac{3\pi}{8}+\cos\frac{5\pi}{8}+\cos\frac{7\pi}{8}\right)$$
$$+\frac{1}{8}\left(\cos\frac{\pi}{4}+\cos\frac{3\pi}{4}+\cos\frac{5\pi}{4}+\cos\frac{7\pi}{4}\right).$$

The sums in the brackets are equal to zero because

$$\cos\frac{\pi}{8}=-\cos\frac{7\pi}{8},\ \cos\frac{3\pi}{8}=-\cos\frac{5\pi}{8}$$

and

$$\cos\frac{\pi}{4}=-\cos\frac{3\pi}{4},\ \cos\frac{5\pi}{4}=-\cos\frac{7\pi}{4}.$$

Consequently, $$S=\frac{3}{2}.$$

554. If in the identity

$$\tan\alpha\tan(60°-\alpha)\tan(60°+\alpha)=\tan 3\alpha \qquad (1)$$

we put $\alpha=20°$ (see Problem 536), then we immediately obtain

$$\tan 20°\tan 40°\tan 80°=\sqrt{3}. \qquad (2)$$

There is another solution in which formula (1) is not used. Let us transform separately the products of sines and cosines. To this end, we apply formulas (13) and (15), page 62, and get

$$\sin 20°\sin 40°\sin 80°=\frac{1}{2}(\cos 20°-\cos 60°)\sin 80°$$
$$=\frac{1}{2}\left(\frac{\sin 100°+\sin 60°}{2}-\frac{1}{2}\sin 80°\right).$$

Noting that $\sin 100°=\sin 80°$, we write

$$\sin 20°\sin 40°\sin 80°=\frac{\sqrt{3}}{8}. \qquad (3)$$

Furthermore, we have

$$\cos 20°\cos 40°\cos 80°=\frac{2\sin 20°\cos 20°\cos 40°\cos 80°}{2\sin 20°}$$
$$=\frac{\sin 40°\cos 40°\cos 80°}{2\sin 20°}$$
$$=\frac{\sin 80°\cos 80°}{4\sin 20°}=\frac{\sin 160°}{8\sin 20°}$$
$$=\frac{\sin 20°}{8\sin 20°}=\frac{1}{8}.$$

Thus, $$\cos 20°\cos 40°\cos 80°=\frac{1}{8}. \qquad (4)$$

Relations (3) and (4) imply (2).

2. Trigonometric Equations and Systems of Equations

555. The equation can be written as

$$4\sin x\cos x(\sin^2 x-\cos^2 x)=1,$$

that is

$$-2\sin 2x\cos 2x=-\sin 4x=1.$$

Answer: $x=-\frac{\pi}{8}+k\cdot\frac{\pi}{2}\ (k=0,\pm 1,\pm 2,\dots)$.

556. The equation makes no sense for $x=\frac{\pi}{2}+k\pi$ and for $x=-\frac{\pi}{4}+k\pi$. For all the other values of x it is equivalent to the equation

$$\frac{\cos x-\sin x}{\cos x+\sin x}=1+\sin 2x.$$

After simple transformations we obtain

$$\sin x(3+\sin 2x+\cos 2x)=0.$$

It is obvious that the equation $\sin 2x+\cos 2x+3=0$ has no solution, and therefore, the original equation is reduced to the equation $\sin x=0$.

Answer: $x=k\pi$.

557. The equation can be written in the following form:

$$(\sin x+\cos x)^2+(\sin x+\cos x)+(\cos^2 x-\sin^2 x)=0,$$

that is $(\sin x+\cos x)(1+2\cos x)=0.$

Equating each of the expressions in the brackets to zero, we find the roots.

Answer: $x_1=-\frac{\pi}{4}+k\pi,\ x_2=\pm\frac{2\pi}{3}+2k\pi.$

558. Rewrite the given equation in the following form :

$$\sin x+1-\cos 2x=\cos x-\cos 3x+\sin 2x.$$

After some simple transformations we obtain

$$\sin x+2\sin^2 x=2\sin 2x\cdot\sin x+\sin 2x$$

and, hence,

$$\sin x(1+2\sin x)(1-2\cos x)=0.$$

Answer: $x_1=k\pi,\ x_2=\frac{\pi}{6}(-1)^{k+1}+k\pi,\ x_3=\pm\frac{\pi}{3}+2k\pi.$

559. Rewrite the equation in the form

$$\left(\frac{1}{2}\sin 2x+\frac{\sqrt{3}}{2}\cos 2x\right)^2-\frac{1}{4}\cos\left(2x-\frac{\pi}{6}\right)-\frac{5}{4}=0,$$

that is

$$4\cos^2\left(2x-\frac{\pi}{6}\right)-\cos\left(2x-\frac{\pi}{6}\right)-5=0. \qquad (1)$$

Solving quadratic equation (1), we find

$$\cos\left(2x-\frac{\pi}{6}\right)=-1, x=\frac{7\pi}{12}+k\pi.$$

The other root of equation (1) is equal to $\frac{5}{4}$ and must be discarded since $|\cos\alpha|\le 1$.

560. Dividing both sides of the equation by 2, we reduce it to the form

$$\sin 17x+\sin\left(5x+\frac{\pi}{3}\right)=0,$$

whence we obtain

$$2\sin\left(11x+\frac{\pi}{6}\right)\cos\left(6x-\frac{\pi}{6}\right)=0.$$

Answer: $x_1=-\frac{\pi}{66}+\frac{k\pi}{11}, x_2=\frac{\pi}{36}+\frac{(2k+1)\pi}{12}$.

561. The given equation makes no sense when $\cos x=0$; therefore we can suppose that $\cos x\neq 0$. Noting that the right-hand member of the equation is equal to $3\sin x\cos x+3\cos^2 x$, and dividing both members by $\cos^2 x$, we obtain

$$\tan^2 x(\tan x+1)=3(\tan x+1),$$

that is $(\tan^2 x-3)(\tan x+1)=0.$

Answer: $x_1=-\frac{\pi}{4}+k\pi, x_2=\frac{\pi}{3}+k\pi, x_3=-\frac{\pi}{3}+k\pi.$

562. Using the formula for the sum of cubes of two members we transform the left-hand side of the equation in the following way :

$$(\sin x+\cos x)(1-\sin x\cos x)=\left(1-\frac{1}{2}\sin 2x\right)(\sin x+\cos x).$$

Hence, the original equation takes the form

$$\left(1-\frac{1}{2}\sin 2x\right)(\sin x+\cos x-1)=0.$$

The expression in the first brackets is different from zero for all x. Therefore it is sufficient to consider the equation $\sin x+\cos x-1=0$. The latter is reduced to the form

$$\sin\left(x+\frac{\pi}{4}\right)=\frac{1}{\sqrt{2}}.$$

Answer: $x_1=2\pi k, x_2=\frac{\pi}{2}+2\pi k.$

563. Using the well-known trigonometric formulas, write the equation in the following way :

$$\csc^2 x - \sec^2 x - \cot^2 x - \tan^2 x - \cos^2 x - \sin^2 x = -3. \qquad (1)$$

Since $\csc^2 x = 1 + \cot^2 x$ and $\sec^2 x = 1 + \tan^2 x$, the above equation is reduced to the form $\tan^2 x = 1$.

Answer: $x = \frac{\pi}{4} + k\frac{\pi}{2}$.

564. Using the identity

$$\sin^4\frac{x}{3} + \cos^4\frac{x}{3} = \left(\sin^2\frac{x}{3} + \cos^2\frac{x}{3}\right)^2 - 2\sin^2\frac{x}{3}\cos^2\frac{x}{3} = 1 - \frac{1}{2}\sin^2\frac{2}{3}x,$$

we transform the equation to the form $\sin^2\frac{2x}{3} = \frac{3}{4}$.

Answer: $x = \frac{3n \pm 1}{2}\pi \ (n = 0, \pm 1, \pm 2, \dots)$.

565. Using the identity obtained in the solution of the preceding problem, we obtain the equation $\sin^2 2x + \sin 2x - 1 = 0$.

Solving it, we get $\quad \sin 2x = \frac{\sqrt{5} - 1}{2}$.

Answer: $x = (-1)^k \frac{1}{2} \arcsin \frac{\sqrt{5} - 1}{2} + \frac{k\pi}{2}$.

566. Let us rewrite the given equation in the form

$$(1 + k)\cos x \cos(2x - \alpha) = \cos(x - \alpha) + k\cos 2x \cos(x - \alpha). \qquad (1)$$

We have

$$\cos x \cos(2x - \alpha) = \frac{1}{2}[\cos(3x - \alpha) + \cos(x - \alpha)]$$

and

$$\cos(x - \alpha)\cos 2x = \frac{1}{2}[\cos(3x - \alpha) + \cos(x + \alpha)],$$

and therefore equation (1) turns into

$$k[\cos(x - \alpha) - \cos(x + \alpha)] = \cos(x - \alpha) - \cos(3x - \alpha),$$

that is

$$k \sin x \sin\alpha = \sin(2x - \alpha)\sin x. \qquad (2)$$

Equation (2) is equivalent to the following two equations :

(a) $\sin x = 0; x = l\pi$ and (b) $\sin(2x - \alpha) = k\sin\alpha$.

Thus,

$$x = \frac{\alpha}{2} + (-1)^n \cdot \frac{1}{2}\arcsin(k\sin\alpha) + \frac{\pi}{2}n.$$

For the last expression to make sense, k and α must satisfy the condition

$$|k\sin\alpha| \le 1.$$

567. Since the numbers a, b, c and d are consecutive terms of an arithmetic progression, we can put $b = a + r$, $c = a + 2r$, $d = a + 3r$ where r is the common difference of the progression. Using the formula

$$\sin\alpha \sin\beta = \frac{1}{2}[\cos(\alpha - \beta) - \cos(\alpha + \beta)],$$

we represent the equation in the form

$$\cos(2a + r)x - \cos(2a + 5r)x = 0$$

or

$$\sin(2a + 3r)x \cdot \sin 2rx = 0,$$

whence

$$x_1 = \frac{k\pi}{2a + 3r}, x_2 = \frac{k\pi}{2r}.$$

These formulas make sense because

$$2a + 3r = b + c > 0 \text{ and } r \neq 0.$$

568. Write the equation in the following form :

$$\cos^2\frac{x}{2} - \sin^2\frac{x}{2} = 2\left(\frac{\sin\frac{x}{2}}{\cos\frac{x}{2}} - 1\right)$$

After some simple transformations it is reduced to the equation

$$\left(\cos\frac{x}{2} - \sin\frac{x}{2}\right)\left(3\cos^2\frac{x}{2} + 2\sin^2\frac{x}{2} + \sin\frac{x}{2}\cos\frac{x}{2}\right) = 0.$$

The equation $3\cos^2\frac{x}{2} + 2\sin^2\frac{x}{2} + \sin\frac{x}{2}\cos\frac{x}{2} = 0$ is equivalent to the equation $2\tan^2\frac{x}{2} + \tan\frac{x}{2} + 3 = 0$ and has no real solutions.

Answer: $x = \frac{\pi}{2} + 2k\pi$.

569. *First solution.* The equation becomes senseless for $x = k\pi$. For all the other values of x it is equivalent to the equation

$$\cos x - \sin x = 2\sin 2x \cdot \sin x. \qquad (1)$$

Replacing the product standing on the right-hand side of (1) by the corresponding sum according to formula (13), page 62, we obtain

$$\cos x - \sin x = \cos x - \cos 3x, \ \sin x = \cos 3x,$$

whence $\sin x = \sin\left(\frac{\pi}{2} - 3x\right)$. Consequently,

$$2\sin\left(2x - \frac{\pi}{4}\right)\cos\left(x - \frac{\pi}{4}\right) = 0.$$

Answer: $x_1 = \frac{\pi}{8} + \frac{k\pi}{2}, x_2 = \frac{3\pi}{4} + k\pi.$ (2)

Second solution. Applying formula (20), page 74, and putting $\tan x = t$, we get the equation

$$t^3 + 3t^2 + t - 1 = 0.$$

Factoring the left member, we obtain

$$(t+1)(t+1-\sqrt{2})(t+1+\sqrt{2}) = 0,$$

whence

$$(\tan x)_1 = -1, (\tan x)_2 = \sqrt{2} - 1, (\tan x)_3 = -1 - \sqrt{2}.$$

Answer: $x_1 = \frac{3\pi}{4} + k\pi;\ x_2 = \text{arc} \tan(\sqrt{2} - 1) + k\pi,$

$$x_3 = -\text{arc} \tan(1 + \sqrt{2}) + k\pi.$$

Note. The above expressions of x_2 and x_3 can be written in the form of one formula (2).

570. Applying formula (14), page 62, to the left-hand side of the equation, we obtain

$$\cos(2x - \beta) + \cos\beta = \cos\beta,$$

whence $\cos(2x - \beta) = 0.$

Consequently, $x = \pm\frac{\pi}{4} + k\pi + \frac{\beta}{2}$ and $\tan x = \tan\left(\frac{\beta}{2} \pm \frac{\pi}{4}\right).$

571. The original equation can be written in the form

$$\sin\alpha + [\sin(2\varphi + \alpha) - \sin(2\varphi - \alpha)] = \sin(\varphi + \alpha) - \sin(\varphi - \alpha),$$

or, after some simple transformations, in the form

$$\sin\alpha + 2\sin\alpha\cos 2\varphi = 2\sin\alpha \cdot \cos\varphi.$$

Assuming $\sin\alpha \neq 0$ (otherwise $\cos\varphi$ becomes indeterminate), we obtain $1 + 2\cos 2\varphi - 2\cos\varphi = 0, 4\cos^2\varphi - 2\cos\varphi - 1 = 0,$

$$\cos\varphi = \frac{1 \pm \sqrt{5}}{4}.$$

The angle φ being in the third quadrant, we have $\cos\varphi < 0$. Hence, $\cos\varphi = \frac{1 - \sqrt{5}}{4}.$

572. Applying the formula $\cos^2\varphi = \frac{1 + \cos 2\varphi}{2}$, write the equation in the form

$$\cos 2(\alpha + x) + \cos 2(\alpha - x) = 2a - 2$$

or $\cos 2\alpha \cos 2x = a - 1,$

whence

$$\cos 2x = \frac{a - 1}{\cos 2\alpha}. \quad (1)$$

On the other hand,

$$\cot x = \pm\sqrt{\frac{1+\cos 2x}{1-\cos 2x}},$$

and therefore from (1) we find

$$\cot x = \pm\sqrt{\frac{a-1+\cos 2\alpha}{1-a+\cos 2\alpha}}.$$

Formula (1) shows that the problem only makes sense if $\cos 2\alpha \neq 0$ and $|\cos 2\alpha| \geq |a-1|$.

573. Using formulas (18) and (19), page 63, we reduce the given relation $\sin\alpha + \cos\alpha = \frac{\sqrt{7}}{2}$ to the form

$$(2+\sqrt{7})\tan^2\frac{\alpha}{2} - 4\tan\frac{\alpha}{2} - (2-\sqrt{7}) = 0.$$

Solving this equation with respect to $\tan\frac{\alpha}{2}$, we obtain

$$\left(\tan\frac{\alpha}{2}\right)_1 = \frac{3}{2+\sqrt{7}} = \sqrt{7}-2 \text{ and } \left(\tan\frac{\alpha}{2}\right)_2 = \frac{\sqrt{7}-2}{3}.$$

Let us verify whether the above values of $\tan\frac{\alpha}{2}$ satisfy the conditions of the problem.

Since $0 < \frac{\alpha}{2} < \frac{\pi}{8}$, we have the condition $0 < \tan\frac{\alpha}{2} < \tan\frac{\pi}{8} = \sqrt{2}-1$.

The value $\left(\tan\frac{\alpha}{2}\right)_2 = \frac{\sqrt{7}-2}{3}$ satisfies this condition because $\frac{\sqrt{7}-2}{3} < \sqrt{2}-1$. The root $\sqrt{7}-2$ should be discarded since

$$\sqrt{7}-2 > \sqrt{2}-1.$$

574. Putting $\sin x - \cos x = t$ and using the identity $(\sin x - \cos x)^2 = 1 - 2\sin x\cos x$, we rewrite the original equation in the form

$$t^2 + 12t - 13 = 0.$$

This equation has the roots $t_1 = -13$ and $t_2 = 1$. But $t = \sin x - \cos x = \sqrt{2}\sin\left(x - \frac{\pi}{4}\right)$, and thus, $|t| \leq \sqrt{2}$. Consequently, the root $t_1 = -13$ must be discarded. Therefore, the original equation is reduced to the equation

$$\sin\left(x-\frac{\pi}{4}\right) = \frac{1}{\sqrt{2}}.$$

Answer: $x_1 = \pi + 2k\pi,\ x_2 = \frac{\pi}{2} + 2k\pi.$

575. Transform the given equation to the form

$$2\cos^2\frac{x}{2}(2+\sin x)+\sin x=0.$$

Using the formula $2\cos^2\frac{x}{2}=1+\cos x$ and opening the brackets, we obtain

$$2+2(\sin x+\cos x)+\sin x\cdot\cos x=0. \qquad (1)$$

This equation is of the same type as in Problem 574. By the substitution $\sin x+\cos x=t$ equation (1) is reduced to the quadratic equation $t^2+4t+3=0$ whose roots are $t_1=-1$ and $t_2=-3$. Since $|\sin x+\cos x|\le\sqrt{2}$, the original equation can only be satisfied by the roots of the equation

$$\sin x+\cos x=-1. \qquad (2)$$

Solving equation (2), we obtain

$$x_1=-\frac{\pi}{2}+2k\pi$$

and

$$x_2=(2k+1)\pi.$$

Here x_2 should be discarded because $\sin x_2=0$, and therefore the original equation makes no sense for $x=x_2$.

Answer: $x=-\frac{\pi}{2}+2k\pi$.

576. The given equation only makes sense for $x\ne k\pi$. For these values of x it can be rewritten in the form

$$\cos^3 x+\cos^2 x=\sin^3 x+\sin^2 x.$$

Transferring all terms to the left-hand side of the equation and factoring it we get

$$(\cos x-\sin x)(\sin^2 x+\cos^2 x+\sin x\cos x+\sin x+\cos x)=0.$$

There are two possible cases here which are considered below.

(a) $\sin x-\cos x=0$, then

$$x_1=\frac{\pi}{4}+k\pi; \qquad (1)$$

(b) $\sin^2 x+\cos^2 x+\sin x\cos x+\sin x+\cos x=0. \qquad (2)$

Equation (2) is analogous to the one considered in Problem 574 and has the solutions

$$x_2=-\frac{\pi}{2}+2k\pi \qquad (3)$$

and

$$x_3=(2k+1)\pi. \qquad (4)$$

But the values of x determined by formula (4) are not roots of the original equations, since the original equation is only considered for $x\ne k\pi$. Consequently, the equation has the roots defined by formulas (1) and (3).

577. Rewrite the equation in the form

$$2\left(\frac{\sin 3x}{\cos 3x}-\frac{\sin 2x}{\cos 2x}\right)=\frac{\sin 2x}{\cos 2x}\left(\frac{\sin 2x}{\cos 2x}\cdot\frac{\sin 3x}{\cos 3x}+1\right).$$

Reducing the fractions to a common denominator and discarding it, we obtain the equation

$$2(\sin 3x\cos 2x-\cos 3x\sin 2x)\cos 2x$$
$$=\sin 2x(\sin 2x\sin 3x+\cos 2x\cos 3x).$$

But the expression in the brackets on the left-hand side is equal to $\sin x$, and the one on the right-hand side is equal to $\cos x$. Therefore, we arrive at the equation

$$2\sin x(\cos 2x-\cos^2 x)=-2\sin^3 x=0,$$

whence $x=k\pi.$

578. The given equation can be rewritten in the form

$$3\left(\frac{\cos 2x}{\sin 2x}-\frac{\cos 3x}{\sin 3x}\right)=\frac{\sin 2x}{\cos 2x}+\frac{\cos 2x}{\sin 2x}$$

or

$$\frac{3\sin x}{\sin 2x\sin 3x}=\frac{1}{\sin 2x\cos 2x}.$$

Note that this equation has sense if the condition

$$\sin 2x\neq 0,\quad \sin 3x\neq 0,\quad \cos 2x\neq 0$$

holds. For the values of x satisfying this condition we have

$$3\sin x\cos 2x=\sin 3x.$$

Transforming the last equation we obtain

$$\sin x(3-4\sin^2 x-3\cos 2x)=0$$

and thus arrive at the equation

$$2\sin^3 x=0,$$

which is equivalent to the equation $\sin x=0$. Hence, due to the above note, the original equation has no solutions.

579. Rewrite the equation in the form

$$6(\tan x+\cot 3x)=\tan 2x+\cot 3x$$

and transform it in the following way :

$$6\left(\frac{\sin x}{\cos x}+\frac{\cos 3x}{\sin 3x}\right)=\frac{\sin 2x}{\cos 2x}+\frac{\cos 3x}{\sin 3x}$$

or

$$\frac{6\cos 2x}{\cos x\sin 3x}=\frac{\cos x}{\cos 2x\sin 3x};$$

$$6\cos^2 2x=\cos^2 x;\ 12\cos^2 2x-\cos 2x-1=0.$$

Solving the last equation, we find

$$\cos 2x=\frac{1\pm 7}{24},$$

whence

(1) $\cos 2x = \frac{1}{3}, x = \pm \frac{1}{2} \text{arc cos} \frac{1}{3} + k\pi;$

(2) $\cos 2x = -\frac{1}{4}, x = \pm \frac{1}{2} \text{arc cos} \left(-\frac{1}{4}\right) + k\pi.$

In the above solution we have multiplied both members of the equation by the product $\cos x \cos 2x \sin 3x$. But it is evident that for neither of the values of x found above this product vanishes. Consequently, all these values of x are the roots of the original equation.

580. Reducing the fractions on the right-hand side of the equation to a common denominator and applying the formula

$$a^5 - b^5 = (a-b)(a^4 + a^3b + a^2b^2 + ab^3 + b^4),$$

we get

$$\sin x \cos x (\sin x - \cos x)(\sin^4 x + \sin^3 x \cos x + \sin^2 x \cos^2 x + \\ + \sin x \cos^3 x + \cos^4 x) = \sin x - \cos x.$$

It follows that either

$$\sin x - \cos x = 0 \qquad (1)$$

or $$\sin x \cos x (\sin^4 x + \sin^3 x \cos x + \sin x \cos^3 x + \cos^4 x + \\ \sin^2 x \cos^2 x) - 1 = 0. \qquad (2)$$

Now, taking advantage of the relations

$\sin^4 x + \cos^4 x = (\sin^2 x + \cos^2 x)^2 - 2\sin^2 x \cos^2 x,$

and $\sin^3 x \cos x + \cos^3 x \sin x = \sin x \cos x,$

we transform equation (2) to the form

$$y^3 - y^2 - y + 1 = 0, \qquad (3)$$

where $y = \sin x \cos x$. Factoring the left member of this equation we obtain

$$(y-1)^2(y+1) = 0.$$

If $y = 1$, i.e. $\sin x \cos x = 1$, then $\sin 2x = 2$ which is impossible, and if $y = -1$, then $\sin 2x = -2$ which is also impossible.

Thus, equation (2) has no roots. Consequently, the roots of the original equation coincide with the roots of equation (1), i.e. $x = \frac{\pi}{4} + \pi n.$

581. The right-hand side of the equation is not determined for $x = k\pi$ and $x = \frac{\pi}{2} + m\pi$, because for $x = 2l\pi$ the function $\cot \frac{x}{2}$ is not defined, for $x = (2l+1)\pi$ the function $\tan \frac{x}{2}$ is not defined and for $x = \frac{\pi}{2} + m\pi$ the denominator of the right member vanishes. For $x \neq k\pi$ we have

$$\tan\frac{x}{2}-\cot\frac{x}{2}=\frac{\sin^2\frac{x}{2}-\cos^2\frac{x}{2}}{\sin\frac{x}{2}\cos\frac{x}{2}}=-\frac{2\cos x}{\sin x}.$$

Hence, for $x\neq k\pi$ and $x\neq\frac{\pi}{2}+m\pi$ (where k and m are arbitrary integers) the right member of the equation is equal to $-2\sin x\cos x$. The left member of the equation has no sense for $x=\frac{\pi}{2}+k\pi$ and $x=\frac{\pi}{4}+l\cdot\frac{\pi}{2}$ $(l=0,\pm 1,\pm 2,\dots)$, and for all the other values of x it is equal to $-\tan x$ because

$$\tan\left(x-\frac{\pi}{4}\right)\tan\left(x+\frac{\pi}{4}\right)=\tan\left(x-\frac{\pi}{4}\right)\cot\left[\frac{\pi}{2}-\left(x+\frac{\pi}{4}\right)\right]$$

$$=-\tan\left(x-\frac{\pi}{4}\right)\cot\left(x-\frac{\pi}{4}\right)=-1.$$

Thus, if $x\neq k\pi$, $x\neq\frac{\pi}{2}+m\pi$ and $x\neq\frac{\pi}{4}+l\frac{\pi}{2}$, then the original equation is reduced to the form

$$\tan x=2\sin x\cos x.$$

This equation has the roots

$$x=k\pi \text{ and } x=\frac{\pi}{4}+l\frac{\pi}{2}.$$

It follows that the original equation has no roots.

582. Multiplying the right member of the equation by $\sin^2 x+\cos^2 x=1$ we reduce it to the form

$$(1-a)\sin^2 x-\sin x\cos x-(a+2)\cos^2 x=0. \quad (1)$$

First let us assume that $a\neq 1$. Then from (1) it follows that $\cos x\neq 0$, since otherwise we have $\sin x=\cos x=0$ which is impossible. Dividing both members of (1) by $\cos^2 x$ and putting $\tan x=t$ we get the equation

$$(1-a)t^2-t-(a+2)=0. \quad (2)$$

Equation (1) is solvable if and only if the roots of equation (2) are real, i.e. if its discriminant is non-negative :

$$D=-4a^2-4a+9\geq 0. \quad (3)$$

Solving inequality (3) we find

$$-\frac{\sqrt{10}+1}{2}\leq a\leq\frac{\sqrt{10}-1}{2}. \quad (4)$$

Let t_1 and t_2 be the roots of equation (2). Then the corresponding solutions of equation (1) have the form

$$x_1=\arctan t_1+k\pi,\quad x_2=\arctan t_2+k\pi.$$

Now let us consider the case $a = 1$.

In this case quation (1) is written in the form

$$\cos x(\sin x + 3\cos x) = 0$$

and has the following solutions :

$$x_1 = \frac{\pi}{2} + k\pi, x_2 = -\text{arc}\tan 3 + k\pi.$$

583. Applying the formulas

$$\sin^4 x = \left(\frac{1-\cos 2x}{2}\right)^2, \cos^2 x = \frac{1+\cos 2x}{2}$$

and putting $\cos 2x = t$ we rewrite the given equation in the form

$$t^2 - 6t + 4a^2 - 3 = 0. \quad (1)$$

The original equation has solutions for a given value of a if and only if, for this value of a, the roots t_1 and t_2 of the equation (1) are real and atleast one of these roots does not exceed unity in its absolute value. Solving equation (1), we find

$$t_1 = 3 - 2\sqrt{3-a^2}, t_2 = 3 + 2\sqrt{3-a^2}.$$

Hence, the roots of equation (1) are real if

$$|a| \le \sqrt{3}. \quad (2)$$

If condition (2) is fulfilled, then $t_2 > 1$ and, therefore, this root can be discarded. Thus, the problem is reduced to finding the values of a satisfying condition (2), for which $|t_1| \le 1$, i.e.

$$-1 \le 3 - 2\sqrt{3-a^2} \le 1. \quad (3)$$

From (3) we find

$$-4 \le -2\sqrt{3-a^2} \le -2,$$

whence

$$2 \ge \sqrt{3-a^2} \ge 1. \quad (4)$$

Since the inequality $2 \ge \sqrt{3-a^2}$ is fulfilled for $|a| \le \sqrt{3}$, the system of inequalities (4) is reduced to the inequality

$$\sqrt{3-a^2} \ge 1,$$

whence we find

$$|a| \le \sqrt{2}.$$

Thus, the original equation is solvable if $|a| \le \sqrt{2}$, and its solutions are

$$x = \pm\frac{1}{2}\text{arc}\cos\left(3 - 2\sqrt{3-a^2}\right) + k\pi.$$

584. Let us transform the given equation by multiplying its both members by $32\sin\frac{\pi x}{31}$. Applying several times the formula

$$\sin\alpha\cos\alpha=\frac{1}{2}\sin 2\alpha, \text{ we get}$$

$$\sin\frac{32}{31}\pi x=\sin\frac{\pi x}{31}$$

or

$$\sin\frac{\pi x}{2}\cos\frac{33}{62}\pi x=0. \qquad (1).$$

Hence, we find the roots

$$x_1=2n, x_2=\frac{31}{33}(2n+1)\ (n=0,\pm 1,\pm 2,\dots).$$

In the above solution of the problem we multiplied both sides of the given equation by the factor $32\sin\frac{\pi x}{31}$ which can turn into zero. Therefore, equation (1) can have extraneous roots. A value of x is an extraneous root if and only if it satisfies the equation

$$\sin\frac{\pi x}{31}=0 \qquad (2)$$

but does not satisfy the original equation.

The roots of equation (2) are given by the formula

$$x=31k\ (k=0,\pm 1,\pm 2,\dots), \qquad (3)$$

and, as is readily seen, they do not satisfy the original equation. Therefore, from the roots of equation (1) found above we should exclude all those of form (3). For the roots expressed by x_1 this leads to the equality $2n=31k$ which is only possible for an even k, i.e. for $k=2l$ and $n=31l\ (l=0,\pm 1,\pm 2,\dots)$. For the roots expressed by x_2 we analogously obtain the equality $\frac{31}{33}(2n+1)=31k$ or $2n+1=33k$, which is only possible for an odd k, i.e. for $k=2l+1$ and

$$n=33l+16\ (l=0,\pm 1,\pm 2,\dots).$$

Thus, the roots of the original equation are

$$\left.\begin{aligned} &x_1=2n, && \text{where } n\neq 31l,\\ &x_2=\frac{31}{33}(2n+1), && \text{where } n\neq 33l+16. \end{aligned}\right\}\ l=0,\pm 1,\pm 2,\dots.$$

585. Rewrite the equation in the form

$$\frac{1}{2}\cos 7x+\frac{\sqrt{3}}{2}\sin 7x=\frac{\sqrt{3}}{2}\cos 5x+\frac{1}{2}\sin 5x$$

or
$$\sin\frac{\pi}{6}\cos 7x+\cos\frac{\pi}{6}\sin 7x=\sin\frac{\pi}{3}\cos 5x+\cos\frac{\pi}{3}\sin 5x,$$

i.e. $$\sin\left(\frac{\pi}{6}+7x\right)=\sin\left(\frac{\pi}{3}+5x\right).$$

But $\sin\alpha=\sin\beta$ if and only if either $\alpha-\beta=2k\pi$ or $\alpha+\beta=(2m+1)\pi$ $(k, m=0, \pm 1, \pm 2, \dots)$. Hence,

$$\frac{\pi}{6}+7x-\frac{\pi}{3}-5x=2k\pi$$

or

$$\frac{\pi}{6}+7x+\frac{\pi}{3}+5x=(2m+1)\pi.$$

Thus, the roots of the equation are

$$\left.\begin{aligned} x&=\frac{\pi}{12}(12k+1),\\ x&=\frac{\pi}{24}(4m+1)\end{aligned}\right\}(k, m=0, \pm 1, \pm 2, \dots).$$

586. The left member of the equation being equal to

$$2-(7+\sin 2x)(\sin^2 x-\sin^4 x)=2-(7+\sin 2x)\sin^2 x\cdot\cos^2 x$$
$$=2-(7+\sin 2x)\frac{1}{4}\sin^2 2x,$$

we can put $t=\sin 2x$ and rewrite the equation in the form

$$t^3+7t^2-8=0. \qquad (1)$$

It is readily seen that equation (1) has the root $t_1=1$. The other two roots are found from the equation

$$t^2+8x+8=0. \qquad (2)$$

Solving this equation we find

$$t=-4+2\sqrt{2} \text{ and } t=-4-2\sqrt{2}.$$

These roots should be discarded because they are greater than unity in their absolute values. Consequently, the roots of the original equation coincide with the roots of the equation $\sin 2x=1$.

Answer: $x=\frac{\pi}{4}+k\pi.$

587. We may suppose that $a^2+b^2\neq 0$, since otherwise the equation attains the form $c=0$, and it is impossible to find $\sin x$ and $\cos x$. As is known, if $a^2+b^2\neq 0$, then there exists an angle $\varphi, 0\le\varphi<2\pi$, such that

$$\sin\varphi=\frac{a}{\sqrt{a^2+b^2}}, \cos\varphi=\frac{b}{\sqrt{a^2+b^2}}. \qquad (1)$$

Dividing the given equation termwise by $\sqrt{a^2+b^2}$ and using (1) we obtain the equivalent equation

$$\sin(x+\varphi)=\frac{c}{\sqrt{a^2+b^2}}. \qquad (2)$$

We always have $|\sin(x+\varphi)| \le 1$, and, hence, this equation is solvable if and only if $|c| \le \sqrt{a^2+b^2}$, i.e. $c^2 \le a^2+b^2$. This is the condition for solvability of the problem. Furthermore, we find

$$\cos(x+\varphi) = \pm\sqrt{1-\sin^2(x+\varphi)} = \pm\frac{\sqrt{a^2+b^2-c^2}}{\sqrt{a^2+b^2}}. \tag{3}$$

Noting that

$$\sin x = \sin(x+\varphi-\varphi) = \sin(x+\varphi)\cos\varphi - \cos(x+\varphi)\sin\varphi$$

and

$$\cos x = \cos(x+\varphi-\varphi) = \cos(x+\varphi)\cos\varphi + \sin(x+\varphi)\sin\varphi,$$

and substituting expressions (1), (2) and (3) into the right-hand side we finally obtain the following two solutions :

(a) $\sin x = \dfrac{bc - a\sqrt{a^2+b^2-c^2}}{a^2+b^2}$, $\cos x = \dfrac{ac + b\sqrt{a^2+b^2-c^2}}{a^2+b^2}$

and

(b) $\sin x = \dfrac{bc + a\sqrt{a^2+b^2-c^2}}{a^2+b^2}$, $\cos x = \dfrac{ac - b\sqrt{a^2+b^2-c^2}}{a^2+b^2}$.

588. Noting that $(b\cos x + a)(b\sin x + a) \ne 0$ (otherwise the equation has no sense), we discard the denominators and get

$$ab\sin^2 x + (a^2+b^2)\sin x + ab = ab\cos^2 x + (a^2+b^2)\cos x + ab,$$

whence

$$(a^2+b^2)(\sin x - \cos x) - ab(\sin^2 x - \cos^2 x) = 0.$$

Therefore, the original equation is reduced to the following two equations :

$$1^\circ \cdot \sin x = \cos x, \text{ whence } x = \frac{\pi}{4} + k\pi,$$

and

$$2^\circ \cdot \sin x + \cos x = \frac{a^2+b^2}{ab}.$$

But the latter equation has no solutions because

$$\frac{a^2+b^2}{|ab|} \ge 2,$$

whereas

$$|\sin x + \cos x| = \sqrt{2}\left|\sin x\cdot\frac{1}{\sqrt{2}} + \cos x\cdot\frac{1}{\sqrt{2}}\right| = \sqrt{2}\left|\sin\left(x+\frac{\pi}{4}\right)\right| \le \sqrt{2}.$$

Answer: $x = \dfrac{\pi}{4} + k\pi$.

589. Using the identity

$$\cos^6 x = \left(\frac{1+\cos 2x}{2}\right)^3 = \frac{1}{8}(1+3\cos 2x+3\cos^2 2x+\cos^3 2x)$$

and the fomula

$$\cos 6x = 4\cos^3 2x - 3\cos 2x \text{ (see (8) page 62)},$$

we reduce the equation to the form

$$4\cos^2 2x + 5\cos 2x + 1 = 0. \quad (1)$$

From (1) we find

$$(\cos 2x)_1 = -1, (\cos 2x)_2 = -\frac{1}{4}.$$

Answer : $x_1 = \left(k+\frac{1}{2}\right)\pi$;

$$x_2 = \pm\frac{1}{2}\text{ arc cos}\left(-\frac{1}{4}\right) + k\pi.$$

590. Applying the formulas

$$\sin^2\alpha = \frac{1-\cos 2\alpha}{2} \text{ and } \cos 2\alpha = 2\cos^2\alpha - 1$$

we rewrite the equation in the form

$$(1-\cos 2x)^3 + 3\cos 2x + 2(2\cos^2 2x - 1) + 1 = 0,$$

or

$$7\cos^2 2x - \cos^3 2x = 0,$$

whence $\cos 2x = 0,\ x = \frac{\pi}{4} + k\frac{\pi}{2}.$

591. From the formulas for $\sin 3x$ and $\cos 3x$ we find

$$\cos^3 x = \frac{\cos 3x + 3\cos x}{4},$$

$$\sin^3 x = \frac{3\sin x - \sin 3x}{4}.$$

Hence, the equation can be rewritten in the form

$$\cos 3x(\cos 3x + 3\cos x) + \sin 3x(3\sin x - \sin 3x) = 0$$

or $3(\cos 3x\cos x + \sin 3x\sin x) + \cos^2 3x - \sin^2 3x = 0,$

that is

$$3\cos 2x + \cos 6x = 0. \quad (1)$$

But, since we have $\cos^3 2x = \frac{\cos 6x + 3\cos 2x}{4}$, equation (1) takes the form

$$4\cos^3 2x = 0,$$

whence

$$\cos 2x = 0, x = \frac{\pi}{4} + \frac{\pi}{2}n.$$

592. Using the identity $(\sin^2 x + \cos^2 x)^2 = 1$ we get

$$\sin^4 x + \cos^4 x = 1 - \frac{1}{2}\sin^2 2x,$$

whence

$$\sin^8 x + \cos^8 x = \left(1 - \frac{1}{2}\sin^2 2x\right)^2 - \frac{1}{8}\sin^4 2x = \frac{17}{32},$$

$$1 - \sin^2 2x + \frac{1}{8}\sin^4 2x = \frac{17}{32}, \quad \sin^4 2x - 8\sin^2 2x + \frac{15}{4} = 0.$$

Solving the biquadratic equation we find

$$\sin^2 2x = 4 \pm \frac{7}{2}, \sin^2 2x = \frac{1}{2}, 2x = \frac{\pi}{4} + k\frac{\pi}{2};$$

whence $$x = \frac{2k+1}{8}\pi.$$

593. Replacing $\sin^2 x$ and $\cos^2 x$, respectively, by $\frac{1 - \cos 2x}{2}$ and $\frac{1 + \cos 2x}{2}$, we rewrite the equation in the form

$$\left(\frac{1 - \cos 2x}{2}\right)^5 + \left(\frac{1 + \cos 2x}{2}\right)^5 = \frac{29}{16}\cos^4 2x$$

or $$(1 - \cos 2x)^5 + (1 + \cos 2x)^5 = 58\cos^4 2x.$$

Putting $\cos 2x = y$, after some simple transformations we obtain the following biquadratic equation with respect to y:

$$24y^4 - 10y^2 - 1 = 0.$$

This equation has only two real roots: $y_{1,2} = \pm\frac{\sqrt{2}}{2}$. Hence, $\cos 2x = \pm\frac{\sqrt{2}}{2}$, whence $x = \frac{\pi}{8}(2k+1)$ where $k = 0, \pm 1, \pm 2, \ldots$.

594. Using the identity obtained in Problem 261 we rewrite the original equation in the form

$$(\sin x + \sin 2x)(\sin 2x + \sin 3x)(\sin x + \sin 3x) = 0.$$

Factoring the sums of sines into products, we arrive at the equation

$$\sin\frac{3x}{2}\sin 2x \sin\frac{5x}{2}\cos x \cos^2\frac{x}{2} = 0.$$

Equating each factor to zero we get the solutions

$$(1)\ x = \frac{2n_1}{3}\pi; \quad (2)\ x = \frac{2n_2}{2}\pi; \quad (3)\ x = \frac{2n_3}{5}\pi;$$

$$(4)\ x = \frac{2n_4 + 1}{2}\pi; \quad (5)\ x = (2n_5 + 1)\pi,$$

where n_1, n_2, n_3, n_4 and n_5 are arbitrary integers.

Noting that the solutions (4) and (5) are contained in (2), we finally obtain the following formulas for the solutions :

(1) $x = \frac{2n_1}{3}\pi$; (2) $x = \frac{2n_2}{2}\pi$; (3) $x = \frac{2n_3}{5}\pi$,

where n_1, n_2 and n_3 are arbitrary integers.

595. *First solution.* For $n = 1$ the equation turns into an identity. If $n > 1$, then, by virtue of the given equation, we drive from the identity

$$1 = (\sin^2 x + \cos^2 x)^n = \sin^{2n} x + C_n^1 \sin^{2(n-1)} x \cos^2 x + \ldots + C_n^{n-1} \sin^2 x \cos^{2(n-1)} x + \cos^{2n} x$$

the equation $C_n^1 \sin^{2(n-1)} x \cos^2 x + \ldots + C_n^{n-1} \sin^2 x \cos^{2(n-1)} x = 0$. All the summands being non-negative, we conclude that either $\sin^2 x = 0$ or $\cos^2 x = 0$ and $x = \frac{\pi}{2} k$.

Second solution. As is obvious, the equation is satisfied if x takes on the values which are integer multiples of $\frac{\pi}{2}$, i.e. if $x = \frac{\pi}{2} k$ (k – integer). Let us show that the equation $\sin^{2n} x + \cos^{2n} x = 1$

has no other roots. Let $x_0 \neq k \cdot \frac{\pi}{2}$; then $\sin^2 x_0 < 1$ and $\cos^2 x_0 < 1$ whence it follows that for $n > 1$ we have $\sin^{2n} x_0 < \sin^2 x_0$ and $\cos^{2n} x_0 < \cos^2 x_0$ and, hence,

$$\sin^{2n} x_0 + \cos^{2n} x_0 < \sin^2 x_0 + \cos^2 x_0 = 1.$$

The proof is thus completed.

596. Put $\frac{3\pi}{10} - \frac{x}{2} = y$, then $\frac{\pi}{10} + \frac{3x}{2} = \pi - 3\left(\frac{3\pi}{10} - \frac{x}{2}\right) = \pi - 3y$, and the equation takes the form

$$\sin 3y = 2 \sin y.$$

With the aid of formula (7), page 73, the last equation can be transformed to the form

$$\sin y\,(4\sin^2 y - 1) = 0. \qquad (1)$$

Equation (1) has the following solutions :

$y_1 = k\pi$, $\quad y_2 = (-1)^k \frac{\pi}{6} + \pi k$, $y_3 = (-1)^{k+1} \cdot \frac{\pi}{6} + \pi k$.

Returning to the argument $x = \frac{3\pi}{5} - 2y$ we finally obtain the solutions of the original equation :

$$x_1 = \frac{3\pi}{5} - 2k\pi, \; x_2 = \frac{3\pi}{5} + (-1)^{k+1} \frac{\pi}{3} - \pi k,$$

$$x_3 = \frac{3\pi}{5} + (-1)^k \cdot \frac{\pi}{3} - \pi k.$$

597. Since $|\cos\alpha| \le 1$ and $\sin\alpha \ge -1$, we have

$|\cos 4x - \cos 2x| \le 2$ and $\sin 3x + 5 \ge 4$.

Thus the left member of the eqution does not exceed 4, the right member being not less than 4. Consequently, we have $|\cos 4x - \cos 2x| = +2$ (and then either $\cos 4x = -1$ and $\cos 2x = 1$, or $\cos 4x = 1$ and $\cos 2x = -1$) and $\sin 3x = -1$. Let us consider all the possible cases.

(a) $\cos 4x = -1$, $x = \left(\frac{n}{2} + \frac{1}{4}\right)\pi$; $\cos 2x = 1, x = \pi k$;

$\sin 3x = -1, x = -\frac{\pi}{6} + \frac{2\pi}{3} l$;

and, hence, in this case there are no common roots.

(b) $\cos 4x = 1,\qquad x = \frac{\pi n}{2}$;

$\cos 2x = -1,\qquad x = \left(k + \frac{1}{2}\right)\pi$;

$\sin 3x = -1, x = -\frac{\pi}{6} - + \frac{2}{3}\pi l = \frac{4l-1}{6}\pi.$

Thus, in this case the common roots are

$$x = \left(2m + \frac{1}{2}\right)\pi, m = 0, \pm 1, \pm 2, \ldots$$

598. Let us transform the euqation to the form

$$\frac{1}{\sqrt{2}}\sin x + \frac{1}{\sqrt{2}}\cos x = \frac{1}{2\sin x\cos x}$$

or

$$\sin\left(x + \frac{\pi}{4}\right) = \frac{1}{\sin 2x},$$

that is

$$\sin\left(x + \frac{\pi}{4}\right)\sin 2x = 1. \qquad (1)$$

We have $|\sin\alpha| \le 1$, and therefore (1) holds if either

$$\sin\left(x + \frac{\pi}{4}\right) = -1 \text{ and } \sin 2x = -1,$$

or

$$\sin\left(x + \frac{\pi}{4}\right) = 1 \quad\text{and}\quad \sin 2x = 1.$$

But the first two equations have no roots in common while the second two equations have the common roots $x = \frac{\pi}{4} + 2k\pi$. Consequently the roots of the given equation are $x = \frac{\pi}{4} + 2k\pi$.

599. Dividing the given equation termwise by 2 and noting that $\frac{1}{2} = \cos\frac{\pi}{3}$ and $\frac{\sqrt{3}}{2} = \sin\frac{\pi}{3}$, we get the equivalent equation

$$\sin\left(x + \frac{\pi}{3}\right)\sin 4x = 1.$$

This equation is satisfied only if $\sin\left(x + \frac{\pi}{3}\right) = \pm 1$ and $\sin 4x = \pm 1$,

whence $$x = -\frac{\pi}{3} \pm \frac{\pi}{2} + 2n\pi \text{ and } x = \frac{1}{4}\left(\pm\frac{\pi}{2} + 2m\pi\right),$$

where n and m are integers. Equating both values and cancelling out π we obtain the equality

$$-\frac{1}{3} \pm \frac{1}{2} + 2n = \pm\frac{1}{8} + \frac{m}{2}.$$

Multiplying by 24 we receive

$$12m - 48n = -8 \pm 9.$$

For any integers m and n, the left member is an even integer, and the right member an odd integer equal to 1 or -17. Thus, the last equation has no integral solutions m and n, and hence the assertion of the problem is proved.

600. *First solution.* The given problem is equivalent to the following problem: what values can the function $\lambda = \sec x + \csc x$ assume if the argument x varies within the range $0 < x < \frac{\pi}{2}$?

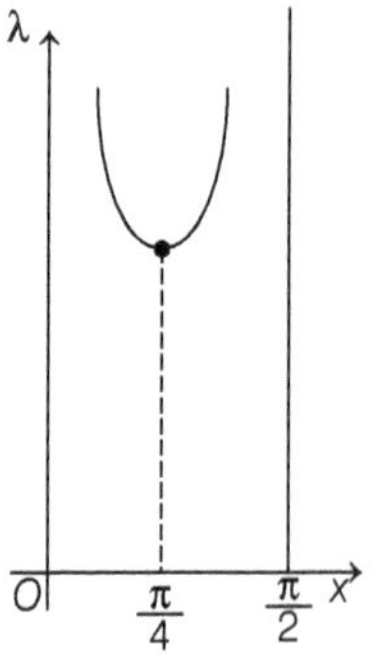

FIG. 248

Consider the function

$$\lambda^2 = (\sec x + \csc x)^2 = \frac{1}{\cos^2 x} + \frac{2}{\sin x \cos x} + \frac{1}{\sin^2 x}$$
$$= \frac{1}{\sin^2 x \cos^2 x} + \frac{2}{\sin x \cos x} = \frac{4}{\sin^2 2x} + \frac{4}{\sin 2x}.$$

As x increases from zero to $\frac{\pi}{2}$, each summand on the right-hand side varies in the following way : it first decreases from $+\infty$ to $4\left(\text{for } 0 < x \le \frac{\pi}{4}\right)$, then increases from 4 to $+\infty$ $\left(\text{for } \frac{\pi}{4} \le x < \frac{\pi}{2}\right)$; for $x = \frac{\pi}{4}$ both summands simultaneously attain their least values and hence, for $x = \frac{\pi}{4}$ the sum takes on the least value as well, and $\lambda^2 = 8$. Therefore, if $0 < x < \frac{\pi}{2}$, then $\lambda^2 \ge 8$, and since $\sec x$ and

$\csc x$ are positive in the first quadrant, we have $\lambda \geq 2\sqrt{2}$. The graph of the function $\lambda(x)$ is shown in Fig. 248.

Second solution. Note that we must confine ourselves to considering only the positive values of λ because for $0 < x < \frac{\pi}{2}$ the functions $\sec x$ and $\csc x$ are positive. Transforming the equation to the form

$$\sin x + \cos x = \lambda \sin x \cos x,$$

we then square both members and obtain

$$1 + 2\sin x \cos x = \lambda^2 \sin^2 x \cos^2 x.$$

Now putting $\sin 2x = z$ we can write

$$\lambda^2 z^2 - 4z - 4 = 0,$$

whence

$$z_{1,2} = \frac{2 \pm \sqrt{4 + 4\lambda^2}}{\lambda^2}. \qquad (1)$$

By the hypothesis, we have $0 < x < \frac{\pi}{2}$, and therefore $z = \sin 2x > 0$. Thus, in equality (1) we must take the plus sign, i.e.

$$z = \frac{2 + \sqrt{4 + 4\lambda^2}}{\lambda^2}.$$

If now we take the values of λ satisfying the inequality

$$\frac{2 + \sqrt{4 + 4\lambda^2}}{\lambda^2} \leq 1, \qquad (2)$$

then the equation $\sin 2x = \frac{2 + \sqrt{4 + 4\lambda^2}}{\lambda^2}$ will have a solution x such that $0 < x < \frac{\pi}{2}$. Obviously, this solution will also satisfy the original equation. But if inequality (2) is not satisfied, the required solution does not exist. We see that the problem is reduced to solving inequality (2). Getting rid of the denominator, we readily find $\lambda \geq 2\sqrt{2}$.

601. From the given system we immediately obtain

$$x + y = k\pi, \quad x - y = l\pi.$$

It follows that

$$x = \frac{k + l}{2}\pi, \; y = \frac{k - l}{2}\pi.$$

By the condition of the problem, we have $0 \leq k + l \leq 2$ and $0 \leq k - l \leq 2$. These inequalities are satisfied by the following five pairs of values of k and l:

(1) $k = 0, l = 0$; (2) $k = 1, l = 0$; (3) $k = 1, l = -1$;
(4) $k = 1, l = 1$; (5) $k = 2, l = 0$.

Answer: $x_1 = 0, y_1 = 0; x_2 = \frac{\pi}{2}, y_2 = \frac{\pi}{2};$

$$x_3 = 0, \quad y_3 = \pi;$$
$$x_4 = \pi, \quad y_4 = 0;$$
$$x_5 = \pi, \; y_5 = \pi.$$

602. Transform the system to the form

$$\left.\begin{aligned} \sin^2 x &= 1 + \sin x \sin y, \\ \cos^2 x &= 1 + \cos x \cos y. \end{aligned}\right\} \qquad (1)$$

Adding together the equations of system (1) and subtracting the first equation from the second we obtain the system

$$\left.\begin{aligned} \cos 2x - \cos(x + y) &= 0, \\ 1 + \cos(x - y) &= 0. \end{aligned}\right\} \qquad (2)$$

The first equation of system (2) can be rewritten as

$$\cos 2x - \cos(x + y) = 2\sin\left(\frac{3x + y}{2}\right)\sin(y - x) = 0.$$

If $\sin(x - y) = 0$, then $x - y = k\pi$. But from the second equation of system (2) we find

$$\cos(x - y) = -1, \quad x - y = (2n + 1)\pi.$$

Consequently, in this case we have an infinitude of solutions : $x - y = (2n + 1)\pi$.

If $\sin\left(\frac{3x + y}{2}\right) = 0$, then $3x + y = 2k\pi$. But $x - y = (2n + 1)\pi$, and, hence,

$$x = \frac{2k + 2n + 1}{4}\pi, \qquad y = \frac{2k - 6n - 3}{4}\pi.$$

603. Squaring both equations, adding them termwise and using the identity

$$\sin^6 x + \cos^6 x = 1 - \frac{3}{4}\sin^2 2x$$

(see Problem 533), we get: $\sin^2 2x = 1$. If $\sin 2x = 1$, then either $x = \frac{\pi}{4} + 2k\pi$ or $x = \frac{\pi}{4} + (2k + 1)\pi$. In the first case from the original system we find $\sin y = \cos y = \frac{1}{\sqrt{2}}$, and in the second case we have $\sin y = \cos y = -\frac{1}{\sqrt{2}}$. The case $\sin 2x = -1$ is treated in a similar way.

Answer: $x_1 = \frac{\pi}{4} + 2k\pi, \; y_1 = \frac{\pi}{4} + 2l\pi; \; x_2 = \frac{\pi}{4} + (2k + 1)\pi,$

$y_2 = \frac{\pi}{4} + (2l + 1)\pi; \; x_3 = \frac{3\pi}{4} + 2k\pi, \quad y_3 = \frac{3\pi}{4} + 2l\pi;$

$x_4 = \frac{3}{4}\pi + (2k + 1)\pi, \quad y_4 = \frac{3}{4}\pi + (2l + 1)\pi.$

604. The first equation can be written in the form

$$\frac{\sin(x+y)}{\cos x \cos y} = 1,$$

whence, by virtue of the second equation, we obtain

$$\sin(x+y) = \cos x \cos y = \frac{\sqrt{2}}{2}.$$

Hence, either

$$x + y = \frac{\pi}{4} + 2k\pi \qquad (1)$$

or

$$x + y = -\frac{\pi}{4} + (2k+1)\pi. \qquad (2)$$

The second equation of the original system can be transformed to the form

$$\cos(x+y) + \cos(x-y) = \sqrt{2}.$$

It follows that

$$\cos(x-y) = \sqrt{2} - \cos(x+y). \qquad (3)$$

If (1) holds, then $\cos(x+y) = \frac{\sqrt{2}}{2}$, and from (3) we find

$$\cos(x-y) = \frac{\sqrt{2}}{2},\ x - y = \pm\frac{\pi}{4} + 2l\pi.$$

If (2) holds, then $\cos(x+y) = -\frac{\sqrt{2}}{2}$ and $\cos(x-y) = \frac{3\sqrt{2}}{2}$ which is impossible.

Thus we have the system of equations

$$\left.\begin{aligned} x + y &= \frac{\pi}{4} + 2k\pi, \\ x - y &= \pm\frac{\pi}{4} + 2l\pi \end{aligned}\right\} \qquad (4)$$

for finding x and y. According to the choice of the sign in the second equation of system (4), we obtain the solutions

$$x_1 = \frac{\pi}{4} + (k+l)\pi,\ y_1 = (k-l)\pi$$

and

$$x_2 = (k+l)\pi, \qquad y_2 = \frac{\pi}{4} + (k-l)\pi.$$

605. Dividing termwise the first equation by the second one we get

$$\cos x \cos y = \frac{3}{4\sqrt{2}}. \qquad (1)$$

Adding this equation to the first one and subtracting the first equation from (1), we obtain the following system equivalent to the original one:

$$\left.\begin{aligned} \cos(x-y) &= \frac{1}{\sqrt{2}}, \\ \cos(x+y) &= \frac{1}{2\sqrt{2}}. \end{aligned}\right\}$$

It follows that

$$\left.\begin{aligned} x-y &= \pm\frac{\pi}{4}+2k\pi, \\ x+y &= \pm\arccos\frac{1}{2\sqrt{2}}+2l\pi. \end{aligned}\right\} \quad (2)$$

According to the choice of the signs in the equations (2) we get the following solutions:

(a) $x_1 = (k+l)\pi + \frac{1}{2}\arccos\frac{1}{2\sqrt{2}} + \frac{\pi}{8}$,

$y_1 = (l-k)\pi + \frac{1}{2}\arccos\frac{1}{2\sqrt{2}} - \frac{\pi}{8}$;

(b) $x_2 = (k+l)\pi + \frac{1}{2}\arccos\frac{1}{2\sqrt{2}} - \frac{\pi}{8}$,

$y_2 = (l-k)\pi + \frac{1}{2}\arccos\frac{1}{2\sqrt{2}} + \frac{\pi}{8}$;

(c) $x_3 = (k+l)\pi - \frac{1}{2}\arccos\frac{1}{2\sqrt{2}} + \frac{\pi}{8}$,

$y_3 = (l-k)\pi - \frac{1}{2}\arccos\frac{1}{2\sqrt{2}} - \frac{\pi}{8}$;

(d) $x_4 = (k+l)\pi - \frac{1}{2}\arccos\frac{1}{2\sqrt{2}} - \frac{\pi}{8}$,

$y_4 = (l-k)\pi - \frac{1}{2}\arccos\frac{1}{2\sqrt{2}} + \frac{\pi}{8}$.

606. Transform the second equation to the form

$$\frac{1}{2}[\cos(x+y)+\cos(x-y)] = a.$$

But, since $x+y=\varphi$, we have $\cos(x-y) = 2a - \cos\varphi$. Thus we obtain the system of equations

$$\left.\begin{aligned} x+y &= \varphi, \\ x-y &= \pm\arccos(2a-\cos\varphi)+k\pi. \end{aligned}\right\}$$

Answer:

$$x = \frac{\varphi}{2} \pm \frac{1}{2}\arccos(2a-\cos\varphi) + k\pi,$$

$$y = \frac{\varphi}{2} \mp \frac{1}{2}\arccos(2a-\cos\varphi) - k\pi,$$

where a and φ must satisfy the relation $|2a-\cos\varphi| \le 1$.

607. The left member of the first equation of the system not exceeding unity, the system is solvable only for $a = 0$. Putting $a = 0$ we obtain the system

$$\left.\begin{aligned} \sin x \cdot \cos 2y = 1, \\ \cos x \cdot \sin 2y = 0. \end{aligned}\right\} \qquad (1)$$

From the second equation of system (1) it follows that either $\cos x = 0$ or $\sin 2y = 0$. If $\cos x = 0$, then for $x_1 = \frac{\pi}{2} + 2m\pi$ we find from the first equation the expression $y_1 = n\pi$, and for $x_2 = -\frac{\pi}{2} + 2k\pi$ we get $y_2 = \left(l + \frac{1}{2}\right)\pi$. The case $\sin 2y = 0$ gives no new solutions. Thus, the system of equations is solvable only for $a = 0$ and has the following solutions:

$$x_1 = \frac{\pi}{2} + 2m\pi, \qquad y_1 = n\pi$$

and

$$x_2 = -\frac{\pi}{2} + 2k\pi, \quad y_2 = \left(l + \frac{1}{2}\right)\pi.$$

608. Note that $\cos y$ cannot be equal to zero. Indeed, if $\cos y = 0$, then $y = \frac{\pi}{2} + k\pi$ and

$$\cos(x - 2y) = \cos(x - \pi) = -\cos x = 0,$$
$$\sin(x - 2y) = \sin(x - \pi) = -\sin x = 0.$$

But $\sin x$ and $\cos x$ cannot vanish simultaneously because $\sin^2 x + \cos^2 x = 1$.

Evidently, we must have $a \neq 0$ since otherwise $\cos(x - 2y) = \sin(x - 2y) = 0$.

Dividing termwise the second equation by the first one (as follows from the above note, the division is permissible), we obtain :

$$\tan(x - 2y) = 1, \qquad x - 2y = \frac{\pi}{4} + k\pi. \qquad \ldots(1)$$

Let us consider the following two possible cases :

(a) k is even. In this case

$$\cos(x - 2y) = \frac{1}{\sqrt{2}} = a \cos^3 y, \cos y = \sqrt[3]{\frac{1}{a\sqrt{2}}} = \lambda,$$

$$y = \pm \text{arc}\cos \lambda + 2m\pi.$$

Substituting this value of y into (1) we get

$$x = \pm 2\, \text{arc}\cos \lambda + (4m + k)\pi + \frac{\pi}{4}.$$

(b) k is odd. Then $\cos(x-2y) = -\frac{1}{\sqrt{2}} = a\cos^3 y$,

$$y = \pm \text{arc}\cos(-\lambda) + 2m\pi.$$

From (1) we find

$$x = \pm 2\,\text{arc}\cos(-\lambda) + (4m+k)\pi + \frac{\pi}{4}.$$

The system is solvable for $a > \frac{1}{\sqrt{2}}$.

609. Squaring the given relations, we obtain

$$\sin^2 x + 2\sin x \sin y + \sin^2 y = a^2, \qquad (1)$$

$$\cos^2 x + 2\cos x\cos y + \cos^2 y = b^2. \qquad (2)$$

Adding and subtracting the equations (1) and (2) termwise, we find

$$2 + 2\cos(x-y) = a^2 + b^2, \qquad (3)$$

$$\cos 2x + \cos 2y + 2\cos(x+y) = b^2 - a^2. \qquad (4)$$

Equation (4) can be transformed to the form

$$2\cos(x+y)\,[\cos(x-y)+1] = b^2 - a^2. \qquad (5)$$

From (3) and (5) we find

$$\cos(x+y) = \frac{b^2-a^2}{a^2+b^2}.$$

610. Using the formula

$$\cos 2x + \cos 2y = 2\cos(x+y)\cos(x-y),$$

we rewrite the second equation of the system in the form

$$4\cos(x-y)\cos(x+y) = 1 + 4\cos^2(x-y).$$

The original system can be replaced by the following equivalent system:

$$\left.\begin{aligned} 4\cos\alpha\cos(x+y) &= 1 + 4\cos^2\alpha, \\ x - y &= \alpha. \end{aligned}\right\} \qquad \begin{matrix}(1)\\(2)\end{matrix}$$

Let us compare the left-hand and right-hand sides of equation (1). We have

$$|4\cos\alpha\cdot\cos(x+y)| \le 4|\cos\alpha|.$$

On the other hand, from the inequality $(1 \pm 2\cos\alpha)^2 \ge 0$ it follows that

$$4|\cos\alpha| \le 1 + 4\cos^2\alpha,$$

the sign of equality appearing only in the case $2|\cos\alpha| = 1$. Consequently, the system of equations (1) and (2) is solvable only if

$$|\cos\alpha| = \frac{1}{2}.$$

Consider the following two possible cases:

(a) $\cos\alpha = \frac{1}{2}$.

From (1) we find that $\cos(x+y)=1$, i.e.

$$x+y=2k\pi. \qquad (3)$$

Solving system (2), (3) we get

$$x_1=\frac{\alpha}{2}+k\pi,\ y_1=k\pi-\frac{\alpha}{2}.$$

(b) $\cos\alpha=-\frac{1}{2}$.

In this case we similarly find

$$x_2=\left(k+\frac{1}{2}\right)\pi+\frac{\alpha}{2},\quad y_2=\left(k+\frac{1}{2}\right)\pi-\frac{\alpha}{2}.$$

611. This problem is analogous to the preceding one. However, we shall demonstrate another method of solution. Applying formula (14), page 62, we represent the first equation of the system in the form $4\cos^2(x-y)+4\cos(x+y)\cos(x-y)+1=0$.

Putting $\cos(x-y)=t$ and taking advantage of the fact that $x+y=\alpha$ we obtain the equation

$$4t^2+4t\cos\alpha+1=0. \qquad (1)$$

This equation has real roots only if $D=16(\cos^2\alpha-1)\geq 0$, i.e. if $|\cos\alpha|=1$. Consider the following two possible cases : $\cos\alpha=1$ and $\cos\alpha=-1$. If $\cos\alpha=1$, then (1) implies that

$$t=\cos(x-y)=-\frac{1}{2}.$$

We obtain the system

$$\left.\begin{aligned} x-y&=\pm\frac{2}{3}\pi+2k\pi,\\ x+y&=\alpha, \end{aligned}\right\}$$

from which we find

$$x_1=\pm\frac{\pi}{3}+k\pi+\frac{\alpha}{2},\quad y_1=\mp\frac{\pi}{3}-k\pi+\frac{\alpha}{2}.$$

If $\cos\alpha=-1$, then we get in like manner the expressions

$$x_2=k\pi+\frac{\alpha}{2}\pm\frac{\pi}{6},\ y_2=\frac{\alpha}{2}-k\pi\mp\frac{\pi}{6}.$$

612. Consider the first equation of the system. By virtue of inequality (1), page 20, we have $\left|\tan x+\frac{1}{\tan x}\right|\geq 2$, the sign of equality taking place only if $\tan x=1$ or $\tan x=-1$. Since the right member of the first equation satisfies the condition $\left|2\sin\left(y+\frac{\pi}{4}\right)\right|\leq 2$, the first equation of the system can only be satisfied in the following cases:

(a) $\left.\begin{aligned} \tan x = 1, \\ \sin\left(y + \frac{\pi}{4}\right) = 1, \end{aligned}\right\}$ (1) (b) $\left.\begin{aligned} \tan x = -1, \\ \sin\left(y + \frac{\pi}{4}\right) = -1. \end{aligned}\right\}$ (2)

System (1) has the solutions

$$x_1 = \frac{\pi}{4} + k\pi, \qquad y_1 = \frac{\pi}{4} + 2l\pi, \tag{3}$$

and system (2) the solutions

$$x_2 = -\frac{\pi}{4} + m\pi, \quad y_2 = -\frac{3\pi}{4} + 2n\pi. \tag{4}$$

It can easily be verified that the solutions determined by formulas (3) do not satisfy the second equation of the original system, and the solutions given by formulas (4) satisfy the second equation (and, hence, the entire system) only for odd values of m. Putting $m = 2k + 1$ in (4), we can write the solutions of the original system in the form

$$x = \frac{3}{4}\pi + 2k\pi, \; y = -\frac{3}{4}\pi + 2n\pi.$$

613. Note that $\cos x \neq 0$ and $\cos y \neq 0$, since otherwise the third equation of the system has no sense. Therefore, the first two equations can be trnasformed to the form

$$(a - 1)\tan^2 x = 1 - b, \tag{1}$$

$$(b - 1)\tan^2 y = 1 - a. \tag{2}$$

But $a \neq 1$, because, if $a = 1$, then from (1) we have $b = 1$, which contradicts the condition $a \neq b$. Similarly, if $b = 1$, then $a = 1$. Consequently, (1) can be divided termwise by (2). Performing the division we obtain

$$\left(\frac{\tan x}{\tan y}\right)^2 = \left(\frac{1-b}{1-a}\right)^2.$$

We now must verify that $a \neq 0$. Indeed, if $a = 0$, then the second equation implies that $\sin y \neq 0$, and the third equation indicates that $b = 0$, i.e. $a = b = 0$ which is impossible.

By virtue of this note, the third equation can be rewritten as

$$\left(\frac{\tan x}{\tan y}\right)^2 = \frac{b^2}{a^2}.$$

Thus,

$$\left(\frac{b}{a}\right)^2 = \left(\frac{1-b}{1-a}\right)^2.$$

If $\frac{b}{a} = \frac{1-b}{1-a}$, then $a = b$, which is impossible. If $\frac{b}{a} = -\frac{1-b}{1-a}$, then $a + b = 2ab$.

Answer: $a + b = 2ab$.

614. By virtue of the first relation, the second one can be rewritten in the form

$$\frac{A\sin\beta}{\cos\alpha}=\frac{B\sin\beta}{\cos\beta}$$

or $$\sin\beta\,(A\cos\beta - B\cos\alpha)=0.$$

The latter relation can be fulfilled either for $\sin\beta = 0$ (and then $\sin\alpha = 0, \cos\beta = \pm 1$ and $\cos\alpha = \pm 1$) or for $A\cos\beta - B\cos\alpha = 0$. In the latter case we obtain the system

$$\left.\begin{aligned}\sin\alpha &= A\sin\beta,\\ A\cos\beta &= B\cos\alpha.\end{aligned}\right\} \qquad ...(1)$$

Squaring each equation and performing substitutions according to the formulas $\sin^2\alpha = 1-\cos^2\alpha$ and $\cos^2\beta = 1-\sin^2\beta$, we get the following system :

$$\left.\begin{aligned}\cos^2\alpha + A^2\sin^2\beta &= 1,\\ B^2\cos^2\alpha + A^2\sin^2\beta &= A^2.\end{aligned}\right\} \qquad (2)$$

It follows that $\cos^2\alpha$ and $\sin^2\beta$ are uniquely specified if and only if $A^2(1-B^2)\neq 0$; in this case

$$\cos\alpha = \pm\sqrt{\frac{1-A^2}{1-B^2}},\ \sin\beta = \pm\frac{1}{A}\sqrt{\frac{A^2-B^2}{1-B^2}}.$$

Consider the singular cases when $A^2(1-B^2)=0$. If $A=0$, then from (1) we obtain $\cos\alpha = \pm 1$ and $B=0$; in this case $\cos\alpha = \pm 1$, $\sin\beta$ remaining indeterminate. If $B^2 = 1$, then from (2) we get $A^2 = 1$, and the given equations do not in fact involve the parameters A and B; therefore the problem of expressing $\cos\alpha$ and $\sin\alpha$ in terms of A and B becomes senseless.

615. From the second equation we conclude that $\sin x = \sin\left(\frac{\pi}{2}-2y\right)$.

and, consequently, either

$$x = \frac{\pi}{2} - 2y + 2k\pi \qquad (1)$$

or $$x = 2y - \frac{\pi}{2} + (2l+1)\pi. \qquad (2)$$

Taking the first equation of the given system, we find in case (1) the relation

$$\cot 2y = \tan^3 y \quad \text{or} \quad \frac{1-\tan^2 y}{2\tan y} = \tan^3 y.$$

Solving the biquadratic equation we obtain $\tan y = \pm\frac{\sqrt{2}}{2}$. In the second case, expressing x from formula (2) and substituting it into the equation $\tan x = \tan^3 y$ we see that there are no real solutions. Thus, we have

$$\tan y = \pm \frac{1}{\sqrt{2}} \text{ and } x = \frac{\pi}{2} - 2y + 2k\pi,$$

whence $$y_1 = \arctan \frac{1}{\sqrt{2}} + n\pi, x_1 = \frac{\pi}{2} + 2k\pi - 2\arctan \frac{1}{\sqrt{2}} - 2\pi n$$

and $$y_2 = -\arctan \frac{1}{\sqrt{2}} + n\pi, x_2 = \frac{\pi}{2} + 2k\pi + 2\arctan \frac{1}{\sqrt{2}} - 2\pi n,$$

which can be written as

$$x_1 = \frac{\pi}{2} - 2\arctan \frac{1}{\sqrt{2}} + 2m\pi,$$

$$y_1 = \arctan \frac{1}{\sqrt{2}} + n\pi \text{ and } x_2 = \frac{\pi}{2} + 2\arctan \frac{1}{\sqrt{2}} + 2m\pi,$$

$$y_2 = -\arctan \frac{1}{\sqrt{2}} + n\pi,$$

where m and n are arbitrary integers.

616. Transforming the left-hand and right-hand sides of the first equation we obtain

$$2\sin \frac{x+y}{2}\left(\cos \frac{x-y}{2} - \cos \frac{x+y}{2}\right) = 0.$$

This equation is satisfied in the following cases:

1°. $x = -y + 2k\pi$ $(k = 0 \pm 1, \ldots)$.

2°. $y = 2l\pi$, x is an arbitrary number $(l = 0, \pm 1, \ldots)$.

3°. $x = 2m\pi$, y is an arbitrary number $(m = 0, \pm 1, \ldots)$.

Relations 1° and the second equation $|x| + |y| = 1$ of the system are only compatible if $k = 0$; indeed, from 1° we derive the inequality

$$|x| + |y| \geq 2|k|\pi,$$

which can hold, under the condition $|x| + |y| = 1$, only if $k = 0$. Now solving the system $x = -y$, $|x| + |y| = 1$, we find two solutions :

$$x_1 = \frac{1}{2},\ y_1 = -\frac{1}{2} \text{ and } x_2 = -\frac{1}{2},\ y_2 = \frac{1}{2}.$$

In cases 2° and 3°, an analogous argument results in four more pairs of solutions :

$x_3 = 1, y_3 = 0;\quad x_4 = -1, y_4 = 0;$

$x_5 = 0, y_5 = 1;\quad x_6 = 0,\quad y_6 = -1.$

Thus, the system under consideration has six solutions.

617. Squaring both members of each equation of the system and adding together the resulting equalities we obtain

$$\sin^2(y-3x)+\cos^2(y-3x)=4(\sin^6 x+\cos^6 x),$$

$$\sin^6 x+\cos^6 x=\frac{1}{4}. \qquad (1)$$

Consider the identity

$$\sin^6 x\cos^6 x=1-\frac{3}{4}\sin^2 2x \qquad (2)$$

proved in Problem 533.

Comparing (1) and (2) we find

$$\sin^2 2x=1, \qquad \sin 2x=\pm 1,$$

$$x=\frac{\pi}{4}(2n+1) \qquad (n=0,\pm 1,\pm 2,\ldots).$$

Multiplying the equations of the given system, we receive

$$\sin(y-3x)\cos(y-3x)=4\sin^3 x\cos^3 x,$$

i.e. $\sin 2(y-3x)=\sin^3 2x.$

But $\sin 2x=\pm 1$, therefore

$$\sin 2(y-3x)=\pm 1,$$

$$y-3x=\frac{\pi}{4}(2m+1)\ (m=0,\pm 1,\pm 2,\ldots).$$

Hence, $$y=\frac{3\pi}{4}(2n+1)+\frac{\pi}{4}(2m+1).$$

In solving the system we multiplied both members of the equation of the expressions dependent on unknowns which can lead to extraneous solutions. Let us verify whether all the pairs of values of x and y found above are solution. We must have

$$\sin\frac{\pi}{4}(2m+1)=2\sin^3\frac{\pi}{4}(2n+1)$$

and $$\cos\frac{\pi}{4}(2m+1)=2\cos^3\frac{\pi}{4}(2n+1).$$

Putting $\sin\frac{\pi}{4}(2m+1)=\frac{1}{\sqrt{2}}\sin\frac{\pi m}{2}+\frac{1}{\sqrt{2}}\cos\frac{\pi m}{2}$

and $$\cos\frac{\pi}{4}(2m+1)=\frac{1}{\sqrt{2}}\cos\frac{\pi m}{2}-\frac{1}{\sqrt{2}}\sin\frac{\pi m}{2},$$

making a similar substitution in the right member and cancelling out the constant factor, we get

$$\sin\frac{\pi m}{2}+\cos\frac{\pi m}{2}=\left(\sin\frac{\pi n}{2}+\cos\frac{\pi n}{2}\right)^3,$$

$$\cos\frac{\pi m}{2}-\sin\frac{\pi m}{2}=\left(\cos\frac{\pi n}{2}-\sin\frac{\pi n}{2}\right)^3.$$

For integral n, the expressions $\sin\frac{\pi n}{2}+\cos\frac{\pi n}{2}$ and $\cos\frac{\pi n}{2}-\sin\frac{\pi n}{2}$ can only assume the values 0, +1, −1, therefore their cubes take on the same values. Therefore,

$$\sin\frac{\pi n}{2}+\cos\frac{\pi n}{2}=\left(\sin\frac{\pi n}{2}+\cos\frac{\pi n}{2}\right)^3$$

and
$$\cos\frac{\pi n}{2}-\sin\frac{\pi n}{2}=\left(\cos\frac{\pi n}{2}-\sin\frac{\pi n}{2}\right)^3;$$

whence we get

$$\sin\frac{\pi}{2}m-\sin\frac{\pi}{2}n=\cos\frac{\pi}{2}n-\cos\frac{\pi}{2}m,$$

$$-\sin\frac{\pi}{2}m+\sin\frac{\pi}{2}n=\cos\frac{\pi}{2}n-\cos\frac{\pi}{2}m.$$

Addition and subtraction of the last relations result in

$$\sin\frac{\pi}{2}m-\sin\frac{\pi}{2}n=0,$$

$$\cos\frac{\pi}{2}n-\cos\frac{\pi}{2}m=0 \qquad (3)$$

or
$$\sin\frac{\pi}{4}(m-n)\cos\frac{\pi}{4}(m+n)=0,$$

$$\sin\frac{\pi}{4}(m-n)\sin\frac{\pi}{4}(m+n)=0.$$

Since $\cos\frac{\pi}{4}(m+n)$ and $\sin\frac{\pi}{4}(m+n)$ cannot vanish simultaneously, the above system is equivalent to the equation $\sin\frac{\pi}{4}(m-n)=0$. Consequently,

$$m-n=4k\,(k=0,\pm 1,\pm 2,\dots). \qquad (4)$$

Thus, the pairs of values of x and y expressed by formulas

$$x=\frac{\pi}{4}(2n+1),\qquad y=\frac{3\pi}{4}(2n+1)+\frac{\pi}{4}(2m+1)$$

are solutions of the system if and only if the integers n and m are connected by relations (4). Hence,

$$x=\frac{\pi}{4}(2n+1).$$

$$y=\frac{\pi}{4}[3(2n+1)+2(n+4k)+1]=\pi\,[2(n+k)+1].$$

But here $n+k$ is an arbitrary integer. Denoting it by p we finally write

$$x=\frac{\pi}{4}(2n+1),\ y=\pi\,(2p+1)\ (n,\ p=0,\pm 1,\pm 2\dots).$$

618. Squaring both members of the first and second equations and leaving the third one unchanged, we obtain the system

$$\left.\begin{aligned}(\sin x+\sin y)^2&=4a^2,\\(\cos x+\cos y)^2&=4b^2,\\\tan x\tan y&=c.\end{aligned}\right\}\qquad(1)$$

Let us derive the conditions on the numbers a, b and c which guarantee the existence of at least one solution of system (1). The given system has been replaced by system (1) which is not equivalent to it, and therefore we have to show that both systems are solvable when a, b and c satisfy the same conditions.

If for some a, b and c the given system has a solution, then, obviously, for the same a, b and c, system (1) is also solvable. The converse is also true : if for some a, b and c system (1) has a solution, then for the same values of a, b and c the given system is also solvable.

Indeed, let x_1, y_1 be a solution of system (1); then there are four possible cases, namely :

(1) $\sin x_1+\sin y_1=2a, \cos x_1+\cos y_1=2b$;

(2) $\sin x_1+\sin y_1=-2a, \cos x_1+\cos y_1=2b$;

(3) $\sin x_1+\sin y_1=-2a, \cos x_1+\cos y_1=-2b$;

(4) $\sin x_1+\sin y_1=2a, \cos x_1+\cos y_1=-2b$.

If the first case takes place, then x_1, y_1 is the solution of the given system; in the second case the given system has, for instance, the solution $-x_1, -y_1$; in the third case it has the solution $\pi+x_1, \pi+y_1$; in the forth case the solution is $\pi-x_1, \pi-y_1$. Consequently, the given system has at least one solution if and only if system (1) has at least one solution. Now let us find out the conditions for solvability of system (1). Adding and subtracting the first and second equations of system (1), we find:

$$\cos(x-y)=2(a^2+b^2)-1,$$
$$\cos 2x+\cos 2y+2\cos(x+y)=4(b^2-a^2)$$

or

$$\cos(x-y)=2(a^2+b^2)-1,$$
$$\cos(x+y)\cos(x-y)+\cos(x+y)=2(b^2-a^2),$$

whence

$$\cos(x-y)=2(a^2+b^2)-1,$$
$$(a^2+b^2)\cos(x+y)=b^2-a^2.$$

Thus, we have the system

$$\left.\begin{aligned}\cos(x-y)&=2(a^2+b^2)-1,\\(a^2+b^2)\cos(x+y)&=b^2-a^2,\\\tan x\tan y&=c,\end{aligned}\right\}$$

which is equivalent to system (1).

If $a^2 + b^2 = 0$, then the second equation is satisfied for any x and y. From the first equation we get $x - y = \pi + 2k\pi$ $(k = 0, \pm 1, \pm 2, \dots)$, the third equation yields $\tan(y + \pi + 2k\pi)\tan y = c$, or $\tan^2 y = c$. The last equation has a solution for any $c \geq 0$. If $a^2 + b^2 \neq 0$, then we have

$$\left.\begin{aligned} \cos(x-y) &= 2(a^2+b^2) - 1, \\ \cos(x+y) &= \frac{b^2-a^2}{a^2+b^2}. \end{aligned}\right\} \tag{2}$$

This system has a solution if and only if

$$|2(a^2+b^2) - 1| \leq 1, \tag{3}$$

$$\left|\frac{b^2-a^2}{a^2+b^2}\right| \leq 1. \tag{4}$$

Inequality (4) is obviously valid if

$$a^2 + b^2 \neq 0,$$

and (3) is equivalent to the inequality

$$0 < a^2 + b^2 \leq 1.$$

Let us represent the left member of the third equation of system (1) in the following way:

$$\tan x \tan y = \frac{\sin x \sin y}{\cos x \cos y}$$

$$= \frac{\frac{1}{2}[\cos(x-y) - \cos(x+y)]}{\frac{1}{2}[\cos(x-y) + \cos(x+y)]}. \tag{5}$$

Now, substituting into (5) the values of $\cos(x+y)$ and $\cos(x-y)$ found from (2) we see that a solution of system (2) satisfies the third equation of the original system if

$$c = \frac{2(a^2+b^2) - 1 - \dfrac{b^2-a^2}{a^2+b^2}}{\dfrac{b^2-a^2}{a^2+b^2} + 2(a^2+b^2) - 1}$$

$$= \frac{(a^2+b^2)^2 - b^2}{(a^2+b^2)^2 - a^2}.$$

Thus, we have arrived at the following result: the given system has at least one solution in the following two cases :

(1) $0 < a^2 + b^2 \leq 1$ and $c = \dfrac{(a^2+b^2)^2 - b^2}{(a^2+b^2)^2 - a^2}$;

(2) $a = b = 0$ and c is an arbitrary non-negative number.

3. Inverse Trigonometric Functions

619. The definition of the principal values of the inverse trigonometric functions implies that

arc cos (cos x) = x if $0 \le x \le \pi$.

To apply this formula we replace, with the aid of the reduction formulas, $\sin\left(-\frac{\pi}{7}\right)$ by the cosine of the corresponding angle contained between 0 and π. We write the equalities

$$\sin\left(-\frac{\pi}{7}\right) = -\sin\frac{\pi}{7} = \cos\left(\frac{\pi}{2}+\frac{\pi}{7}\right) = \cos\frac{9\pi}{14}$$

and finally obtain

$$\text{arc cos}\left[\sin\left(-\frac{\pi}{7}\right)\right] = \text{arc cos}\left(\cos\frac{9\pi}{14}\right) = \frac{9\pi}{14}.$$

620. By analogy with the solution of the foregoing problem, we have

$$\cos\frac{33}{5}\pi = \cos\left(6\pi+\frac{3}{5}\pi\right) = \cos\frac{3}{5}\pi = \sin\left(\frac{\pi}{2}-\frac{3}{5}\pi\right) = \sin\left(-\frac{\pi}{10}\right).$$

Hence,

$$\text{arc sin}\left(\cos\frac{33}{5}\pi\right) = \text{arc sin}\left[\sin\left(-\frac{\pi}{10}\right)\right] = -\frac{\pi}{10}.$$

621. Let arc tan $\frac{1}{3} = \alpha_1$, arc tan $\frac{1}{5} = \alpha_2$, arc tan $\frac{1}{7} = \alpha_3$ and arc tan $\frac{1}{8} = \alpha_4$.

Obviously, $0 < \alpha_i < \frac{\pi}{4}$, $i = 1, 2, 3, 4$. Therefore,

$$0 < \alpha_1 + \alpha_2 + \alpha_3 + \alpha_4 < \pi.$$

To prove the identity it is sufficient to establish that

$$\tan(\alpha_1 + \alpha_2 + \alpha_3 + \alpha_4) = 1.$$

Since $\tan(\alpha_1 + \alpha_2) = \frac{4}{7}$, and $\tan(\alpha_3 + \alpha_4) = \frac{3}{11}$, we have

$$\tan(\alpha_1 + \alpha_2 + \alpha_3 + \alpha_4) = \frac{\tan(\alpha_1+\alpha_2) + \tan(\alpha_3+\alpha_4)}{1 - \tan(\alpha_1+\alpha_2)\tan(\alpha_3+\alpha_4)} = 1.$$

622. Putting arc sin $x = \alpha$ and arc cos $x = \beta$, we obtain

$$x = \sin\alpha \text{ and } x = \cos\beta = \sin\left(\frac{\pi}{2}-\beta\right).$$

By the definition of the principal values, we have $-\frac{\pi}{2} \le \alpha \le \frac{\pi}{2}$ and $0 \le \beta \le \pi$. The last inequality implies the inequality

$$-\frac{\pi}{2} \le \frac{\pi}{2} - \beta \le \frac{\pi}{2}.$$

Hence, $\alpha = \frac{\pi}{2} - \beta$, because the angles α and $\frac{\pi}{2} - \beta$ lie between $-\frac{\pi}{2}$ and $\frac{\pi}{2}$, and the sines of these angles are equal. Thus the formula is proved.

623. Taking advantage of the relation arc sin x + arc cos $x = \frac{\pi}{2}$ (see the solution of Problem 622) we transform the equation to the form

$$12\pi t^2 - 6\pi^2 t + (1 - 8\alpha)\pi^3 = 0, \qquad (1)$$

where t = arc sin x. For $\alpha < \frac{1}{32}$ the discriminant of this equation satisfies the inequality

$$D = 36\pi^4 - 48\pi^4(1 - 8\alpha) < 0.$$

Consequently, the roots of equation (1) are nonreal, and therefore the original equation has no solutions for $\alpha < \frac{1}{32}$.

624. Put arc cos $x = \alpha$ and arc sin $\sqrt{1 - x^2} = \beta$.

(a) If $0 \le x \le 1$, then $0 \le \alpha \le \frac{\pi}{2}$ and $0 \le \beta \le \frac{\pi}{2}$ (because $0 \le \sqrt{1 - x^2} \le 1$).

Thus, we must only verify that $\sin\alpha = \sin\beta$. But, by virtue of the inequality $0 \le \alpha \le \frac{\pi}{2}$, we have in fact $\sin\alpha = +\sqrt{1 - x^2}$.

On the other hand, for all y ($|y| \le 1$) we have sin arc sin $y = y$; in particular, $\sin\beta = \sin \arcsin \sqrt{1 - x^2} = \sqrt{1 - x^2}$. Hence, for $0 \le x \le 1$, the formula arc cos x = arc sin $\sqrt{1 - x^2}$ holds true.

(b) If $-1 \le x \le 0$, then $\frac{\pi}{2} \le \alpha \le \pi, 0 \le \beta \le \frac{\pi}{2}$ and $\frac{\pi}{2} \le \pi - \beta \le \pi$.

Besides, we have $\sin\alpha = \sqrt{1 - x^2}$ and $\sin(\pi - \beta) = \sin\beta = \sqrt{1 - x^2}$, and therefore $\alpha = \pi - \beta$, i.e. for $-1 \le x \le 0$ the formula arc cos $x = \pi$ − arc sin $\sqrt{1 - x^2}$ holds true.

625. We shall prove that arc sin $(-x) = -$ arc sin x. Put arc sin $(-x) = \alpha$; then $-x = \sin\alpha$ and, by the definition of the principal values, we have

$$-\frac{\pi}{2} \le \alpha \le \frac{\pi}{2}. \qquad (1)$$

Since $\sin(-\alpha) = -\sin\alpha = x$ and inequality (1) implies the inequality $-\frac{\pi}{2} \le -\alpha \le \frac{\pi}{2}$, we can write $-\alpha = \arcsin x$, whence $\alpha = -$ arc sin x, i.e. arc sin $(-x) = -$arc sin x.

The formula arc cos $(-x) = \pi$ − arc cos x is proved in a similar way.

626. The definition of the principal values of the inverse trigonometric functions implies that $\arcsin(\sin\alpha)=\alpha$ if $-\frac{\pi}{2}\le\alpha\le\frac{\pi}{2}$. If

$$-\frac{\pi}{2}+2k\pi\le x\le\frac{\pi}{2}+2k\pi, \text{ then } -\frac{\pi}{2}\le x-2k\pi\le\frac{\pi}{2}.$$

But then $\arcsin(\sin x)=\arcsin[\sin(x-2k\pi)]=x-2k\pi$.

627. By the hypothesis, we have

$$\tan\frac{\alpha}{2}=\frac{1+x}{1-x}. \qquad (1)$$

Using the formula $\sin\alpha=\dfrac{2\tan\frac{\alpha}{2}}{1+\tan^2\frac{\alpha}{2}}$, we obtain, by virtue of (1), the

expression $$\sin\alpha=\frac{1-x^2}{1+x^2},$$

whence $$y=\arcsin(\sin\alpha)=\arcsin\frac{1-x^2}{1+x^2}=\beta. \qquad (2)$$

Since $0<x<1$, we have $\frac{\pi}{4}<\arctan\frac{1+x}{1-x}<\frac{\pi}{2}$ and $\frac{\pi}{2}<\alpha<\pi$. Then

and $$-\frac{\pi}{2}<\alpha-\pi\le 0$$

$\arcsin[\sin(\alpha-\pi)]=\arcsin(-\sin\alpha)=-\arcsin(\sin\alpha)=-y$.

But the angle $\alpha-\pi$ lies within the range of the principal value $\arcsin x$. Hence,

$$y=\arcsin(\sin\alpha)=\pi-\alpha. \qquad (3)$$

From (2) and (3) we obtain $\alpha+\beta=\pi$.

628. In the expressions arc sin cos arc sin x and arc cos sin arc cos x we take the principal values of the inverse trigonometric functions. Let us consider cos arc sin x. This is the cosine of an arc whose sine is equal to x. Hence, $\cos\arcsin x=+\sqrt{1-x^2}$, where $-1\le x\le 1$.

Of course, it is essential here that $-\frac{\pi}{2}\le\arcsin x\le\frac{\pi}{2}$. Analogously,

$\sin\arccos x=+\sqrt{1-x^2}$, where $-1\le x\le 1$.

Let $y=+\sqrt{1-x^2}$; then $0\le y\le 1$.

Thus, it is necessary to find the relation between $\arcsin y$ and $\arccos y$ for $0\le y\le 1$. These are two complimentary angles (see the solution of Problem 622). Thus,

$$\arcsin\cos\arcsin x+\arccos\sin\arccos x=\frac{\pi}{2}.$$

4. Trigonometric Inequalities

629. The given inequality is equivalent to the inequality

$$\sin^2 x + \sin x - 1 > 0. \qquad (1)$$

Factoring the quadratic trinomial on the left-hand side of (1), we get

$$\left(\sin x + \frac{1+\sqrt{5}}{2}\right)\left(\sin x - \frac{\sqrt{5}-1}{2}\right) > 0. \qquad (2)$$

But $\frac{1+\sqrt{5}}{2} > 1$, and, therefore, $\sin x + \frac{1+\sqrt{5}}{2} > 0$. Consequently, the original inequality is equivalent to $\sin x > \frac{\sqrt{5}-1}{2}$ and has the following solutions: $2k\pi + \varphi < x < \pi - \varphi + 2k\pi$ where $\varphi = \arcsin \frac{\sqrt{5}-1}{2}$ $(k = 0, \pm 1, \pm 2, \dots)$.

630. The expression under consideration only makes sense for $x \neq \frac{\pi}{2} + \pi n$. For these values of x we multiply both members of the inequality by $\cos^2 x$ and arrive at the equivalent inequality

$$(\sin 2x)^2 + \frac{3}{2}\sin 2x - 2 > 0.$$

Solving the above quadratic inequality we find that either $\sin 2x < \frac{-3-\sqrt{41}}{4}$ or $\sin 2x > \frac{\sqrt{41}-3}{4}$. The former cannot be fulfilled. Hence,

$$k\pi + \frac{1}{2}\arcsin\frac{\sqrt{41}-3}{4} < x < \frac{\pi}{2} - \frac{1}{2}\arcsin\frac{\sqrt{41}-3}{4} + k\pi.$$

631. Transforming the product of sines into the sum, we replace the given inequality by the equivalent inequality

$$\cos 3x > \cos 7x \text{ or } \sin 5x \sin 2x > 0.$$

But for $0 < x < \frac{\pi}{2}$ we have $\sin 2x > 0$ and, consequently, the original inequality is equivalent to $\sin 5x > 0$.

Answer: $0 < x < \frac{\pi}{5}$ and $\frac{2}{5}\pi < x < \frac{\pi}{2}$.

632. The denominator of the left member of the inequality is positive because $|\sin x + \cos x| = \left|\sqrt{2}\sin\left(x + \frac{\pi}{4}\right)\right| \leq \sqrt{2}$. Therefore, the given inequality is equivalent to $\sin^2 x > \frac{1}{4}$ or $|\sin x| > \frac{1}{2}$.

Answer: $\frac{\pi}{6} + k\pi < x < \frac{5}{6}\pi + k\pi$.

633. Let us write the inequality in the form

$$(\cos x - \sin x)\,[1 - (\cos x + \sin x)]$$
$$= 2\sin\frac{x}{2}\left(\sin\frac{x}{2} - \cos\frac{x}{2}\right)(\cos x - \sin x) > 0. \qquad (1)$$

But $\sin\frac{x}{2} > 0$, since $0 < x < 2\pi$. Let us consider the following two possible cases when inequality (1) is fulfilled :

Case 1.
$$\left.\begin{array}{r}\cos x - \sin x > 0,\\ \sin\frac{x}{2} - \cos\frac{x}{2} > 0.\end{array}\right\} \qquad (2)$$

By the hypothesis, we have $0 < x < 2\pi$. Taking this into account, we find from (2) that the first inequality is fulfilled if $0 < x < \frac{\pi}{4}$ or $\frac{5}{4}\pi < x < 2\pi$ and the second if $\frac{\pi}{2} < x < 2\pi$. Hence, in this case

$$\frac{5}{4}\pi < x < 2\pi.$$

Case 2.
$$\left.\begin{array}{r}\cos x - \sin x < 0,\\ \sin\frac{x}{2} - \cos\frac{x}{2} < 0.\end{array}\right\} \qquad (3)$$

Taking into consideration that $0 < x < 2\pi$, we see that system (3) is satisfied if
$$\frac{\pi}{4} < x < \frac{\pi}{2}.$$

Answer: $\frac{\pi}{4} < x < \frac{\pi}{2}$ and $\frac{5}{4}\pi < x < 2\pi$.

634. Put $\tan\frac{x}{2} = t$. Then the inequality takes the form $t > \frac{2t - 2 + 2t^2}{2t + 2 - 2t^2}$

or
$$\frac{(t-1)(t^2+t+1)}{t^2 - t - 1} > 0. \qquad (1)$$

Since $t^2 + t + 1 > 0$ for all real values of t, inequality (1) is equivalent to the inequality

$$\frac{t-1}{t^2 - t - 1} > 0. \qquad (2)$$

The trinomial $t^2 - t - 1$ has the roots $\frac{1-\sqrt{5}}{2}$ and $\frac{1+\sqrt{5}}{2}$. Solving (2), we find that either $\tan\frac{x}{2} > \frac{1+\sqrt{5}}{2}$ or $\frac{1-\sqrt{5}}{2} < \tan\frac{x}{2} < 1$

Answer: (a) $2k\pi + 2\,\text{arc tan}\,\frac{1+\sqrt{5}}{2} < x < \pi + 2k\pi$.

(b) $2k\pi - 2\,\text{arc tan}\,\frac{\sqrt{5}-1}{2} < x < \frac{\pi}{2} + 2k\pi$.

635. From the formulas for $\sin 3x$ and $\cos 3x$ given on page 62 we find

$$\cos^3 x = \frac{\cos 3x + 3\cos x}{4}, \sin^3 x = \frac{3\sin x - \sin 3x}{4}.$$

Using these formulas, we rewrite the given inequality in the form

$$(\cos 3x + 3\cos x)\cos 3x - (3\sin x - \sin 3x)\sin 3x > \frac{5}{2}$$

or

$$\sin^2 3x + \cos^2 3x + 3(\cos 3x \cos x - \sin 3x \sin x) > \frac{5}{2},$$

i.e.

$$\cos 4x > \frac{1}{2}, \text{ whence } -\frac{\pi}{3} + 2\pi n < 4x < \frac{\pi}{3} + 2\pi n$$

or

$$-\frac{\pi}{12} + \frac{1}{2}\pi n < x < \frac{\pi}{12} + \frac{1}{2}\pi n \ (n = 0, \pm 1, \pm 2, \dots).$$

636. The inequality to be proved can be written in the form

$$\cot\frac{\varphi}{2} > \frac{\cos^2\frac{\varphi}{2} - \sin^2\frac{\varphi}{2} + \sin\varphi}{\sin\varphi}. \qquad \dots(1)$$

But $\sin\varphi > 0$ for $0 < \varphi < \frac{\pi}{2}$, and therefore, multiplying both members of inequality (1) by $\sin\varphi$, we get the equivalent inequality

$$2\cos^2\frac{\varphi}{2} > \cos^2\frac{\varphi}{2} - \sin^2\frac{\varphi}{2} + \sin\varphi,$$

i.e. $1 > \sin\varphi$. The last inequality is fulfilled for $0 < \varphi < \frac{\pi}{2}$, and, hence, the original inequality is also valid.

637. Putting $\tan x = t$ we obtain

$$\tan 2x = \frac{2t}{1 - t^2}, \quad \tan 3x = \frac{\tan x + \tan 2x}{1 - \tan 2x \tan x} = \frac{3t - t^3}{1 - 3t^2}.$$

The left member is not determined for the values of x satisfying the relations $t^2 = 1$ and $t^2 = \frac{1}{3}$. For all the other values of x the left member of the inequality is equal to $t^4 + 2t^2 + 1$ and, hence, assumes positive values.

638. By virtue of the relations

$$\cot^2 x - 1 = \frac{\cos 2x}{\sin^2 x}, \qquad 3\cot^2 x - 1 = \frac{3\cos^2 x - \sin^2 x}{\sin^2 x}$$

and

$$\cot 3x \tan 2x - 1 = \frac{\cos 3x \sin 2x - \sin 3x \cos 2x}{\sin 3x \cos 2x} = -\frac{\sin x}{\sin 3x \cos 2x},$$

the left member of the inequality can be rewritten in the form

$$-\frac{\sin x\,(3\cos^2 x-\sin^2 x)}{\sin^4 x\sin 3x}.$$

But

$\sin 3x=\sin(x+2x)=\sin x\cos 2x+\cos x\sin 2x=\sin x\,(3\cos^2 x-\sin^2 x)$,
and, therefore, the given inequality is reduced to the inequality

$$-\frac{1}{\sin^4 x}\le -1,$$

which obviously holds.

639. Using the formula $\tan(\theta-\varphi)=\dfrac{\tan\theta-\tan\varphi}{1+\tan\theta\tan\varphi}$ and the condition $\tan\theta=n\tan\varphi$ we get

$$\tan^2(\theta-\varphi)=\frac{(n-1)^2\tan^2\varphi}{(1+n\tan^2\varphi)^2}=\frac{(n-1)^2}{(\cot\varphi+n\tan\varphi)^2}.$$

We now must prove that

$$(\cot\varphi+n\tan\varphi)^2\ge 4n \text{ or } (1+n\tan^2\varphi)^2\ge 4n\tan^2\varphi.$$

Thus, we arrive at the inequality

$$(1-n\tan^2\varphi)^2\ge 0,$$

which obviously holds.

640. The given inequality can be rewritten in the form

$$\frac{1}{2}+\frac{1-\sin x}{2-\sin x}-\frac{2-\sin x}{3-\sin x}\ge 0.$$

Multiplying it by $2(2-\sin x)(3-\sin x)>0$ we replace it by the equivalent inequality $\sin^2 x-5\sin x+4\ge 0$, i.e.

$$(4-\sin x)(1-\sin x)\ge 0. \qquad (1)$$

From (1) we conclude that the last inequality, and, consequently, the original one, is fulfilled for all x, the sign of equality appearing for $x=\dfrac{\pi}{2}+2k\pi$.

641. Let us first establish that

$$|\sin x|\le|x|.$$

Consider the unit circle shown in Fig. 249. Let x be the radian measure of a positive or negative angle AOM. For any position of the point M we have

$$\overset{\smile}{AM}=|x|\cdot OA=|x|,$$
$$|BM|=|\sin x|.$$

Since $|BM| \le \overset{\frown}{AM}$, we have $|\sin x| \le |x|$ (the sign of equality appears here for $x=0$). Now we conclude that if $0 \le \varphi \le \frac{\pi}{2}$, i.e. if $0 \le \cos\varphi \le 1 < \frac{\pi}{2}$, then $\sin\cos\varphi < \cos\varphi$. But $0 \le \sin\varphi \le \varphi \le \frac{\pi}{2}$ and, therefore, $\cos\varphi \le \cos\sin\varphi$. We finally obtain $\cos\sin\varphi \ge \cos\varphi > \sin\cos\varphi$. The inequality has been proved.

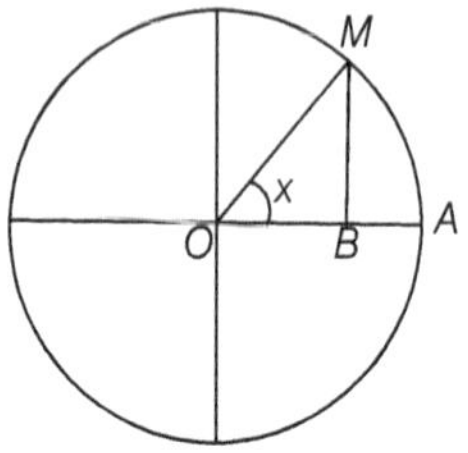

Fig. 249

642. We shall apply the method of complete induction. Let $n=2$, then $0 < \alpha < \frac{\pi}{4}$. Hence,

$$\tan 2\alpha = \frac{2\tan\alpha}{1-\tan^2\alpha} > 2\tan\alpha,$$

because $0 < 1 - \tan^2\alpha < 1$. Suppose that

$$\tan n\alpha > n\tan\alpha \tag{1}$$

for

$$0 < \alpha < \frac{\pi}{4(n-1)}. \tag{2}$$

We shall prove that $\tan(n+1)\alpha > (n+1)\tan\alpha$, if $0 < \alpha < \frac{\pi}{4n}$.

Let us use the formula

$$\tan(n+1)\alpha = \frac{\tan n\alpha + \tan\alpha}{1-\tan\alpha\tan n\alpha}. \tag{3}$$

Since inequality (1) is fulfilled under condition (2), it automatically holds for $0 < \alpha < \frac{\pi}{4n}$. But we have

$$0 < \tan\alpha < 1, \tag{4}$$

and, since $0 < n\alpha < \frac{\pi}{4}$, we obtain

$$0 < \tan n\alpha < 1. \tag{5}$$

Now inequalities (4) and (5) imply

$$0 < 1 - \tan\alpha\tan n\alpha < 1. \tag{6}$$

From (6) and (3) it follows that $\tan(n+1)\alpha > (n+1)\tan\alpha$, i.e. we have obtained what we set out to prove.

643. Since to a greater angle in the first quadrant there corresponds a greater value of the tangent, we can write

$$\tan\alpha_1 < \tan\alpha_i < \tan\alpha_n \qquad (1)$$

for $i = 1, 2, \dots, n$. Besides, $\cos\alpha_i > 0$ $(i = 1, 2, \dots n)$. Therefore, inequalities (1) can be rewritten in the form

$$\tan\alpha_1 \cos\alpha_i < \sin\alpha_i < \tan\alpha_n \cos\alpha_i. \qquad (2)$$

Let us make i in inequality (2) assume the values 1, 2, ..., n and add together all the inequalities thus obtained. This results in

$$\tan\alpha_1(\cos\alpha_1 + \dots + \cos\alpha_n) < \sin\alpha_1 + \dots$$
$$+ \sin\alpha_n < \tan\alpha_n(\cos\alpha_1 + \dots + \cos\alpha_n). \qquad (3)$$

Dividing all the members of inequalities (3) by $\cos\alpha_1 + \dots + \cos\alpha_n$ (which is permissible since $\cos\alpha_1 + \dots + \cos\alpha_n > 0$) we obtain

$$\tan\alpha_1 < \frac{\sin\alpha_1 + \dots + \sin\alpha_n}{\cos\alpha_1 + \dots + \cos\alpha_n} < \tan\alpha_n.$$

644. Denote the left-hand side of the inequality by t. Then

$$t = \frac{1}{2}\left(\cos\frac{A-B}{2} - \cos\frac{A+B}{2}\right)\cos\frac{A+B}{2},$$

because $\sin\frac{C}{2} = \cos\frac{A+B}{2}$.

Putting $\cos\frac{A+B}{2} = x$,

after obvious transformations we obtain

$$t = -\frac{1}{2}\left(x^2 - 2x\frac{1}{2}\cos\frac{A-B}{2} + \frac{1}{4}\cos^2\frac{A-B}{2}\right) +$$
$$+ \frac{1}{8}\cos^2\frac{A-B}{2} = \frac{1}{8}\cos^2\frac{A-B}{2} - \frac{1}{2}\left(x - \frac{1}{2}\cos\frac{A-B}{2}\right)^2$$

Consequently,

$$t \le \frac{1}{8}\cos^2\frac{A-B}{2} \le \frac{1}{8}.$$

645. Transform the left member of the given inequality in the following way:

$$\frac{\cos x}{\sin^2 x(\cos x - \sin x)} = \frac{1}{\sin^2 x(1 - \tan x)}$$
$$= \frac{\frac{1}{\cos^2 x}}{\tan^2 x(1 - \tan x)}$$
$$= \frac{1 + \tan^2 x}{\tan x} \cdot \frac{1}{\tan x(1 - \tan x)}.$$

For brevity, let us put $\tan x = t$. Since $0 < x < \frac{\pi}{4}$, we have

$$0 < t < 1. \qquad (1)$$

Thus, the problem is reduced to proving the inequality

$$\frac{1+t^2}{t} \cdot \frac{1}{t(1-t)} > 8$$

for $0 < t < 1$. By virtue of inequality (1), page 13, we have $\frac{1+t^2}{t} > 2$.

Furthermore, $t(1-t) = \frac{1}{4} - \left(\frac{1}{2} - t\right)^2 \le \frac{1}{4}$. Hence,

$\frac{1+t^2}{t} \cdot \frac{1}{t(1-t)} > 2 \cdot \frac{1}{\frac{1}{4}} = 8$ which is what we set out to prove.

5. Miscellaneous Problems

646. Put $\arctan \frac{1}{5} = \alpha$, $\arctan \frac{5}{12} = \beta$ and consider $\tan(2\alpha - \beta)$. Using the formula for the tangent of the difference of two angles, we get

$$\tan(2\alpha - \beta) = \frac{\tan 2\alpha - \tan\beta}{1 + \tan 2\alpha \tan\beta}. \qquad (1)$$

But, since $\tan\alpha = \frac{1}{5}$, we have $\tan 2\alpha = \frac{2\tan\alpha}{1-\tan^2\alpha} = \frac{5}{12}$. Substituting $\tan 2\alpha$ and $\tan\beta$ into formula (1) we find $\tan(2\alpha - \beta) = 0$. Thus,

$$\sin(2\alpha - \beta) = \sin\left(2\arctan\frac{1}{5} - \arctan\frac{5}{12}\right) = 0.$$

647. Let us prove that $\tan(\alpha + 2\beta) = 1$. To compute $\tan(\alpha + 2\beta)$ we use the formula

$$\tan(\alpha + 2\beta) = \frac{\tan\alpha + \tan 2\beta}{1 - \tan\alpha \tan 2\beta}. \qquad (1)$$

We first compute $\tan 2\beta$ by the formula

$$\tan 2\beta = \frac{\sin 2\beta}{\cos 2\beta} = \frac{2\sin\beta\cos\beta}{\cos 2\beta}.$$

Now we must find $\cos\beta$ and $\cos 2\beta$. But $\cos\beta = +\sqrt{1-\sin^2\beta} = \frac{3}{\sqrt{10}}$ (because β is an angle in the first quadrant)

and $\cos 2\beta = \cos^2\beta - \sin^2\beta = \frac{4}{5}$. Hence, $\tan 2\beta = \frac{3}{4}$. Substituting the found value of $\tan 2\beta$ into (1) we get

$$\tan(\alpha + 2\beta) = 1.$$

Now we can prove that $\alpha + 2\beta = \frac{\pi}{4}$.

Since $\tan\alpha = \frac{1}{7}$, $\tan\beta = \frac{\sin\beta}{\cos\beta} = \frac{1}{3}$ and, besides, by the condition of the problem, α and β are angles in the first quadrant, we have $0 < \alpha < \frac{\pi}{4}$ and $0 < \beta < \frac{\pi}{4}$. Hence, we find that $0 < \alpha + 2\beta < \frac{3}{4}\pi$. But the only angle lying between 0 and $\frac{3}{4}\pi$ whose tangent is equal to 1 is $\frac{\pi}{4}$. Thus, $\alpha + 2\beta = \frac{\pi}{4}$.

648. We must have $\cos x \neq 0$, $\sin x \neq 0$ and $\sin x \neq -1$, and therefore $x \neq \frac{k\pi}{2}$ where k is an integer. For all the values of x other than $x = \frac{k\pi}{2}$, y has sense, and

$$y = \frac{\sin x\left(1 + \frac{1}{\cos x}\right)}{\cos x\left(1 + \frac{1}{\sin x}\right)} = \frac{\sin^2 x\,(1 + \cos x)}{\cos^2 x\,(1 + \sin x)}. \qquad (1)$$

Relation (1) implies that $y > 0$ because for $x \neq \frac{k\pi}{2}$ we have

$$\cos x < 1 \text{ and } \sin x < 1.$$

649. Transforming the product $\sin\alpha \cdot \sin 2\alpha \cdot \sin 3\alpha$ into a sum by formula (13), page 62, we obtain

$$\sin\alpha \cdot \sin 2\alpha \cdot \sin 3\alpha = \frac{1}{2}\sin 2\alpha\,(\cos 2\alpha - \cos 4\alpha)$$

$$= \frac{1}{4}\sin 4\alpha - \frac{1}{2}\sin 2\alpha \cdot \cos 4\alpha \leq \frac{1}{4} + \frac{1}{2} < \frac{4}{5}.$$

650. We have $\sin 5x = \sin 3x\cos 2x + \cos 3x\sin 2x$, and therefore, using formulas (5) to (8), page 62, after simple computations, we find

$$\sin 5x = 5\sin x - 20\sin^3 x + 16\sin^5 x. \qquad (1)$$

Putting $x = 36°$ in formula (1) we obtain the equation $16t^5 - 20t^3 + 5t = 0$ for determining $\sin 36°$. This equation has the roots

$$t_1 = 0,\ t_2 = +\sqrt{\frac{5+\sqrt{5}}{8}},\ t_3 = -\sqrt{\frac{5+\sqrt{5}}{8}},$$

$$t_4 = +\sqrt{\frac{5-\sqrt{5}}{8}} \text{ and } t_5 = -\sqrt{\frac{5-\sqrt{5}}{8}},$$

among which only t_2 and t_4 are positive. But $\sin 36° \neq t_2$ because $\frac{5+\sqrt{5}}{8} > \frac{1}{2}$, and, hence, $t_2 > \frac{1}{\sqrt{2}}$. Thus,

$$\sin 36° = t_4 = \frac{1}{2}\sqrt{\frac{5-\sqrt{5}}{2}}.$$

651. Using the identity proved in Problem 533, we get $\varphi(x) = \frac{1+3\cos^2 2x}{4}$, whence it follows that the greatest value of $\varphi(x)$ is equal to 1, and the least to $\frac{1}{4}$.

652. Performing simple transformations we obtain

$$y = 1 - \cos 2x + 2(1+\cos 2x) + 3\sin 2x = 3 + 3\sin 2x + \cos 2x.$$

Introducing the auxiliary angle $\varphi = \arctan\frac{1}{3}$, we can write

$$y = 3 + \sqrt{10}\left(\frac{3}{\sqrt{10}}\sin 2x + \frac{1}{\sqrt{10}}\cos 2x\right)$$
$$= 3 + \sqrt{10}\sin(2x+\varphi).$$

Hence, the greatest value of y is equal to $3+\sqrt{10}$, and the least to $3-\sqrt{10}$.

653. If n is an integer satisfying the condition of the problem, we have for all x the relation

$$\cos n(x+3\pi)\cdot\sin\frac{5}{n}(x+3\pi) = \cos nx\cdot\sin\frac{5}{n}x. \qquad (1)$$

In particular, putting $x = 0$, we conclude from (1) that n must satisfy the equation $\sin\frac{15\pi}{n} = 0$. This equation is only satisfied by the integers which are the divisors of the number 15, i.e.

$$n = \pm 1, \pm 3, \pm 5, \pm 15. \qquad (2)$$

The direct substitution shows that for each of these values the function $\cos nx \cdot \sin\frac{5}{n}x$ is periodic with period 3π. Formula (2) exhausts all the required values of n.

654. Since the sum under consideration is equal to zero for $x = x_1$, we have

$$a_1\cos(\alpha_1 + x_1) + \dots + a_n\cos(\alpha_n + x_1)$$
$$= (a_1\cos\alpha_1 + \dots + a_n\cos\alpha_n)\cos x_1 -$$
$$-(a_1\sin\alpha_1 + \dots + a_n\sin\alpha_n)\sin x_1 = 0. \qquad (1)$$

But, by the condition of the problem,

$$a_1\cos\alpha_1 + \dots + a_n\cos\alpha_n = 0. \qquad (2)$$

Besides, $\sin x_1 \neq 0$ because $x_1 \neq k\pi$. From (1) and (2) we get

$$a_1\sin\alpha_1 + \dots + a_n\sin\alpha_n = 0. \qquad (3)$$

Let now x be an arbitrary number. Then we have

$$a_1 \cos(\alpha_1 + x) + \ldots + a_n \cos(\alpha_n + x)$$
$$= (a_1 \cos\alpha_1 + \ldots + a_n \cos\alpha_n)\cos x$$
$$- (a_1 \sin\alpha_1 + \ldots + a_n \sin\alpha_n)\sin x = 0,$$

since, by virtue of (2) and (3), the sums in the brackets are equal to zero.

655. Suppose the contrary, i.e. assume that there exists $T \neq 0$ such that for all $x \geq 0$ we have

$$\cos\sqrt{x+T} = \cos\sqrt{x} \qquad (1)$$

(the condition $x \geq 0$ must hold because the radical $\sqrt{x}$ is imaginary for $x < 0$). Let us first put $x = 0$ in formula (1); then

$$\cos\sqrt{T} = \cos 0 = 1 \qquad (1)$$

and, consequently,

$$\sqrt{T} = 2k\pi. \qquad (3)$$

Now we substitute the value $x = T$ into (1). According to (1) and (2) we obviously obtain $\cos\sqrt{2T} = \cos\sqrt{T} = 1$, whence

$$\sqrt{2T} = 2l\pi.$$

By the hypothesis, we have $T \neq 0$, and therefore, dividing (4) by (3), we get $\sqrt{2} = \frac{l}{k}$ where l and k are integers which is impossible.

656. *First solution.* Let us consider the sum

$$S = (\cos x + i\sin x) + (\cos 2x + i\sin 2x) + \ldots + (\cos nx + i\sin nx).$$

Applying De Moivre's formula $(\cos x + i\sin x)^n = \cos nx + i\sin nx$ we compute S as the sum of a geometric progression. We thus obtain

$$S = \frac{(\cos x + i\sin x)^{n+1} - (\cos x + i\sin x)}{\cos x + i\sin x - 1}$$

The sought-for sum $\sin x + \sin 2x + \ldots + \sin nx$ is equal to the imaginary part of S.

Second solution. Multiplying the left-hand side by $2\sin\frac{x}{2}$ and applying formula (13), page 62, we get

$$\left(\cos\frac{x}{2} - \cos\frac{3}{2}x\right) + \left(\cos\frac{3}{2}x - \cos\frac{5}{2}x\right) + \ldots$$
$$\ldots + \left(\cos\frac{2n-1}{2}x - \cos\frac{2n+1}{2}x\right) = \cos\frac{x}{2} - \cos\frac{2n+1}{2}x =$$
$$= 2\sin\frac{nx}{2}\cdot\sin\frac{n+1}{2}x,$$

which results in the required formula.

657. Denote the required sum by A and add the sum

$$B = \frac{\sin\frac{\pi}{4}}{2} + \frac{\sin\frac{2\pi}{4}}{2^2} + \ldots + \frac{\sin\frac{\pi n}{4}}{2^n}$$

multiplied by i to it. This results in

$$A + Bi = \frac{1}{2}\left(\cos\frac{\pi}{4} + i\sin\frac{\pi}{4}\right) + \frac{1}{2^2}\left(\cos\frac{2\pi}{4} + i\sin\frac{2\pi}{4}\right) + \\ + \ldots + \frac{1}{2^n}\left(\cos n\frac{\pi}{4} + i\sin n\frac{\pi}{4}\right).$$

Applying De Moivre's formula, we find

$$A + Bi = \frac{1}{2}\left(\cos\frac{\pi}{4} + i\sin\frac{\pi}{4}\right) + \ldots + \frac{1}{2^n}\left(\cos\frac{\pi}{4} + i\sin\frac{\pi}{4}\right)^n =$$

$$= \frac{1}{2}\left(\cos\frac{\pi}{4} + i\sin\frac{\pi}{4}\right)\frac{1 - \frac{1}{2^n}\left(\cos\frac{\pi}{4} + i\sin\frac{\pi}{4}\right)^n}{1 - \frac{1}{2}\left(\cos\frac{\pi}{4} + i\sin\frac{\pi}{4}\right)}.$$

When deriving the last expression, we have used the formula for the sum of terms of a geometric progression. The sought-for sum A can be found as the real part of this expression. Noting that

$$\cos\frac{\pi}{4} = \sin\frac{\pi}{4} = \frac{1}{\sqrt{2}},$$

we write $$A + Bi = \frac{1}{2}\left(\cos\frac{\pi}{4} + i\sin\frac{\pi}{4}\right)\frac{1 - \frac{1}{2^n}\left(\cos\frac{\pi}{4} + i\sin\frac{\pi}{4}\right)^n}{1 - \frac{1}{2}\left(\cos\frac{\pi}{4} + i\sin\frac{\pi}{4}\right)} =$$

$$= \frac{1}{2\sqrt{2}}(1 + i)\frac{1 - \frac{1}{2^n}\left(\cos n\frac{\pi}{4} + i\sin n\frac{\pi}{4}\right)}{1 - \frac{1}{2\sqrt{2}} - \frac{i}{2\sqrt{2}}}$$

$$= \frac{(1 + i)\left[\left(2^n - \cos n\frac{\pi}{4}\right) - i\sin n\frac{\pi}{4}\right]}{2^n\left[(2\sqrt{2} - 1) - i\right]}$$

$$= \frac{(1 + i)(2\sqrt{2} - 1 + i)}{2^n\left[(2\sqrt{2} - 1)^2 + 1\right]}\left[\left(2^n - \cos n\frac{\pi}{4}\right) - i\sin n\frac{\pi}{4}\right] =$$

$$= \frac{\left[(2\sqrt{2} - 2) + 2n\sqrt{2}\right]\left[\left(2^n - \cos n\frac{\pi}{4}\right) - i\sin n\frac{\pi}{4}\right]}{2^n(10 - 4\sqrt{2})}$$

Taking the real part, we get

$$A = \frac{(\sqrt{2}-1)\left(2^n - \cos n\frac{\pi}{4}\right) + \sqrt{2}\sin n\frac{\pi}{4}}{2^n(5-2\sqrt{2})}.$$

658. The assertion will be proved if we establish that $A = B = 0$. Let $A^2 + B^2 \neq 0$, i.e. at least one of the number A, B is other than zero. Then

$$f(x) = \left(\frac{A}{\sqrt{A^2+B^2}}\cos x + \frac{B}{\sqrt{A^2+B^2}}\sin x\right)$$

$$\sqrt{A^2+B^2} = \sqrt{A^2+B^2}\sin(x+\varphi),$$

where

$$\sin\varphi = \frac{A}{\sqrt{A^2+B^2}}, \cos\varphi = \frac{B}{A^2+B^2}.$$

Let now x_1 and x_2 be the two values of the argument indicated in the problem; then $f(x_1) = f(x_2) = 0$ and, since $\sqrt{A^2+B^2} \neq 0$, we have $\sin(x_1+\varphi) = \sin(x_2+\varphi) = 0$. It follows that $x_1 + \varphi = m\pi$, $x_2 + \varphi = n\pi$ and, hence, $x_1 - x_2 = k\pi$ at an integer k. This equality leads to a contradiction, because, by the hypothesis, we must have $x_1 - x_2 \neq k\pi$. Consequently, $A^2 + B^2 = 0$, whence $A = B = 0$.

Ingram Content Group UK Ltd.
Milton Keynes UK
UKHW020738070623
423023UK00014B/651

9 789324 191748